Springer Series on
Wave Phenomena 17

Edited by L. M. Brekhovskikh

Springer
Berlin
Heidelberg
New York
Barcelona
Hong Kong
London
Milan
Paris
Singapore
Tokyo

Springer Series on

Wave Phenomena

Editors: L.M. Brekhovskikh L.B. Felsen H.A. Haus

Managing Editor: H.K.V. Lotsch

Volume 1
Mechanics of Continua and Wave Dynamics
2nd Edition
By L. M. Brekhovskikh, V. Goncharov

Volume 2
Rayleigh-Wave Theory and Application
Editors: E. A. Ash, E. G. S. Paige

Volume 3
Electromagnetic Surface Excitations
Editors: R. F. Wallis, G. I. Stegeman

Volume 4
Short-Wavelength Diffraction Theory
Asymptotic Methods
By V. M. Babic, V. S. Buldyrev

Volume 5
Acoustics of Layered Media I
Plane and Quasi-Plane Waves
By L. M. Brekhovskikh, O. A. Godin

Volume 6
Geometrical Optics of Inhomogeneous Media
By Yu. A. Kravtsov, Yu. I. Orlov

Volume 7
Recent Developments in Surface Acoustic Waves
Editors: D. F. Parker, G. A. Maugin

Volume 8
Fundamentals of Ocean Acoustics
2nd Edition
By L. M. Brekhovskikh, Yu. P. Lysanov

Volume 9
Nonlinear Optics in Solids
Editor: O. Keller

Volume 10
Acoustics of Layered Media II
Point Sources and Bounded Beams
By L. M. Brekhovskikh, O. A. Godin

Volume 11
Resonance Acoustic Spectroscopy
By N. D. Veksler

Volume 12
Scalar Wave Theory
Green's Functions and Applications
By J. A. De Santo

Volume 13
Radar Target Imaging
Editors: W.-M. Boerner, H. Überall

Volume 14
Random Media and Boundaries
Unified Theory, Two-Scale Method, and Applications
By K. Furutsu

Volume 15
Caustics, Catastrophes, and Wave Fields
By Yu. A. Kravtsov, Yu. I. Orlov

Volume 16
Electromagnetic Pulse Propagation in Causal Dielectrics
By K. E. Oughstun, G. C. Sherman

Volume 17
Wave Scattering from Rough Surfaces
2nd Edition
By A. G. Voronovich

Volume 18
Electromagnetic Wave Propagation in Turbulence
Evaluation and Application of Mellin Transforms
By R. J. Sasiela

Volume 20
Surface Acoustic Waves in Inhomogeneous Media
By S.V. Biryukov, Y. V. Gulyaev, V. V. Krylov and V. P. Plessky

Alexander G. Voronovich

Wave Scattering from Rough Surfaces

Second, Updated Edition
With 22 Figures

Professor Alexander Voronovich
National Oceanic and Atmospheric Administration
Environmental Technology Laboratory
325 Broadway
Boulder CO 80303-3328
USA

ISSN 0931-7252
ISBN 3-540-64673-6 2nd Edition Springer-Verlag Berlin Heidelberg New York
ISBN 3-540-57439-5 1st Edition Springer-Verlag Berlin Heidelberg New York

Library of Congress Cataloging-in-Publication Data. Voronovich, Alexander G., 1949 – Wave scattering from rough surfaces / Alexander G. Voronovich. -- 2nd ed. p. cm. -- (Springer series on wave phenomena ; 17) Includes bibliographical references and index. Includes bibliographical references and index. ISBN 3-540-64673-6 (hardcover : alk. paper) 1. Waves -- Mathematics. 2. Scattering (Physics) -- Mathematics. 3. Surfaces (Physics) -- Mathematics. I. Title. II. Series. QC157.V67 1998 531'.1133 -- 98-41639

SPIN: 10683389 54/3144 - 5 4 3 2 1 0 - Printed on acid-free paper

Preface

Since the first edition of this book was published in the 1994, the theory of wave scattering from rough surfaces has continued to develop intensively. The community of researchers working in this area keeps growing, which provides justification for issuing this second edition.

In preparing the second edition, I was challenged by the problem of selecting new material from the many important results obtained recently. Eventually, a new section was added to the central Chap. 6 of this book. This section describes the operator expansion technique put forward by M. Milder, which conforms well with the general approach adopted in the book and which to my mind is one of the most promising.

Remote sensing of the terrain and ocean surface represents one of the most important and interesting challenges to the theory of wave scattering from rough surfaces. Rapid progress in electronics results in sensors with new capabilities. New powerful computers and data communication systems allow more sophisticated data processing techniques. What information about soil or air-sea interaction processes can be obtained from gigaflops of data streaming from air- or space-borne radars? To use this information efficiently, one cannot rely entirely on heuristic approaches and needs adequate theory. I hope that this book will contribute to progress in this important area.

Boulder, September 1998 *A. Voronovich*

Preface to the First Edition

Diverse areas of physics and its applications deal with wave scattering at rough surfaces: optics, acoustics, remote sensing, radio astronomy, physics of solids, diffraction theory, radio and wave-propagation techniques, etc. The mathematical description of relevant problems is rather universal and theoretical methods analysing this phenomenon form a definite area of mathematical physics. It was founded by Lord Rayleigh at the beginning of the 20th century and has been intensively developed since 1950 in response to practical needs. Modern theory involves, along with the classical methods, some new approaches with much more extensive possibilities.

The theoretical methods describing wave scattering at rough surfaces represent the subject of this book. Having studied this area over many years, the author came to the conclusion that most of the results found in this theory can easily be obtained and comprehended in the framework of a rather universal scheme and this became the motive for writing the present monograph.

The first half of the book deals with the classical results. However, several new problems are considered in connection with the methods used here (for example, some applications of the Rayleigh hypothesis and its relation to the Lippman argument). The second half is fully devoted to recent results presented in papers but not yet in books. For this reason the author hopes that the present book will be of interest to both newcomers and experts in this area.

To read this book, knowledge of common mathematical methods of modern physics is sufficient. In the few cases where the means applied are not quite standard, necessary information is given in the text (for example, calculation of functions of operators).

The list of references was intended to be concise. It includes either the basic works or recent ones directly related to the questions under consideration. The detailed bibliography on a particular problem is easily available from references in current papers. This saves the reader's time but has the shortcoming that some works which played a remarkable part earlier are not cited.

The author is sincerely grateful to academician L. M. Brekhovskikh for useful remarks and significant support to the idea of writing this book, and to all his colleagues in the Ocean Acoustics Department, P. P. Shirshov Institute of Oceanology, Moscow, for discussing the problems concerned here. I am deeply indebted to my wife V. V. Vavilova for the help in translation and preparing the manuscript.

Moscow, Russia, June 1993 *A. Voronovich*

Contents

1. Introduction

Wave scattering at rough surfaces can be principally considered as a limiting case of wave propagation in inhomogeneous media whose properties change by jumps. Since wave motion is characteristic for all continua often possessing well defined boundaries, then, generally speaking, these scattering processes must be encountered nearly as often as waves themselves. In fact, wave scattering at rough surfaces is of importance in various physical situations and more and more new areas of physics deal with it by responding to practical needs. Here are some instances.

The interaction potential of a molecular beam with a monocrystal surface can be approximated with an infinitely high bended potential barrier. In studying the appropriate scattering process, a classical diffraction problem arises whose solution enables us to reconstruct the shape of the potential barrier [1.1]. Physics of solids considers phonon scattering at crystal boundaries significant for different kinetic phenomena under some conditions [1.2–4].

The problem of scattering at rough surfaces is often tackled in optics (in the diffraction gratings theory, for example [1.5]). Recently, it was treated in RX optics when endeavouring to create mirrors out of layered structures [1.6].

Scattering at rough surfaces is of interest also in radioastronomy in radar experiments on the Moon and other planets [1.7].

It remains actual in radiophysics [1.8–10] and underwater acoustics [1.11] where it is set up as a special field of investigation.

The relationship between roughness properties and scattered field is of practical importance in remote sensing [1.12].

This enumeration does not pretend to be complete and just illustrates how widely this problem is encountered in physics.

Wave scattering at rough surfaces was examined for the first time as a problem of mathematical physics in 1907 by *Rayleigh* [1.13], and has been intensively developed since the 50*th*.

The papers on this problem can be divided into some directions. First, the works which apply the known theoretical approaches to concrete physical problems and experimental results. Development of numerical methods of calculating the diffraction wave fields resulted from scattering at the surface of a given shape is the second direction. It considers in fact different approaches of constructing the set of appropriate linear equations and solving them with computers. This problem is not trivial, since such equations can be written in many ways which lead to systems with different possibilities. Recently, using supercomputers allowed numerical calculations even for statistical ensembles of roughness, e.g., see *Thorsos* [1.14] and *Berman* [1.15].

Numerical modelling is of great significance also because experimental verification of various theoretical approaches to the reference diffraction problem can be now wholly transferred to the numerical experiment (at least for two-dimensional problems). In reality many additional factors should be taken into account while comparing the theoretical and experimental results. For instance, in studying the scattering of underwater acoustic signals from the rough sea surface, under certain conditions air bubbles in the surface layer should be considered. Furthermore, statistical properties of roughness are not sufficiently known. Consideration of experimental conditions is complicated even in the laboratory. Therefore, numerical experiment is the "purest" if applicability of wave equation is not an issue and the form of boundary conditions is reliably established. However numerical modelling does not answer all questions. Moreover, when dealing with the three-dimensional problems, dimensions of appropriate sets of linear equations can be too large.

Finally, the third direction concerns development of theoretical methods describing wave scattering both in deterministic and statistical cases. This area is rather important since the results obtained here serve as a base for solving different concrete practical problems and elaborating numerical approaches. The problems considered in the present book are practically completely referred to this third direction.

The scattering problems can be classified as well by the type of rough surface—compact (i.e., enclosing some finite volume) or incompact. This book deals exclusively with scattering at incompact and averagely plane in some sense surfaces.

The well known and widely applied in theoretical physics language of S-matrix is universal, and in many respects most fitting to describe the wave scattering at rough surfaces. Matrix element $S(\boldsymbol{k}, \boldsymbol{k}_0)$ of this matrix is in the present case an amplitude of the plane wave with a horizontal projection of wave vector equal to $\boldsymbol{k}$ which resulted from scattering at the rough boundary of some plane wave with a horizontal projection of wave vector equal to $\boldsymbol{k}_0$. Application of S-matrix for describing the wave scattering at rough surfaces is well known. The appropriate language is systematically used in the T-matrix approach of *Watermann* [1.16], for example.

The Green's function $G(\boldsymbol{R}, \boldsymbol{R}_0)$ is principally the most exhaustive characteristic of the given problem. It presents the field at the point $\boldsymbol{R}$ generated by the point source situated at $\boldsymbol{R}_0$. However, it is clear that scattering amplitude (SA) also allows us to determine the Green's function using a representation of the source radiated field as a superposition of plane waves and subsequent summation of the fields of all scattered waves. Calculation of the corresponding quadrature is rather important practically, but presents principally a technical problem.

In theoretical investigations, SA is more convenient than the Green's function because formulas are much simpler with the former. In fact, to derive SA from the Green's function the source and receiver positions should be assumed at infinity which simplifies considerations as compared to the general case. Moreover, SA $S(\boldsymbol{k}, \boldsymbol{k}_0)$ depends on four arguments, while the Green's function $G(\boldsymbol{R}, \boldsymbol{R}_0)$ on six. Finally, we describe scattering in terms of SA of plane waves using the basis related to the symmetry of "average" boundary which additionally simplifies formulas in the statistical case.

Such properties as unitarity and reciprocity theorem are easily formulated in terms of SA for waves of different nature. Finally, experimentally measured quantities—the mean reflection coefficient and the scattering cross section—are immediately related with statistical moments of SA.

Description of scattering processes in terms of scattering amplitude is constantly used in this book.

The reference problem of scattering for roughness of more or less general type cannot be solved exactly. Therefore only approximate analytic approaches to this problem exist. Two of them are used most widely and are of a special significance. These are classical methods of handling the problems in physics: the perturbation theory and quasi-classical approximation. In the present case the latter corresponds to the tangent plane method also known as the Kirchhoff approximation.

The perturbation theory was first applied by *Rayleigh* in his classical work [1.13]. A small parameter of this theory is the Rayleigh parameter—the value proportional to the ratio of roughness heights to the incident wave length. Thus this approach allows us to obtain the low frequency asymptotic of solution. In the lowest order the Bragg's scattering at spectral components of roughness essentially takes place here.

Quasi-classical approximation for the given problem was built in the work by *Brekhovskikh* [1.17] and in application to the statistical problem first considered by *Isakovich* [1.18]. This theory assumes the local character of the interaction of radiation with roughness, that is the boundary can be approximated at each point by a tangent plane. This approximation involves some small parameters, however, from physical considerations it is clear that the short wave asymptotic of solution is dealt anyway.

The applicability ranges of "classical" approaches are schematically shown in Fig. 1.1. Their results agree only near the coordinates' origin. Note that in the general case the small slope of roughness should be required anyway.

In practice such situations often take place which cannot be considered in the framework of only one "classical" approach. Such, for example, is an important problem of acoustic or electromagnetic waves scattering at a rough sea surface. In these cases the so-called two-scale model is used which is built as a combination of the two above mentioned classical methods. The two-scale model was for the first time suggested by *Kur'yanov* [1.19]. In this approach the boundary roughness is divided into two classes: large- and small-scale. The scattering processes at large-scale components are described in the tangent plane approximation and small-scale roughness is accounted for as a perturbation. The parameter separating large- and small-scale roughness is chosen arbi-

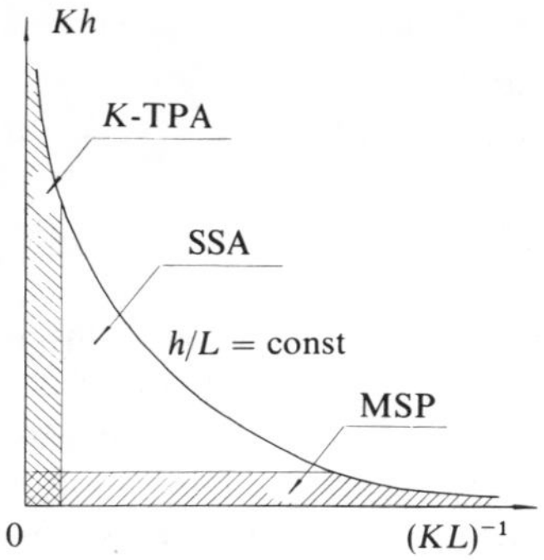

Fig. 1.1. Regions of validity of "classical" approaches (shaded): small perturbation method and Kirchhoff – tangent plane approximation. The area below hyperbola corresponds to applicability of small slope approximation (see the text below)

trarily in certain limits. The circumstance is inconvenient, complicates calculating corrections and hinders studying the inverse problem: determining the roughness spectrum by characteristics of the scattered field. The inverse problem simply solvable within the small perturbation method is a poorly examined problem for the two-scale situation. In particular, its solution was analyzed by *Fuks* [1.20] who demonstrated the great role of the ambiguously determined parameter separating scales.

For this reason some other theories were suggested uniting both "classical" approaches and giving in corresponding limiting cases the results of the perturbation theory and quasi-classical approximation. Chapter 6 describes six different theoretical approaches gaining this aim.

Now we present a brief content of the book. In Chap. 2 the scattering amplitde (SA) of plane waves is introduced. SA (which is generally matrix-valued) can be built uniformly for the waves of different nature. A set of SA forms the so-called *S*-matrix. Formulation of the reciprocity theorem and unitarity condition for the *S*-matrix are discussed in Sect. 2.3 and the relation of statistical moments of SA with scattering cross sections and the mean reflection coefficient is established in Sect. 2.4.

Chapter 3 is devoted to the Helmholtz formula and the Rayleigh hypothesis. Section 3.1 suggests a simple derivation of the Helmholtz formula when it is a direct consequence of the rule of differentiation of discontinuous functions. From this derivation easily follows that its formulation is ambiguous. Since this formula is often used in analytic approaches, this ambiguity should be taken into account. The Helmholtz formula helps to obtain integral equations for the unknown density of surface sources generating the scattered field. These equations are considered in Sect. 3.1 as well. The above mentioned ambiguity manifests in that the scattered field can be assumed to be generated by either monopole or dipole surface sources.

The Rayleigh hypothesis is discussed in Sects. 3.2–4. The question at issue is whether representation for the scattered field in the form of outgoing waves can be used at all points including those at the boundary? The *Lippmann* [1.21] argument and results of *Petit* and *Cadilhac* [1.22] negate this, while the results of *Marsh* [1.23] and *Millar* [1.24] confirm this. This problem, besides its theoretical interest, is of large practical importance, since the application of the Rayleigh hypothesis enables us to write down immediately an equation for SA which is conveniently solved with the perturbation theory. Sections 3.2–4 and the introduction to Chap. 4 show that this procedure is right, the series expansion of SA is determined correct and the Lippman argument is invalid. This result is constantly used in Chap. 4 which is devoted to the perturbation theory. SA in the perturbation theory is represented as an integral-power expansion in elevations. Various boundary problems differ only in the coefficient functions of these expansions which are readily found by a unique scheme based on the Rayleigh equation. In the lower order, the Bragg's wave scattering at spectral components of undulation is in fact considered, and in the general case it can be described by a dimensionless roughness independent function $B(\boldsymbol{k}, \boldsymbol{k}_0)$ (matrix-valued in the case of multicomponental fields) where $\boldsymbol{k}_0$ and $\boldsymbol{k}$ are horizontal components of wave vectors of the incident and scattered wave, respectively. Chapter 4 considers both simple scalar Dirichlet and Neumann problems and multicomponental problems concerning electromagnetic wave scattering. We examined both perfect boundary conditions with no energy flux through the boundary, as well as the cases when waves can penetrate into the other medium.

Section 4.2 is devoted to the direct verification of the statement that the application of the Rayleigh hypothesis in calculating the coefficient functions of integral-power expansions of SA give the correct result. For this the reference integral-power expansion for the Dirichlet problem is built in another "fair" way also, i.e., proceeding from the exact integral equation for dipole surface sources. In such case, the Rayleigh hypothesis is not involved at all. It is proved that these two expansions – obtained by the "fair" way and by using the Rayleigh hypothesis – coincide in all orders.

For all problems considered the *apriori* limitations imposed on SA are obtained by a unique scheme. For this some antisymmetric bilinear form expressed through asymptotic values of fields and vanishing at exact solution of the problem, is introduced. Substitution into this bilinear form the fields corresponding to scattering of two different plane waves yields the reciprocity theorem. When, for one of the solutions a "time-reversed solution" is taken, then the same bilinear form gives the law of energy conservation.

Chapter 5 is devoted to the Kirchhoff-tangent plane approximation correspondent to the quasi-classical approximation. It can be built in two ways. The first is based on the assumption that wave reflection at each point of the boundary is as if the boundary coincides with the tangent plane at each point. Then the value of the field and its normal derivative at the boundary are easily expressed through the incident field, and the scattered one is determined using the Helmholtz formula. This idea enables us to find readily SA for any problem that is shown in Sect. 5.1 in several examples.

However, in the framework of this approach it is difficult to calculate corrections and to obtain applicability criteria of the theory. But the same results can be obtained proceeding from the integral equation, for the surface source's density when solving it with iterations. The first iteration corresponds to the Kirchhoff approximation and one more iteration is needed to calculate the diffraction corrections. Besides that, in the geometric-optical limit multiple reflections and shadowing of the surface areas by other ones can arise. To describe these phenomena many iterations of integral equation should be done as demonstrated by *Liszka* and *McCoy* [1.25]. All these questions are considered in Sects. 5.2, 3.

In addition, Sect. 5.4 considers corrections to the Kirchhoff approximation with respect to statistical problems. In this case one can attempt to proceed further than in the deterministic situation, since the terms of previously mentioned iteration series are easily represented in the form containing roughness of the boundary only in the exponents. Hence this series can be easily averaged especially for Gauss statistics of roughness. It is convenient to represent the result in diagram notation. Further, it is possible to perform a partial summation of the series and this was done by *Zipfel* and *DeSanto* [1.26] and *DeSanto* [1.27]. But corresponding results are not reproduced in this book because of the comprehensive review paper by *DeSanto* and *Brown* [1.28] (which deals also with some other interesting approaches to the problem). Section 5.4 presents some other diagram approach due to *Voronovich* [1.29] which is convenient for evaluation of higher order diagrams using expansions with respect to some small parameters. Analysis of the first non-trivial diagram shows that there are two of them: the ratio of the rms slope of undulation to the grazing angle of the incident wave and some other parameter which requires that the horizontal scale of roughness be large enough. Both restrictions become harder as the grazing angle diminishes. It is of interest that this approach allows us to separate contributions of geometric optics and diffractional effects – the former are frequency independent and the latter vanish as $K \to \infty$.

Section 5.5 treats the two-scale model mentioned above.

Chapter 6 is completely devoted to "nonclassical" approaches uniting the perturbation theory and the Kirchhoff approximation.

One uniting approach results from the iterative solution of an integral equation for surface dipole density. In this case two first iterations give the result which transforms in corresponding limiting cases into formulas of the perturbation theory and the Kirchhoff approximation, respectively. The practical disadvantage of this approach is in integrations of rather high order in the resulting formulas.

Sections 6.1, 2 outline the small slope approximation having a relationship to all approaches pointed out below. This approach was suggested in the work by *Voronovich* [1.30] and is based on using the transformation properties of SA with respect to shifts of rough boundary in space. SA is represented in the form of product where the first "Kirchhoff-type" factor ensures fulfillment of required transformation properties and the second factor is sought in the form of an integral-power series. As a result, the small parameter of this expansion appears to be the slope of roughness. The last assertion cannot be proved in general cases and should be checked for any specific boundary conditions. Nevertheless, this property seems very plausible and is, in fact, discovered in a number of problems.

The small parameter – roughness slope – principally differs from those used in classical approaches since it is independent of wave length. This circumstance is rather unexpected and to clarify the validity of the built approximation, the results of calculating scattering at an echelette grating surface was compared to the exact solution of the problem. This comparison demonstrated fairly good agreement for both small and large Rayleigh parameters according to the theory predictions. An echelette grating surface was chosen because scattering at this surface cannot be considered with the Kirchhoff approximation due to the points with infinite curvature.

The small roughness slopes should be required anyway in the other approximation theories as well. However, in the present theory this small parameter claims to be the only one (Fig. 1.1).

This point should be strongly emphasized because of its importance. More detailed consideration shows the necessity of the requirement that the incident and scattering angles should be large compared to the angles of roughness slope, i.e., the geometric shadowing should be absent. The same requirement must be satisfied for virtual intermediate waves arising at multiple scattering. For a periodic surface it is necessary for the grazing spectra to be weakly excited.

Resonances with grazing spectra for an acoustically rigid surface result in Wood's anomalies (*Hessel* and *Oliner* [1.31]), and this phenomenon cannot be considered within the framework of small-slope approximation.

Small-slope approximation is a uniting approach. For Dirichlet and Neumann problems the second order expression for SA in the high-frequency limit transforms exactly to the tangent-plane's formula. On the other hand, for the low-frequency case the perturbational results are automatically reproduced.

The formulas for SA in small-slope approximation are reciprocal. Moreover, the first order expression of SA appeared to be related with the unitarity of scattering.

The coefficient functions of the previously mentioned integral-power expansion can be calculated in different ways. In particular, it is sufficient to know SA expansion into the ordinary perturbative series in elevations. The appropriate procedure is especially simple for the first order of small-slope approximation. In this case, the formula for SA splits into the product of the "Kirchhoff" factor of purely geometric origin (which is boundary conditions-independent) and the Bragg's factor $B(\boldsymbol{k}, \boldsymbol{k}_0)$.

Section 6.3 presents the phase perturbation technique due to *Winebrenner* and *Ishimaru* [1.32, 33]. This technique is analogous to Rytov's method in wave propagation theory for inhomogeneous media. Its essense is that the density of monopole sources at the boundary is expressed as an exponential of some new function sought in the form of expansion in powers of elevations. The first approximation formulas for a horizontal plane situated at an arbitrary level gives an exact solution of the problems. Hence, the Rayleigh parameter cannot be a small parameter of the theory and for gently undulating surfaces with large horizontal scale, the results correspond to those obtained in the Kirchhoff approximation (even including corrections). Moreover, the method assures the correct limiting transfer to the perturbation theory. High efficiency of the method was demonstrated by numerical experiments including the cases for rather grazing incident (not scattering) angles as well (*Broschat* et al. [1.34]). Generally speaking, formulas for SA originating in this method do not satisfy the reciprocity theorem which can even be used in calculating scattering in grazing directions (reversed phase perturbation technique).

The phase operator method which is a matrix analog of the phase perturbation technique was independently put forward by *Voronovich* [1.35]. It is discussed in Sect. 6.4. The set of scattering amplitudes considered at all arguments forms an S-matrix. Solving the scattering problem by almost all methods is in essence one or another way of approximating the calculating of the various matrix elements independently. In such a case introducing the S-matrix is a matter of terminology. On the contrary, the main idea of the phase operator method is to consider instead of the S-matrix its logarithm – naturally called the phase operator – and to seek in the form of series expansion not $\hat{S}$ itself but matrix $\hat{H} = \log \hat{S}$. The present procedure also changes the small parameter of expansion which is proportional to the slope irregularities. A situation of interest is when in the roughness spectrum there are no small scale components and excitation of inhomogeneous waves can be ignored. In this case, the phase operator $\hat{H}$ becomes Hermitian and the scattering matrix becomes unitary. Hence, under these conditions the law of conservation of energy is exactly fulfilled. There exists also such modification of the present approach when unitarity of scattering is exactly ensured for any accuracy of calculating the phase operator.

The phase operator technique is one of the ways to take into account multiple scattering. It cannot be referred to analytic approaches completely because it gives explicit expressions only for the phase operator, and to find the S-matrix the corresponding matrix exponent should be calculated. However, the last operation is readily realized with a computer. On the other hand, matrix elements of the phase operator are immediately expressed through the Fourier coefficients of irregularities and in contrast to other numerical methods do not require complicated quadratures.

The phase operator possesses a curious property: its expansion into the integral-power series of elevations consists only of odd powers.

Section 6.5 presents the approach independently suggested by *Meecham* [1.36] and *Lysanov* [1.37] in 1956 which consists in that the kernel of the integral equation for the surface monopole density is under certain conditions approximated by a difference kernel and then this equation is easily solved by a Fourier transform.

In Sect. 6.6, comparison between small-slope approximation and the full-wave approach due to *Bahar* [1.38] is performed. The latter can be considered as a generalization of the reference waveguide approach. It is demonstrated that for a two-dimensional Dirichlet problem both perturbative schemes in the lowest order lead to coinciding results.

Chapter 7 of the book deals with the practically important problem of determining the correlation function of the scalar field in the horizontally layered refractive waveguide bounded from above by the rough surface. This problem is also considered using the concept of SA.

The total correlation function is built of coherent (related to mean-field) and noncoherent parts. The mean-field evaluation can be exactly reduced to the problem and wave propagation in a layered waveguide with an impedance boundary condition. The impedance is related to both the waveguide's and the roughness's parameters.

Determination of the noncoherent component of correlation function is a more complicated problem. However, it is possible to progress seriously in it by using the following consideration. The wave field at some horizon can be divided into up- and downgoing plane waves. The upgoing waves incident upon the rough boundary and scatter at it. This process was considered in Chap. 6 in detail. Downgoing waves refract in the stratified lower medium and return again to the boundary. By that the waves undergo some horizontal shift which in high-frequency limit is related to the skip distance of the corresponding ray. In many practical cases this shift exceeds much of the radius of correlation of roughness. Based on a related small parameter one can promote seriously in this problem without concretizing characteristics of the scattering process at the rough boundary. The appropriate approach will be referred to as an approximation of noncorrelated successive reflections. In particular, for the mean-field this approximation leads in the above mentioned impedance boundary condition to the replacement of the unperturbed reflection coefficient by a mean reflection coefficient. The latter is closely related to the first statistical moment of SA.

The problem of calculating the noncoherent component of the correlation function is reduced to the solution of some integral equation. After space-averaging of this function over some horizontal area this integral equation transforms to the radiative transport equation. The kernel of the corresponding integral term is closely related to the second statistical moment of SA at the rough boundary – scattering cross-section. It should be stressed that the applicability of geometric optics to the description of wave-fields is not supposed; necessary approximations can be substantiated for the expense of the above mentioned space averaging without passing to a high-frequency limit.

Additional simplification of radiative transport equations can be achieved when considering the problem of a noise field generated by the surface sources uniformly distributed over a surface. In this case the resulting equation describes the angle distribution of the noises. The transition to the diffusion approximation by changing the integral equation to a differential one is also considered.

The theory proposed in Chap. 7 does not use modal expansion and deals with the quantities allowing direct local measurements.

2. Scattering Amplitude and Its Properties

The problem of calculating the wave scattering from the rough and plane in average boundary is formulated in the present chapter. Such a boundary either divides two homogeneous half-spaces or bounds only one of them. The waves propagating in these media can be scalar or multicomponent (possessing polarization). The most natural language describing scattering processes is an amplitude of scattering (SA) the plane waves into each other. Using SA instead of Green's function is more preferable in many relations. Applications of SA in the problems of wave scattering from rough surfaces is well known; *Waterman* [2.1] was one of the first to apply it systematically.

SA is introduced in Sect. 2.2 and is quite analogous to SA used in quantum mechanics [2.2] and has similar properties. Two main general relations which SA must satisfy – the reciprocity theorem and unitarity – are discussed in Sect. 2.3. And last, Sect. 2.4 considers statistical properties of SA and establishes a simple relation between the first and second statistical moments of SA and experimentally measured values – the mean reflection coefficient and the scattering cross-section (or the scatterng coefficient), respectively.

2.1 Formulation of Problem and Some Auxiliary Relations

The main problem to consider in this book is the following. Let Ω and Ω' be two homogeneous half-spaces with an interface between them running along a certain rough surface Σ. The wave fields Ψ and Ψ' can propagate in both half-spaces. Certain sources of the field Q are located in the region Ω below the boundary (Fig. 2.1). We assume the boundary as motionless and radiation from the sources as monochromatic with frequency ω: a temporal factor is taken to be $\exp(-\mathrm{i}\omega t)$ and is omitted everywhere in what follows. The source generated field will be reflected when bouncing at the boundary and will penetrate area Ω'. This process will be accompanied by wave diffraction at boundary irregularities. For brevity we name the process of interaction of wave fields with boundary roughness as scattering. The problem is to calculate the resulting field in regions Ω and Ω'.

The reference problem can also be considered as statistical. That is, if the reference rough boundary belongs to some statistical ensemble, then the task is to calculate statistical moments of the resulting field.

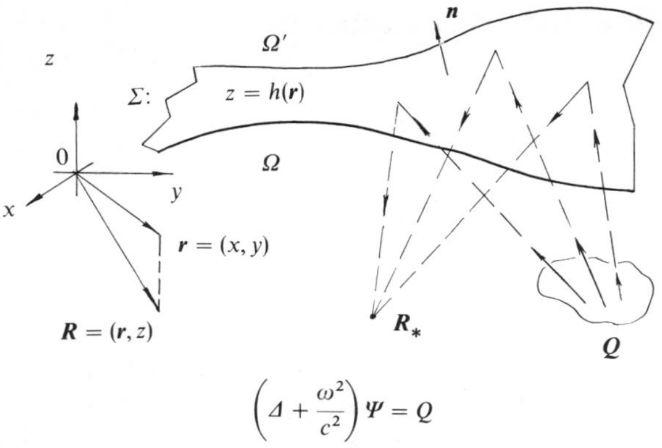

Fig. 2.1. Geometry of the problem and notation

We shall use everywhere the right Cartesian coordinates with z-axis directed upward. We take vector $\boldsymbol{R}$ for the three-dimensional radius vector and $\boldsymbol{r}$ for its projection on the horizontal plane: The boundary equation is written in the form

$$z = h(\boldsymbol{r}) \; . \tag{2.1.1}$$

The present book is devoted exclusively to the case of incompact boundaries which are assumed to be plane on average in a sense. Namely, the assumption is that the boundary lies completely in a horizontal layer, $h_{\min} < h(\boldsymbol{r}) < h_{\max}$. In some practically important situations radiation cannot penetrate region Ω' and is perfectly scattered backward into area Ω. For example, such a case takes place when a sound wave is incident on the interface with a vacuum, or when an electromagnetic wave is incident on a perfectly conducting surface. Then the field in region Ω' need not be considered, and only boundary conditions at Σ are imposed.

Wave propagation is described mathematically by the Helmholtz equation

$$(\Delta + K^2)\Psi = Q \; , \tag{2.1.2}$$

where $\Delta = \partial_x^2 + \partial_y^2 + \partial_z^2$ and $K = \omega/c = \text{const}$ is a wave number.

For sound waves the field Ψ may be identified with the velocity potential, so that the formula for local velocity takes the form

$$\boldsymbol{v} = \nabla\Psi \; . \tag{2.1.3}$$

It follows from Newton's law $\varrho_0 \partial_t \boldsymbol{v} = -\nabla p$ that acoustical pressure for monochromatic waves is expressed by the formula

$$p = \mathrm{i}\omega\varrho_0 \Psi \tag{2.1.4}$$

with ϱ_0 being medium density.

Let $\boldsymbol{I}$ denote the energy flux vector averaged over the wave period:

$$\boldsymbol{I} = \frac{\omega}{2\pi} \int_0^{2\pi/\omega} p\boldsymbol{v}\, dt \ .$$

It is obvious that in terms of the complex field Ψ, vector $\boldsymbol{I}$ is expressed as follows:

$$\boldsymbol{I} = (\mathrm{i}\omega\varrho_0/4)(\Psi\nabla\Psi^* - \Psi^*\nabla\Psi) \tag{2.1.5}$$

(here and below the asterisk denotes the complex conjugation).

In the general case the wave field can consist of several components $\Psi = \{\Psi_\sigma\}$, $\sigma = 1, 2, \ldots, n$. For example, in the electromagnetic case the waves of two polarizations ($n = 2$) occur and the wave field in an isotropic solid consists of three components: two polarizations of transverse waves and a longitudinal wave. Generally speaking, various wave components differ in propagation velocities. However, when using the scattering amplitude to describe scattering, then multicomponental fields will be treated similarly to the scalar case. The scattering amplitude is introduced in Sect. 2.2, and here we limit ourselves to a simple scalar case.

The field Ψ at the rough boundary Σ must satisfy certain boundary conditions which in the general case can be formulated as

$$\hat{B}\Psi|_{z=h(\boldsymbol{r})} = 0 \ , \tag{2.1.6}$$

where $\hat{B}$ is a linear operator. This operator often appears to be nonlocal. That is, its effect depends not only on the field values and some of its spatial derivatives at the point $\boldsymbol{R} = [\boldsymbol{r}, h(\boldsymbol{r})]$, but as well on the field in the vicinity of the point $\boldsymbol{R}$ (with certain reservations one can say that the operator depends on field derivatives of all orders at point $\boldsymbol{R}$). Without loss of generality one can assume that field derivatives, with respect to coordinate z of the order not higher than unity, enter into the boundary condition (2.1.6). Indeed, the derivatives with respect to the second and higher orders can be expressed in terms of derivatives with respect to x and y by the homogeneous Helmholtz equation (2.1.2). Thus, in the general case the boundary condition (2.1.6) is written as

$$\int d\boldsymbol{r}' \cdot \left[K_1(\boldsymbol{r}, \boldsymbol{r}')\Psi(\boldsymbol{r}', z) + K(\boldsymbol{r}, \boldsymbol{r}') \frac{\partial \Psi(\boldsymbol{r}', z)}{\partial z} \right]\Bigg|_{z=h(\boldsymbol{r}')} = 0 \ . \tag{2.1.7}$$

Assuming that the kernel $K(\boldsymbol{r}, \boldsymbol{r}')$ is not degenerated, i.e., such function $K^{-1}(\boldsymbol{r}', \boldsymbol{r})$ exists that

$$\int K^{-1}(\boldsymbol{r}_1, \boldsymbol{r}') \cdot K(\boldsymbol{r}', \boldsymbol{r}_2)\, d\boldsymbol{r}' = \delta(\boldsymbol{r}_1 - \boldsymbol{r}_2)$$

then (2.1.7) can be solved for $\partial\Psi/\partial z$ and written in the form

$$\frac{\partial \Psi}{\partial z}\bigg|_{z=h(\boldsymbol{r})} = \int M_1(\boldsymbol{r}, \boldsymbol{r}') \cdot \Psi(\boldsymbol{r}', h(\boldsymbol{r}'))\, d\boldsymbol{r}' \ , \tag{2.1.8}$$

where

$$M_1(\boldsymbol{r},\boldsymbol{r}') = -\int K^{-1}(\boldsymbol{r},\boldsymbol{r}'')\cdot K_1(\boldsymbol{r}'',\boldsymbol{r}')\,d\boldsymbol{r}'' \;.$$

The boundary condition (2.1.8) can be given in another form. The unit vector of external normal $\boldsymbol{n}$ to surface Σ is

$$\boldsymbol{n} = [1 + (\nabla h)^2]^{-1/2}(\boldsymbol{N} - \nabla h) \;, \tag{2.1.9}$$

where $\nabla = (\partial_x, \partial_y)$ is the horizontal gradient operator and $\boldsymbol{N} = \boldsymbol{e}_z$ is the unit vector directed along z-axis. Correspondingly, the operator of differentiation with respect to external normal is

$$\frac{\partial \Psi}{\partial \boldsymbol{n}} = [1 + (\nabla h)^2]^{-1/2}\left(\frac{\partial \Psi}{\partial z} - \nabla h\cdot\nabla\Psi\right) \;. \tag{2.1.10}$$

The area of surface element $d\Sigma$ whose projection on a horizontal plane corresponds to element $d\boldsymbol{r} = dx\,dy$ is

$$d\Sigma = [1 + (\nabla h)^2]^{1/2}\,d\boldsymbol{r} \;. \tag{2.1.11}$$

Using formulas (2.1.10, 11), it is easy to rearrange the boundary condition (2.1.8) in the following form:

$$\left.\frac{\partial \Psi}{\partial \boldsymbol{n}}\right|_{\boldsymbol{R}\in\Sigma} = \int L_1(\boldsymbol{R},\boldsymbol{R}')\Psi(\boldsymbol{R}')\,d\Sigma' \;, \tag{2.1.12}$$

where the kernel L_1 could be easily expressed in terms of M_1.

To find Ψ we rearrange (2.1.7) similarly to the previous equation, under the condition that kernel K_1 is not degenerated, then as a result we get the relation

$$\Psi(\boldsymbol{R})|_{\boldsymbol{R}\in\Sigma} = \int L_2(\boldsymbol{R},\boldsymbol{R}')\left.\frac{\partial \Psi}{\partial \boldsymbol{n}'}\right|_{\boldsymbol{R}'\in\Sigma} d\Sigma' \;. \tag{2.1.13}$$

In application such a case often occurs so that kernel L_1 in (2.1.12) vanishes and the boundary condition becomes

$$\left.\frac{\partial \Psi}{\partial \boldsymbol{n}}\right|_{\boldsymbol{R}\in\Sigma} = 0 \;. \tag{2.1.14}$$

This boundary condition is referred to as the Neumann boundary condition. It describes, for example, scattering of sound waves at those irregularities where normal to the boundary the component of the velocity turns to zero, see (2.1.3). Such boundaries are called acoustically rigid. Another example is the scattering of vertically polarized electromagnetic waves at a two-dimensional perfectly conducting surface. This case will be considered in Sect. 4.4.

Boundary condition (2.1.13) with $L_2 = 0$ is the Dirichlet boundary condition:

$$\Psi|_{R \in \Sigma} = 0 \ . \qquad (2.1.15)$$

It describes scattering of sound waves at the boundary where acoustic pressure becomes zero, see (2.1.4). The interface between a medium and vacuum represents the sort of boundary which is called acoustically soft or pressure-release. Scattering of horizontally polarized electromagnetic waves at the two-dimensional roughness of a perfect conductor obeys the Dirichlet boundary condition as well. Quantum particles scattering at an infinitely high potential barrier also obey this type of boundary condition. This approximation is also used for considering the interaction of molecular beams with crystal boundaries.

Theoretical studies very often deal with Neumann and Dirichlet boundary conditions.

Suppose now that operator L_1 is local and takes the following form:

$$L_1(\boldsymbol{R}, \boldsymbol{R}') = \mathrm{i}KZ(\boldsymbol{R})\delta_\Sigma(\boldsymbol{R} - \boldsymbol{R}') \ .$$

The nondimensional complex quantity Z is called impedance. Thus, in this case, the boundary condition (2.1.12) is reduced to the impedance boundary condition:

$$\frac{\partial \Psi(\boldsymbol{R})}{\partial \boldsymbol{n}} = \mathrm{i}KZ(\boldsymbol{R})\Psi(\boldsymbol{R}) \ , \qquad \boldsymbol{R} \in \Sigma \ . \qquad (2.1.16)$$

The quantity $Y = Z^{-1}$ is the admittance. Using it, we can also write the boundary condition (2.1.16) as

$$\Psi(\boldsymbol{R}) = (\mathrm{i}K)^{-1} Y(\boldsymbol{R}) \frac{\partial \Psi}{\partial \boldsymbol{n}} \ , \qquad \boldsymbol{R} \in \Sigma \ .$$

Let us calculate the average energy flux through the impedance boundary. According to (2.1.5),

$$I_n = \boldsymbol{I} \cdot \boldsymbol{n} = \frac{\mathrm{i}\omega\varrho_0}{4}\left(\Psi \frac{\partial \Psi^*}{\partial \boldsymbol{n}} - \Psi^* \frac{\partial \Psi}{\partial \boldsymbol{n}}\right) .$$

Using formula (2.1.16) we find

$$I_n = \frac{\varrho_0 \omega^2}{2c} \cdot \mathrm{Re}\{Z\} \cdot |\Psi|^2 \ .$$

In the case of purely imaginary impedance the energy flux through the boundary is equal to zero. For $\mathrm{Re}\{Z\} > 0$, energy will leave the region Ω through the boundary Σ.

The boundary condition (2.1.16) should be completed with the radiation condition at infinity, which implies that the wave field for $z \to -\infty$ is bound to consist only of outgoing waves. The simplest and most applicable formulation of a radiation condition is the limiting absorption principle: by introducing small losses the field will decay while going away from the boundary:

$$\text{if } \operatorname{Im}\{K\} = \varepsilon > 0 \quad \text{then} \quad \Psi \to 0,\ z \to -\infty \ . \tag{2.1.17}$$

In what follows we shall often use the Green's function for free space satisfying the Helmholtz equation (2.1.2)

$$(\Delta + K^2)G_0 = \delta(\boldsymbol{R}) \ . \tag{2.1.18}$$

Applying the Fourier transform one can easily rewrite the solution of (2.1.18) in the following form:

$$G_0(\boldsymbol{R}) = (2\pi)^{-3} \int (K^2 - \xi^2)^{-1} \exp(\mathrm{i}\xi \cdot \boldsymbol{R})\, d^3\xi \ . \tag{2.1.19}$$

Validity of the formula is clearly seen when substituting (2.1.19) into (2.1.18) and using the following integral representation of the δ-function which will be widely applied further:

$$(2\pi)^{-n} \int \exp(\mathrm{i}\xi \cdot \boldsymbol{R})\, d^n\xi = \delta(\boldsymbol{R}), \tag{2.1.20}$$

where n is space dimension ($n = 3$ in the present case). For purely real K the integrand in (2.1.19) becomes infinity at $|\xi| = K$. To avoid arising difficulties we assume that, according to (2.1.17), K contains a small positive imaginary part. In this case there are no infinities in formula (2.1.19).

The integral in (2.1.19) can be easily calculated in spherical coordinates with the polar axis directed along $\boldsymbol{R}$. After straightforward calculations one gets

$$G_0(R) = -\frac{\exp(\mathrm{i}KR)}{4\pi R} \tag{2.1.21}$$

with $R = |\boldsymbol{R}|$. It is obvious that the limiting absorption principle (2.1.17) for the Green's function (2.1.21) is fulfilled.

The Green's function for the Helmholtz equation in the two-dimensional case is of interest also,

$$(\partial_x^2 + \partial_z^2 + K^2)G_0^{(2)} = \delta(x)\delta(z) \ . \tag{2.1.22}$$

This equation can be solved using the Fourier transform in a similar way as previously. However, it may be noted that (2.1.22) is obtained from (2.1.18) when the latter is integrated over the coordinate y in infinite limits. Therefore,

$$G_0^{(2)}(\varrho) = -\frac{1}{4\pi} \int_{-\infty}^{+\infty} \frac{\exp[\mathrm{i}K(\varrho^2 + y^2)^{1/2}]}{(\varrho^2 + y^2)^{1/2}}\, dy \ , \tag{2.1.23a}$$

where $\varrho = (x^2 + z^2)^{1/2}$. This relation is, in fact, the integral representation of the Hankel function, so that

$$G_0^{(2)}(\varrho) = -\frac{\mathrm{i}}{4} H_0^{(1)}(K\varrho) \ . \tag{2.1.23b}$$

Another important representation of the Green's function is obtained after integrating over ξ_z in (2.1.19) at fixed $(\xi_x, \xi_y) = \boldsymbol{k}$. This integral is easily calculated with the help of the residues theorem that yields

$$G_0(R) = -\frac{i}{8\pi^2} \int \exp[i\boldsymbol{k}\cdot\boldsymbol{r} + iq(k)\cdot|z|]q^{-1}(k)\,d\boldsymbol{k} \ . \tag{2.1.24}$$

Here and below

$$q(k) = q_k = (K^2 - k^2)^{1/2} \ . \tag{2.1.25}$$

In this formula one should set $q(k) > 0$ for $|\boldsymbol{k}| < K$ and $q(k) = +i|K^2 - k^2|^{1/2}$ for $|\boldsymbol{k}| > K$, i.e., $\mathrm{Im}\{q\} \geqslant 0$. Let us assume that $q \stackrel{\mathrm{def}}{=} q(k)$. It is easily seen that the limiting absorption principle (2.1.17) is fulfilled due to this choice of square root branch in (2.1.25). Formula (2.1.24) is the Weyl formula. It gives the Green's function representation in the form of superposition of plane waves propagating upward at $z > 0$ and downward at $z < 0$. For $|\boldsymbol{k}| > K$ these waves exponentially decay when going far away from the source level.

Each elementary plane wave

$$\Psi = \Psi_0 \exp[i\boldsymbol{k}\cdot\boldsymbol{r} \pm iq(k)z] \tag{2.1.26}$$

entering into expansion (2.1.24) satisfies the homogeneous Helmholtz equation

$$(\Delta + K^2)\Psi = 0 \ . \tag{2.1.27}$$

Let us calculate the z-component of the energy flux in a plane wave of the kind given by (2.1.26). According to (2.1.5) for homogeneous waves $|\boldsymbol{k}| < K$ (q is real) we have

$$I_z = \pm\omega\varrho_0 q|\Psi_0|^2/2 \tag{2.1.28}$$

($I_z = 0$ for inhomogeneous waves with $|\boldsymbol{k}| > K$). In what follows it is convenient to choose the plane wave amplitude so that the energy flux in the waves along the z-axis is constant and $\boldsymbol{k}$-independent. Namely, according to (2.1.28) we choose the basis of plane waves in the form

$$\Psi_{\boldsymbol{k}} = q_k^{-1/2} \exp(i\boldsymbol{k}\cdot\boldsymbol{r} + iq_k z) \tag{2.1.29}$$

for incident waves, and in the form

$$\Psi_{\boldsymbol{k}} = q_k^{-1/2} \exp(i\boldsymbol{k}\cdot\boldsymbol{r} - iq_k z) \tag{2.1.30}$$

for scattered waves.

2.2 Scattering Amplitude

We start this section by considering the case of only one media. It was shown in Sect. 2.1 that the scattering problem is described mathematically by the Helmholtz equation

$$(\Delta + K^2)\Psi = Q \tag{2.2.1}$$

and boundary condition

$$\hat{B}\Psi|_{z=h(\boldsymbol{r})} = 0 \ . \tag{2.2.2}$$

Moreover, the solution for $z \to -\infty$ must consist of outgoing waves only. For simplicity, we shall start with scalar fields.

It is clear that it is sufficient to consider the problem for the case of a point source at an arbitrary point $\boldsymbol{R}_0$:

$$Q(\boldsymbol{R}) = \delta(\boldsymbol{R} - \boldsymbol{R}_0) \ .$$

In this case, the corresponding solution presents, by definition, the Green's function of boundary problem (2.2.1, 2)

$$\Psi = G(\boldsymbol{R}, \boldsymbol{R}_0) \ .$$

The problem for arbitrary distribution of sources is solved according to the superposition principle:

$$\Psi(\boldsymbol{R}) = \int G(\boldsymbol{R}, \boldsymbol{R}')Q(\boldsymbol{R}')\,d\boldsymbol{R}' \ . \tag{2.2.3}$$

Further, we shall consider the practically important case for which field sources are below the lowest point of the boundary, i.e., $Q(\boldsymbol{R}) \neq 0$ only for $z < \min h(\boldsymbol{r})$.

We shall designate by ψ_{in} (referring to incident) the solution of (2.2.1) corresponding to the case in which boundary Σ is absent and the whole space is filled with a homogeneous medium. This field is the initial radiation from the sources Q and incident onto the rough boundary. It is obvious that in the general case

$$\Psi_{\text{in}}(\boldsymbol{R}) = \int G_0(\boldsymbol{R} - \boldsymbol{R}')Q(\boldsymbol{R}')\,d\boldsymbol{R}' \ , \tag{2.2.4}$$

where G_0 is the Green's function of homogeneous space (2.1.21). Substituting into (2.2.4) the representation of G_0 in the form of superposition of plane waves (2.1.24), we have

$$\Psi_{\text{in}} = \int a(\boldsymbol{k})q^{-1/2}(k)\exp[\mathrm{i}\boldsymbol{k}\cdot\boldsymbol{r} + \mathrm{i}q(k)z]\,d\boldsymbol{k} \ , \tag{2.2.5}$$

where

$$a(\boldsymbol{k}) = -(\mathrm{i}/8\pi^2)\int q^{-1/2}(k)\exp[-\mathrm{i}\boldsymbol{k}\cdot\boldsymbol{r}' - \mathrm{i}q(k)z']Q(\boldsymbol{r}', z')\,d\boldsymbol{r}'\,dz' \ . \tag{2.2.6}$$

In these formulas the normalizations (2.1.29–30) are used for plane waves. Note that while substituting (2.1.24) above, we set $|z - z'| = z - z'$, since the field Ψ_{in} generated by sources Q and incident at the surface Σ should be calculated for $z' < z$.

Hence for the field sources located below the boundary the field incident onto the boundary can be represented by superposition of plane waves of the kind (2.2.5). Therefore, it is sufficient to consider the case of a single plane wave which is incident onto the boundary Σ and has an arbitrary horizontal component of wave vector $\boldsymbol{k}_0$:

$$\Psi_{\text{in}} = q^{-1/2}(k_0)\exp[\mathrm{i}\boldsymbol{k}_0 \cdot \boldsymbol{r} + \mathrm{i}q(k_0)z]. \tag{2.2.7}$$

As a result of interaction with the boundary this field generates a scattered field Ψ_{sc}, so that the total field will be the sum

$$\Psi = \Psi_{\text{in}} + \Psi_{\text{sc}} \ . \tag{2.2.8}$$

It is true that the scattered field for $z \to -\infty$ must also satisfy the radiation condition and consist only of outgoing waves. We again consider the area below the boundary: $z < \min h(\boldsymbol{r})$. The general solution of the Hemlholtz homogeneous equation (2.1.27) consisting of waves propagating or decaying in z-direction takes the following form:

$$\Psi_{\text{sc}}(\boldsymbol{R}) = \int b(\boldsymbol{k})q^{-1/2}(k)\exp[\mathrm{i}\boldsymbol{k} \cdot \boldsymbol{r} - \mathrm{i}q(k)z]\,d\boldsymbol{k} \ . \tag{2.2.9}$$

Here normalization (2.1.30) is again used for outgoing waves. Amplitude $b(\boldsymbol{k})$ in (2.2.9) is surely dependent on the incident field Ψ_{in}. Let us take for Ψ_{in} the plane wave of the form (2.2.7) and denote by $S(\boldsymbol{k}, \boldsymbol{k}_0)$ the corresponding $b(\boldsymbol{k})$. Then, according to (2.2.7–9), the solution of the primary boundary problem (2.2.1, 2) can be given in the form

$$\Psi = q_0^{-1/2}\exp(\mathrm{i}\boldsymbol{k}_0 \cdot \boldsymbol{r} + \mathrm{i}q_0 z) + \int S(\boldsymbol{k}, \boldsymbol{k}_0)q^{-1/2}\exp(\mathrm{i}\boldsymbol{k} \cdot \boldsymbol{r} - \mathrm{i}qz)\,d\boldsymbol{k} \ . \tag{2.2.10}$$

Here and below, while writing formulas we keep in mind that the argument not indicated in quantity q equals automatically to k: $q \overset{\text{def}}{=} q(k)$, similarly $q_0 \overset{\text{def}}{=} q(k_0)$. Quantity $S(\boldsymbol{k}, \boldsymbol{k}_0)$ introduced in (2.2.10) is referred to as *scattering amplitude* (SA) of wave $\boldsymbol{k}_0$ to wave $\boldsymbol{k}$. In this case it is implied that $\boldsymbol{k}_0$ and $\boldsymbol{k}$ are related to the certainly normalized plane waves of the form as given in (2.1.29, 30). Since we shall use exclusively the basis of plane waves further, this will not be mentioned.

Note that formula (2.2.10) is obtained under assumption $z < \min h(\boldsymbol{r})$. However, since only the field far from roughness is often of interest, this limitation is not of practical significance in most cases. Moreover, for smooth undulations with small enough slopes, formula (2.2.10) gives values of field at each point of area Ω including the boundary $z = h(\boldsymbol{r})$. Extracting information from (2.2.11) on the field immediately near the boundary for $\min h(\boldsymbol{r}) < z < h(\boldsymbol{r})$ is closely connected to the so-called Rayleigh hypothesis and is discussed in Sect. 3.4.

It should be emphasized that the horizontal components of wave vectors in incident and scattered waves $\boldsymbol{k}_0$ and $\boldsymbol{k}$, respectively, take all real values. At $|\boldsymbol{k}_0| < K$ a homogeneous wave is incident at the surface, at $|\boldsymbol{k}_0| > K$ an inhomogeneous one in which amplitude is exponentially attenuated to the boundary. Analogously at $|\boldsymbol{k}| < K$ the scattered wave is an ordinary plane wave and at $|\boldsymbol{k}| > K$ an evanescent one exponentially decaying for $z \to -\infty$. The scattering amplitude $S(\boldsymbol{k}, \boldsymbol{k}_0)$ describes all types of scattering processes dependent on $\boldsymbol{k}$ and $\boldsymbol{k}_0$: the interaction between homogeneous and between inhomogeneous waves, and between inhomogeneous and homogeneous ones, and vice versa.

For the known scattering amplitude the expression for the Green's function $G(\boldsymbol{R}, \boldsymbol{R}_0)$ can be easily obtained. In fact, setting in (2.2.6)

$$Q(\boldsymbol{r}', z') = \delta(\boldsymbol{r}' - \boldsymbol{r}_0)\delta(z' - z_0) \ ,$$

we show that the point source field is a superposition of the plane waves of the form (2.2.5):

$$\Psi_{\text{in}} = \int a(\boldsymbol{k}_0) \cdot q_0^{-1/2} \exp(\mathrm{i}\boldsymbol{k}_0 \cdot \boldsymbol{r} + \mathrm{i}q_0 z)\, d\boldsymbol{k}_0 \ ,$$

where

$$a(\boldsymbol{k}_0) = -\frac{\mathrm{i}}{8\pi^2 q_0^{1/2}} \exp(-\mathrm{i}\boldsymbol{k}_0 \cdot \boldsymbol{r}_0 - \mathrm{i}q_0 z_0) \ . \tag{2.2.11}$$

Therefore, using the superposition principle and proceeding from (2.2.10) we immediately obtain

$$G(\boldsymbol{R}, \boldsymbol{R}_0) = G_0(\boldsymbol{R} - \boldsymbol{R}_0) - \frac{\mathrm{i}}{8\pi^2} \int q^{-1/2} \exp(\mathrm{i}\boldsymbol{k} \cdot \boldsymbol{r} - \mathrm{i}qz)$$
$$\times\ S(\boldsymbol{k}, \boldsymbol{k}_0) q_0^{-1/2} \exp(-\mathrm{i}\boldsymbol{k}_0 \cdot \boldsymbol{r}_0 - \mathrm{i}q_0 z_0)\, d\boldsymbol{k}\, d\boldsymbol{k}_0 \ . \tag{2.2.12}$$

Function G determined by formula (2.2.12) for any scattering amplitude S is a solution of equation

$$(\Delta + K^2)G = \delta(\boldsymbol{R} - \boldsymbol{R}_0) \tag{2.2.13}$$

and satisfies the radiation condition for $z \to -\infty$. Similarly, the representation of the scattered field Ψ_{sc} in the form

$$\Psi_{\text{sc}} = \int S(\boldsymbol{k}, \boldsymbol{k}_0) q^{-1/2} \exp(\mathrm{i}\boldsymbol{k} \cdot \boldsymbol{r} - \mathrm{i}qz)\, d\boldsymbol{k} \tag{2.2.14}$$

automatically satisfies the radiation condition and the homogeneous Helmholtz equation. As we were convinced above, the description of scattering in terms of SA and the Green's function are principally equivalent while neglecting the problem of calculating the field in the mere vicinity of the boundary. However, the description in terms of SA gives practical advantages. First of all in the general case SA depends on four arguments, while the Green's function on six (and in a two-dimensional case on two and four arguments, respectively). Further, in those cases where we need to consider the spatially homogeneous statistical ensemble of surface roughness $h(\boldsymbol{r})$, we obtain a description of scattering in the basis related to the translational symmetry of the statistical ensemble that leads to some simplifications. In particular, it appears that the scattering parameters measured experimentally are easily expressed through statistical moments of SA, see Sect. 2.4.

Now suppose that the wave field is multicomponental: $\Psi(R) = \Psi_i(R)$, $i = 1, \dots, N$. Components of electric and magnetic fields and/or displacements of particles may appear as Ψ_i. In this case several kinds of plane waves can propagate in the medium. We shall denote them by the integer index $\sigma = 1, 2, \dots, m$. Generally speaking, each of these waves possesses its own propagation velocity $c = c_\sigma$. Therefore, elementary solutions in the form of plane waves satisfy equations

$$(\Delta + K_\sigma^2)\Psi_i^\sigma = 0 \ , \tag{2.2.15}$$

where $K_\sigma = \omega/c_\sigma$. These elementary solutions take the following form

$$\Psi_i^\sigma(\boldsymbol{R}) = {}^{\uparrow\downarrow}A_i^\sigma(\boldsymbol{k}) q_\sigma^{-1/2}(k) \exp[\mathrm{i}\boldsymbol{k}\cdot\boldsymbol{r} \pm q_\sigma(k) z] \ , \tag{2.2.16}$$

where $q_\sigma(k) = (K^2 - k^2)^{1/2}$. The arrows at the factor A denote the direction of wave propagation with respect to the z-axes: $\uparrow$ and $\downarrow$ correspond to the incident and the scattered fields. We shall refer the wave corresponding to given σ as a wave of σth polarization. Coefficients ${}^{\downarrow}A_i^\sigma$ (or ${}^{\uparrow}A_i^\sigma$) for fixed σ and various i are related to each other and are determined to an accuracy of an arbitrary constant factor. It is convenient to choose this factor in a way similar to (2.1.29, 30), so that in the plane wave of the form (2.2.16) the energy flux in the Z-direction is equal to unity.

In infinite medium, plane waves of the form (2.2.16) do not interact and can propagate independently. However, while arriving at the rough surface a certain incident wave with $\sigma = \sigma_0$ and $\boldsymbol{k} = \boldsymbol{k}_0$ will produce the scattered waves with all possible σ and $\boldsymbol{k}$. As a result, the scattered field will be

$$\Psi_i^{(\mathrm{sc})} = \sum_{\sigma=1}^{m} \int d\boldsymbol{k}\, S_{\sigma\sigma_0}(\boldsymbol{k}, \boldsymbol{k}_0) {}^{\downarrow}A_i^\sigma(\boldsymbol{k}) q_\sigma^{-1/2} \exp(\mathrm{i}\boldsymbol{k}\cdot\boldsymbol{r} - \mathrm{i}q_\sigma z) \ . \tag{2.2.17}$$

Then in the case of multicomponental fields the scattering amplitude depends on discrete indices σ and σ_0, along with $\boldsymbol{k}$ and $\boldsymbol{k}_0$. SA $S_{\sigma\sigma_0}(\boldsymbol{k}, \boldsymbol{k}_0)$ describes the amplitude of scattering process of the wave $(\sigma_0, \boldsymbol{k}_0)$ into the wave $(\sigma, \boldsymbol{k})$.

We can naturally arrange the following square matrix out of components of SA $S_{\sigma\sigma_0}$:

$$\hat{S}(\boldsymbol{k}, \boldsymbol{k}_0) = \begin{bmatrix} S_{11}(\boldsymbol{k}, \boldsymbol{k}_0) & \dots & S_{1m}(\boldsymbol{k}, \boldsymbol{k}_0) \\ \dots & \dots & \dots \\ S_{m1}(\boldsymbol{k}, \boldsymbol{k}_0) & \dots & S_{mm}(\boldsymbol{k}, \boldsymbol{k}_0) \end{bmatrix} \ . \tag{2.2.18}$$

The Green's function for the multicomponental problem can be constructed quite similarly to the case of a scalar field. The point source located below all boundary irregularities at $\boldsymbol{R}_0 = (\boldsymbol{r}_0, z_0)$ generates the field incident at the surface Σ and consists of wave superposition of the form (2.2.16) with various σ_0 and $\boldsymbol{k}_0$:

$$\Psi_i^{(\text{in})}(\boldsymbol{R}) = \sum_{\sigma_0=1}^{m} \int d\boldsymbol{k}_0 {}^{\uparrow}A_i^{\sigma_0}(\boldsymbol{k}_0) q_{\sigma_0}^{-1/2}(k_0) \exp(\mathrm{i}\boldsymbol{k}_0 \cdot \boldsymbol{r} + \mathrm{i}q\sigma_0 z) a_{\sigma_0}(\boldsymbol{k}_0) \ . \quad (2.2.19)$$

In the general case the amplitudes of these waves can be presented analogously to (2.2.11) in the form

$$a_{\sigma_0}(\boldsymbol{k}_0) = -(\mathrm{i}/8\pi^2)\alpha^{\sigma_0}(\boldsymbol{k}_0) q_{\sigma_0}^{-1/2}(k_0) \exp[-\mathrm{i}\boldsymbol{k}_0 \cdot \boldsymbol{r}_0 - \mathrm{i}q_{\sigma_0}(k_0) z_0] \ ,$$

where coefficients $\alpha^{\sigma_0}(\boldsymbol{k}_0)$ depend on the type of point sources. Each elementary plane wave participating in superposition (2.2.19) is scattered, according to (2.2.17). Summing up the fields of all scattered waves at the point $\boldsymbol{R} = (\boldsymbol{r}, z)$, we obtain

$$G_i^{(\text{sc})}(\boldsymbol{R}, \boldsymbol{R}_0) = -\frac{\mathrm{i}}{8\pi^2} \sum_{\sigma, \sigma_0} \int d\boldsymbol{k}\, d\boldsymbol{k}_0 {}^{\downarrow}A_i^{\sigma}(\boldsymbol{k}) q_{\sigma}^{-1/2}(k) \exp[\mathrm{i}\boldsymbol{k} \cdot \boldsymbol{r} - \mathrm{i}q_{\sigma}(k) z]$$
$$\times S_{\sigma\sigma_0}(\boldsymbol{k}, \boldsymbol{k}_0) \alpha^{\sigma_0}(\boldsymbol{k}_0) q_{\sigma_0}^{-1/2}(k_0) \exp[-\mathrm{i}\boldsymbol{k}_0 \cdot \boldsymbol{r}_0 - \mathrm{i}q_{\sigma_0}(k_0) z_0] \ . \quad (2.2.20)$$

In this formula we wrote the Green's function component related only to the scattered field; the total Green's function includes the direct field which is also produced by the reference point source in the boundless homogeneous space.

Now consider the case in which the interface Σ is between two homogeneous half-spaces and we have the notation for them: Ω_1 and Ω_2, see Fig. 2.2. For simplicity we suppose that the wave fields in the media Ω_1 and Ω_2 are scalar. Then an elementary plane wave incident onto the boundary from below (i.e., from the half-space Ω_1) takes the following form:

$$\Psi_{\text{in}}^{(1)} = \exp(\mathrm{i}\boldsymbol{k}_0 \cdot \boldsymbol{r} + \mathrm{i}q_0^{(1)} z) \cdot \mathscr{D}_0^{(1)} / \sqrt{q_0^{(1)}} \ , \quad (2.2.21)$$

where $q_0^{(1)} \stackrel{\text{def}}{=} q^{(1)}(k_0) = (\omega^2/c_1^2 - k_0^2)^{1/2}$. The root branch is chosen analogously to (2.1.25).

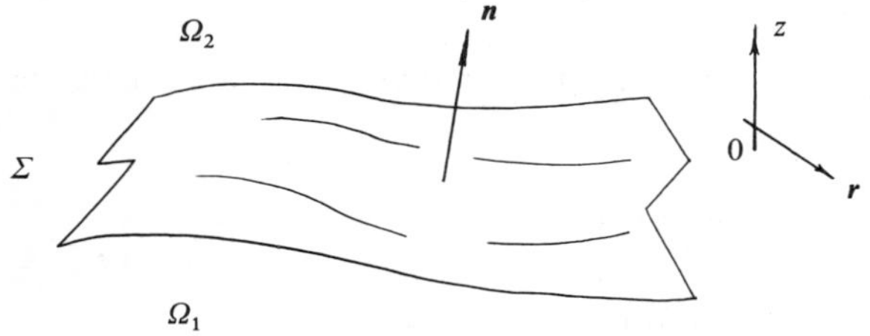

Fig. 2.2. The case of interface between two homogeneous half-spaces

Equation (2.2.21) is different from (2.2.7) by the factor $\mathscr{D}^{(1)}(k_0)$ which is introduced in formula (2.2.21). It fixes the normalization of plane wave amplitudes and should be chosen as convenient. We shall choose this factor so that the z-component of the energy flux in the wave is equal to the same constant (unit, for example) in the both media. In particular, for two liquid half-spaces considered in Sect. 3.5, we set $\mathscr{D}^{(1)} = \varrho_1^{-1/2}$ and $\mathscr{D}^{(2)} = \varrho_2^{-1/2}$, see (2.2.23), according to (2.1.28), where ϱ_1 and ϱ_2 are the densities of these media.

Elementary plane waves propagating off the boundary (i.e., satisfying the radiation condition) are written as follows:

$$\Psi_{\mathrm{sc}}^{(1)} = \exp(\mathrm{i}\boldsymbol{k}\cdot\boldsymbol{r} - \mathrm{i}q^{(1)}z)\mathscr{D}^{(1)}/\sqrt{q^{(1)}} \ , \qquad z < h(\boldsymbol{r}) \tag{2.2.22}$$

for the first medium, $(\boldsymbol{r}, z) \in \Omega_1$, and

$$\Psi_{\mathrm{sc}}^{(2)} = \exp(\mathrm{i}\boldsymbol{k}\cdot\boldsymbol{r} + \mathrm{i}q^{(2)}z)\mathscr{D}^{(2)}/\sqrt{q^{(2)}} \ , \qquad z > h(\boldsymbol{r}) \tag{2.2.23}$$

for the second medium, $(\boldsymbol{r}, z) \in \Omega_2$. In formula (2.2.23)

$$q^{(2)} \overset{\mathrm{def}}{=} q^{(2)}(k) = (\omega^2/c_2^2 - k^2)$$

and the sign of the square root is determined as it was previously. As a result, the scattered field outside a layer $\min h(\boldsymbol{r}) < z < \max h(\boldsymbol{r})$, including the whole boundary Σ, can be represented in the following form:

$$\Psi_{\mathrm{sc}} = \begin{cases} \int \exp(\mathrm{i}\boldsymbol{k}\cdot\boldsymbol{r} + \mathrm{i}q^{(2)}z)\dfrac{\mathscr{D}^{(2)}}{\sqrt{q^{(2)}}}\cdot S^{21}(\boldsymbol{k},\boldsymbol{k}_0)\,d\boldsymbol{k} \ , & z > \max h(\boldsymbol{r}) \quad (2.2.24\mathrm{a}) \\ \int \exp(\mathrm{i}\boldsymbol{k}\cdot\boldsymbol{r} - \mathrm{i}q^{(1)}z)\dfrac{\mathscr{D}^{(1)}}{\sqrt{q^{(1)}}}\cdot S^{11}(\boldsymbol{k},\boldsymbol{k}_0)\,d\boldsymbol{k} \ , & z < \min h(\boldsymbol{r}) \ . \quad (2.2.24\mathrm{b}) \end{cases}$$

A pair of indices (2, 1) and (1, 1) which appear in this formula indicate that the wave incident from the first medium is scattered into the second and the first media: $(2,1) = (2 \leftarrow 1)$, $(1,1) = (1 \leftarrow 1)$. Similarly, the plane wave of the form

$$\Psi_{\mathrm{in}}^{(2)} = \exp(\mathrm{i}\boldsymbol{k}_0\cdot\boldsymbol{r} - \mathrm{i}q_0^{(2)}z)\cdot\mathscr{D}_0^{(2)}/\sqrt{q_0^{(2)}} \tag{2.2.25}$$

incident from the upper half-space Ω_2 at the boundary give rise to the following scattered field:

$$\Psi_{\mathrm{sc}} = \begin{cases} \int \exp(\mathrm{i}\boldsymbol{k}\cdot\boldsymbol{r} + \mathrm{i}q^{(2)}z)\dfrac{\mathscr{D}^{(2)}}{\sqrt{q^{(2)}}}\cdot S^{22}(\boldsymbol{k},\boldsymbol{k}_0)\,d\boldsymbol{k} \ , & z > \max h(\boldsymbol{r}) \quad (2.2.26\mathrm{a}) \\ \int \exp(\mathrm{i}\boldsymbol{k}\cdot\boldsymbol{r} - \mathrm{i}q^{(1)}z)\dfrac{\mathscr{D}^{(1)}}{\sqrt{q^{(1)}}}\cdot S^{12}(\boldsymbol{k},\boldsymbol{k}_0)\,d\boldsymbol{k} \ , & z < \min h(\boldsymbol{r}) \ , \quad (2.2.26\mathrm{b}) \end{cases}$$

where the superscripts (2, 2) and (1, 2) referred to the medium where the wave incident from the second medium is scattered, $(2,2) = (2 \leftarrow 2)$ and $(1,2) = (1 \leftarrow 2)$.

Finally, when multicomponental fields occur in the half-spaces Ω_1 and Ω_2, the scattering amplitude acquires indices σ and σ_0 and takes the form $S^{NN_0}_{\sigma\sigma_0}(\boldsymbol{k}, \boldsymbol{k}_0)$ $(N, N_0 = 1, 2)$. This quantity characterizes the scattering amplitude for the wave of polarization σ_0, and horizontal projection of the wave vector equal to $\boldsymbol{k}_0$ and propagating in the N_0th medium $(N_0 = 1, 2)$ into the wave with corresponding parameters $(\sigma, \boldsymbol{k})$ which propagates in the Nth medium $(N = 1, 2)$. The reference quantity is introduced according to formulas (2.2.17, 24, 26).

For fixed indices N and N_0 the set of quantities $S^{NN_0}_{\sigma\sigma_0}$ with various σ and σ_0 forms a certain matrix which is not, generally speaking, squared for $N \neq N_0$ (since the number of waves of different polarization in the first and second media is different). At last, a complex of four such matrices $\hat{S}^{NN_0}$ forms one matrix of block structure

$$\hat{S} = \begin{pmatrix} \hat{S}^{11} & \hat{S}^{12} \\ \hat{S}^{21} & \hat{S}^{22} \end{pmatrix} . \tag{2.2.27}$$

The dimension of such a matrix is equal to the total number of waves in the first and second media.

Hence the aggregate of scattering amplitudes $S^{NN_0}_{\sigma\sigma_0}$ (or, that is the same, matrix $\hat{S}$) completely describes wave scattering at the rough boundary.

Now we consider a more general situation. That is, when the scattering matrix $\hat{S}$ is known for the boundary between two media, it is easily transfered to the case of wave scattering at some multilayered structure, see Fig. 2.3.

Suppose that the problem of scattering from arbitrary medium in region $\tilde{\Omega}$ with boundary $\tilde{\Sigma}$ is solved. This implies that SA $\tilde{S}_{\sigma\sigma_0}(\boldsymbol{k}, \boldsymbol{k}_0)$ is known which connects the amplitudes of incident and scattered waves at the boundary $\tilde{\Sigma}$:

$$a^{(2)}_\sigma(\boldsymbol{k}) = \sum_{\sigma'} \int \tilde{S}_{\sigma\sigma'}(\boldsymbol{k}, \boldsymbol{k}') b^{(2)}_{\sigma'}(\boldsymbol{k}')\, d\boldsymbol{k}' . \tag{2.2.28}$$

We assume that superposition of plane waves with amplitudes $a^{(1)}_{\sigma_0}(\boldsymbol{k}_0)$ is incident from the medium 1 onto the boundary Σ. This superposition finally generates all the fields. Due to scattering at the boundary $\tilde{\Sigma}$, the waves with amplitudes $a^{(2)}_\sigma(\boldsymbol{k})$ are also incident at the boundary Σ from above. According to the definition of SA $S^{NN_0}_{\sigma\sigma_0}$, we can write for scattering at the interface between the

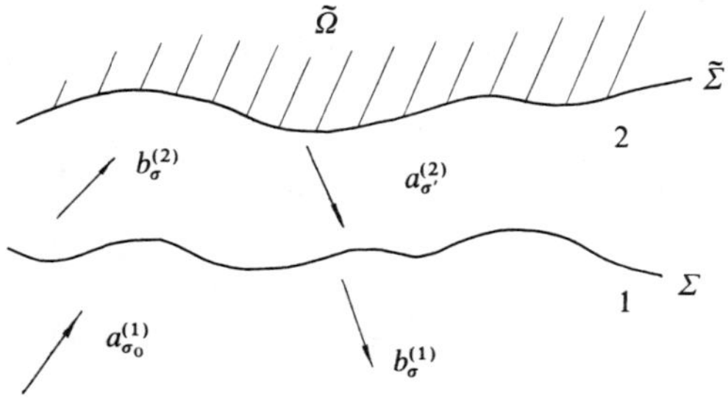

Fig. 2.3. Recounting of SA for the boundary $\tilde{\Sigma}$ to that for the boundary Σ

first and the second media the following relations between the amplitudes of different waves:

$$b_\sigma^{(1)}(\boldsymbol{k}) = \sum_{\sigma_0} \int S_{\sigma\sigma_0}^{11}(\boldsymbol{k}, \boldsymbol{k}_0) a_{\sigma_0}^{(1)}(\boldsymbol{k}_0)\, d\boldsymbol{k}_0 + \sum_{\sigma'} \int S_{\sigma\sigma'}^{12}(\boldsymbol{k}, \boldsymbol{k}') a_{\sigma'}^{(2)}(\boldsymbol{k}')\, d\boldsymbol{k}' \ , \tag{2.2.29}$$

$$b_\sigma^{(2)}(\boldsymbol{k}) = \sum_{\sigma_0} \int S_{\sigma\sigma_0}^{21}(\boldsymbol{k}, \boldsymbol{k}_0) a_{\sigma_0}^{(1)}(\boldsymbol{k}_0)\, d\boldsymbol{k}_0 + \sum_{\sigma'} \int S_{\sigma\sigma'}^{22}(\boldsymbol{k}, \boldsymbol{k}') a_{\sigma'}^{(2)}(\boldsymbol{k}')\, d\boldsymbol{k}' \ . \tag{2.2.30}$$

Our task is to find the coefficients of the linear relation between the amplitudes $b_\sigma^{(1)}(\boldsymbol{k})$ and $a_{\sigma_0}^{(1)}(\boldsymbol{k}_0)$ which compose a set of some new scattering amplitudes $\tilde{S}_{\sigma\sigma_0}^{(\mathrm{New})}(\boldsymbol{k}, \boldsymbol{k}_0)$, quite similar to the quantities already used in (2.2.28). Scattering amplitudes $\tilde{S}^{(\mathrm{New})}$ will be referred to the inhomogeneous medium including an additional layer, $\tilde{\Omega}^{(\mathrm{New})} = \tilde{\Omega}$ +2nd medium (Fig. 2.3).

It is quite natural to use matrix notation at this stage. Let us introduce multiindex $\mu = (\sigma, \boldsymbol{k})$ instead of arguments σ and $\boldsymbol{k}_0$, characterizing plane waves. In this case SA $S_{\sigma\sigma_0}(\boldsymbol{k}, \boldsymbol{k}_0)$ is written as

$$S_{\sigma\sigma_0}(\boldsymbol{k}, \boldsymbol{k}_0) = S_{\mu\mu_0} \ , \qquad \mu = (\sigma, \boldsymbol{k}) \ , \qquad \mu_0 = (\sigma_0, \boldsymbol{k}_0) \ . \tag{2.2.31}$$

Assume also

$$\sum_\mu \stackrel{\mathrm{def}}{=} \sum_\sigma \int d\boldsymbol{k} \ . \tag{2.2.32}$$

Quantity $S_{\mu\mu_0}$ may be considered as a matrix (more precisely, as the kernel of the linear integral operator), which affects the vector a_{μ_0} according to the standard rule

$$b_\mu = \sum_{\mu_0} S_{\mu\mu_0} a_{\mu_0} \ .$$

Here we determine the product of two matrices as usual:

$$C = A \cdot B \quad \text{means} \quad C_{\mu\mu_0} = \sum_{\mu'} A_{\mu\mu'} \cdot B_{\mu'\mu_0} \ .$$

The last formula in usual notation is

$$C_{\sigma\sigma_0}(\boldsymbol{k}, \boldsymbol{k}_0) = \sum_{\sigma'} \int dk' \cdot A_{\sigma\sigma'}(\boldsymbol{k}, \boldsymbol{k}') \cdot B_{\sigma'\sigma_0}(\boldsymbol{k}', \boldsymbol{k}_0) \ .$$

One may extend the value of multi-index μ and include in it index N referred to the number of medium ($N = 1, 2$) as well:

$$v = (N, \mu) = (N, \sigma, \boldsymbol{k}) \tag{2.2.33}$$

with

$$\sum_v \stackrel{\mathrm{def}}{=} \sum_{N=1,2} \sum_\sigma \int d\boldsymbol{k} \ .$$

In this notation

$$S_{\sigma\sigma_0}^{NN_0}(\boldsymbol{k}, \boldsymbol{k}_0) = S_{\nu\nu_0} .$$

The aggregate of quantities $S_{\nu\nu_0}$ considered from this point of view is named the *scattering matrix* or *S-matrix*. In the cases where we can limit ourselves to considering the propagation only in one medium (for example, in the first: $N = N_0 = 1$) we refer the S-matrix also to the complex of quantities $S_{\mu\mu_0}^{11}$.

In matrix notation, (2.2.28–30) are written in the form

$$a^{(2)} = \hat{\hat{S}} b^{(2)} , \tag{2.2.34a}$$

$$b^{(1)} = \hat{S}^{11} a^{(1)} + \hat{S}^{12} a^{(2)} , \tag{2.2.34b}$$

$$b^{(2)} = \hat{S}^{21} a^{(1)} + \hat{S}^{22} a^{(2)} . \tag{2.2.34c}$$

Substituting (2.2.34c) into (2.2.34a) we find

$$a^{(2)} = \hat{\hat{S}} \cdot \hat{S}^{21} a^{(1)} + \hat{\hat{S}} \cdot \hat{S}^{22} \cdot a^{(2)}$$

whence

$$a^{(2)} = (1 - \hat{\hat{S}} \cdot \hat{S}^{22})^{-1} \cdot \hat{\hat{S}} \hat{S}^{21} \cdot a^{(1)} .$$

Now (2.2.34b) yields

$$b^{(1)} = \hat{S}^{11} \cdot a^{(1)} + \hat{S}^{12} \cdot (1 - \hat{\hat{S}} \cdot \hat{S}^{22})^{-1} \cdot \hat{\hat{S}} \cdot \hat{S}^{21} \cdot a^{(1)}$$

or

$$\hat{\hat{S}}^{(\mathrm{New})} = \hat{S}^{11} + \hat{S}^{12} \cdot (1 - \hat{\hat{S}} \cdot \hat{S}^{22})^{-1} \cdot \hat{\hat{S}} \cdot \hat{S}^{21} .$$

Thus, we can calculate wave scattering from a certain multilayered structure shown in Fig. 2.4, according to the following recurrence scheme:

$$\hat{\hat{S}}_{(N)} = \hat{S}^{N,N} , \tag{2.2.35a}$$

$$\hat{\hat{S}}_{(n-1)} = \hat{S}^{n-1,n-1} + \hat{S}^{n-1,n} \cdot (1 - \hat{\hat{S}}_{(n)} \cdot \hat{S}^{n,n})^{-1} \cdot \hat{\hat{S}}_{(n)} \cdot \hat{S}^{n,n-1} , \tag{2.2.35b}$$

for $n = N, N-1, \ldots, 2$. Formulas (2.2.35a) and (2.2.35b) correspond to scattering at the boundaries between the N and $N+1$ media, and between $(n-1)$ and n media ($\sum_{n-1,n}$ in Fig. 2.4).

Under the proper conditions continuous medium can be approximated by a large number of thin homogeneous layers. Therefore, the skill to calculate scattering at the interface between two homogeneous media practically enables us to calculate the wave propagation in inhomogeneous media as well. Hence, the problem of wave scattering at a rough interface between two homogeneous media formulated in Sect. 2.1 is of more interest than would have been primarily thought.

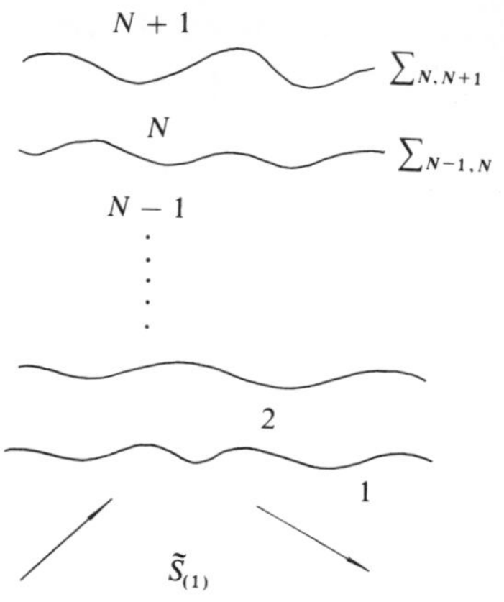

Fig. 2.4. On calculating of scattering at multilayered structure

Apropos of this the following notes should be taken into consideration.

1) It appears (Sect. 4.1) that if surface roughness is periodical, then waves are scattered only in several discrete directions. In this case the continuous index in (2.2.31) become discrete and integration in (2.2.32) is replaced by summation, $S_{\mu\mu_0}$ being the actual matrix, and the same is referred to quantities $\hat{\hat{S}}$ and $\hat{S}$ in (2.2.35). These matrices are, generally speaking, infinite. However, if the scattering processes including the waves with very large $\boldsymbol{k}$ can be neglected, then all matrices from (2.2.35) become finite.

2) When realizing algorithm (2.2.35), it may appear that at some step

$$\det(1 - \hat{\hat{S}}_{(n)} \cdot \hat{S}^{n,n}) = 0 \ . \tag{2.2.36}$$

The latter implies that some normal mode arose [2.3], i.e., such wave motion that is bounded in the vertical direction and is not accompanied by energy input or output. In this case calculation by formula (2.2.35b) is impossible and the description of scattering should be changed. The case of waveguide propagation will be examined in Chap. 7, and yet we assume that condition (2.2.36) is not fulfilled.

3) If media parameters in the $(n-1)$th and nth layers are close to one another, then the scattering matrix $\hat{S}^{n,n}$ in (2.2.35b) will be small, and in calculating the inverse matrix the following approximation will be sufficient:

$$(1 - \hat{\hat{S}}_{(n)} \cdot \hat{S}^{n,n})^{-1} \approx 1 + \hat{\hat{S}}_{(n)} \cdot \hat{S}^{n,n} \ .$$

2.3 Reciprocity Theorem and Unitarity Condition

In the present section we consider several relations which SA should satisfy in connection with the energy conservation and the reciprocity theorem.

Let us examine the case of a scalar field which obeys the Helmholtz equation (2.1.27). Take two different solutions Ψ_1 and Ψ_2 of this equation

$$(\Delta + K^2)\Psi_1 = 0 \ ,$$

$$(\Delta + K^2)\Psi_2 = 0 \ .$$

We multiply the first and the second equations by Ψ_2 and Ψ_1, respectively, and estimate the difference between the results. Then we obtain

$$\Psi_2 \cdot \Delta\Psi_1 - \Psi_1 \Delta\Psi_2 = 0$$

or, that which is the same,

$$\operatorname{div}(\Psi_2 \cdot \nabla\Psi_1 - \Psi_1 \nabla\Psi_2) = 0 \ . \qquad (2.3.1)$$

Let us integrate (2.3.1) over volume Ω_0 included between the boundary Σ and the horizontal plane at the level $z = z_0$ below all boundary irregularities (Fig. 2.5). Using the Gauss formula we transform the reference volume integral into a surface one

$$\int \operatorname{div} \boldsymbol{A}\, d\Omega_0 = \int_{\partial\Omega_0} \boldsymbol{A}\, d\boldsymbol{\Sigma}_0 = \int_{\partial\Omega_0} \boldsymbol{A}\cdot\boldsymbol{n}\, d\Sigma_0 \ . \qquad (2.3.2)$$

From this it follows that

$$\int_{z=z_0} \left(\Psi_2 \frac{\partial\Psi_1}{\partial z} - \psi_1 \frac{\partial\Psi_2}{\partial z}\right) d\boldsymbol{r} = \int \left(\Psi_2 \frac{\partial\Psi_1}{\partial \boldsymbol{n}} - \psi_1 \frac{\partial\Psi_2}{\partial \boldsymbol{n}}\right) d\Sigma \ . \qquad (2.3.3)$$

Relation (2.3.3) results directly from the Helmholtz equation.

While transferring from (2.3.2) to (2.3.3) we omitted the integral over the infinitely far side surface supposing that both incident and scattered fields decay in respective directions. Physically this assumption implies that the surface waves cannot propagate along the infinitely far part of boundary Σ (which may be assumed for convenience to be a plane).

Now suppose that the boundary condition at Σ takes the form of (2.1.12) and the kernel L_1 is symmetrical. It is obvious in this case that the right-hand side of equality (2.3.3) becomes zero. The same follows when considering the boundary condition (2.1.13) and assuming the L_2-kernel symmetry. Thus under the reference assumption the relation

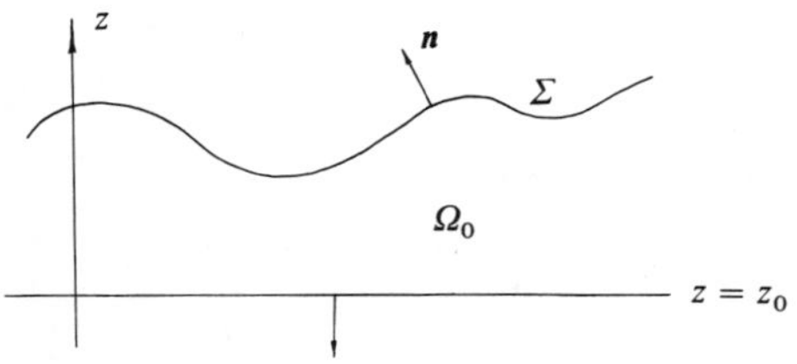

Fig. 2.5. On derivation of relation (2.3.3)

$$\int_{z=z_0} \left(\Psi_2 \frac{\partial \Psi_1}{\partial z} - \Psi_1 \frac{\partial \Psi_2}{\partial z} \right) d\boldsymbol{r} = 0 \tag{2.3.4}$$

should be fulfilled for two arbitrary solutions of the homogeneous Helmholtz equation. Note that (2.3.4) is fulfilled in particular for the impedance boundary condition (2.1.16).

Now for Ψ_1 and Ψ_2 we take the solutions of the form (2.2.10) which describe the scattering of two waves with horizontal wave vectors $\boldsymbol{k}_1$ and $\boldsymbol{k}_2$:

$$\Psi_1 = q_1^{-1/2} \cdot \exp(\mathrm{i}\boldsymbol{k}_1 \cdot \boldsymbol{r} + \mathrm{i}q_1 z) + \int q^{-1/2} \exp(\mathrm{i}\boldsymbol{k} \cdot \boldsymbol{r} - \mathrm{i}qz) S(\boldsymbol{k}, \boldsymbol{k}_1)\, d\boldsymbol{k} \ , \tag{2.3.5a}$$

$$\Psi_2 = q_2^{-1/2} \exp(\mathrm{i}\boldsymbol{k}_2 \cdot \boldsymbol{r} + \mathrm{i}q_2 z) + \int q^{-1/2} \exp(\mathrm{i}\boldsymbol{k} \cdot \boldsymbol{r} - \mathrm{i}qz) S(\boldsymbol{k}, \boldsymbol{k}_2)\, d\boldsymbol{k} \ . \tag{2.3.5b}$$

Substituting (2.3.5) into (2.3.4) and integrating over $\boldsymbol{r}$ we easily find by formula (2.1.20) that

$$S(-\boldsymbol{k}_1, \boldsymbol{k}_2) = S(-\boldsymbol{k}_2, \boldsymbol{k}_1) \ .$$

Since $\boldsymbol{k}_1$ and $\boldsymbol{k}_2$ are arbitrary, the last relation can also be written in the form

$$S(\boldsymbol{k}_1, \boldsymbol{k}_2) = S(-\boldsymbol{k}_2, -\boldsymbol{k}_1) \ . \tag{2.3.6}$$

The latter is a formulation of the reciprocity theorem in terms of SA. In fact, we replace variables $\boldsymbol{k} \to -\boldsymbol{k}_0$ and $\boldsymbol{k}_0 \to -\boldsymbol{k}$ in formula (2.2.12) for the Green's function and find

$$G(\boldsymbol{R}, \boldsymbol{R}_0) = G_0(\boldsymbol{R} - \boldsymbol{R}_0) - \frac{\mathrm{i}}{8\pi^2} \int q_0^{-1/2} \exp(-\mathrm{i}\boldsymbol{k}_0 \cdot \boldsymbol{r} - \mathrm{i}q_0 z) \cdot S(-\boldsymbol{k}_0, -\boldsymbol{k})$$
$$\times\, q^{-1/2} \exp(\mathrm{i}\boldsymbol{k} \cdot \boldsymbol{r}_0 - \mathrm{i}qz_0)\, d\boldsymbol{k}\, d\boldsymbol{k}_0 \ .$$

Comparing this formula with (2.2.12) and taking into account (2.3.6), we see that the Green's function is symmetric

$$G(\boldsymbol{R}, \boldsymbol{R}_0) = G(\boldsymbol{R}_0, \boldsymbol{R}) \ . \tag{2.3.7}$$

Thus, (2.3.7) and (2.3.6) are equivalent.

Now we pass to the law of conservation of energy. Suppose that there are no losses in the medium: $\mathrm{Im}\{K\} = 0$. Applying complex conjugation, we see that Ψ_2^* (asterisk denotes complex conjugation) can be also used as a solution for the Helmholtz equation. Then, relation (2.3.4) takes the form

$$\int_{z=z_0} \left(\Psi_2^* \frac{\partial \Psi_1}{\partial z} - \psi_1 \frac{\partial \Psi_2^*}{\partial z} \right) d\boldsymbol{r} = \int_{\Sigma} \left(\Psi_2^* \frac{\partial \Psi_1}{\partial \boldsymbol{n}} - \Psi_1 \frac{\partial \Psi_2^*}{\partial \boldsymbol{n}} \right) d\Sigma \ . \tag{2.3.8}$$

Assume that the kernel in the boundary condition (2.1.12) is Hermitian: $L_1(\boldsymbol{R}, \boldsymbol{R}') = L_1^*(\boldsymbol{R}', \boldsymbol{R})$. It is obvious in this case that the right-hand side in (2.3.8) becomes zero. We may examine the case of Hermitian kernel L_2 in a similar manner. For the impedance boundary condition Hermitianity of the kernel implies that impedance and admittance are purely imaginary. Thus, under the above assumptions the following relation is fulfilled.

$$\int_{z=z_0} \left(\Psi_2^* \frac{\partial \Psi_1}{\partial z} - \Psi_1 \frac{\partial \Psi_2^*}{\partial z} \right) d\boldsymbol{r} = 0 \ . \tag{2.3.9}$$

We substitute again for Ψ_1 and Ψ_2 in (2.3.9) solutions of the form (2.3.5). Direct calculation by formula (2.1.20) yields

$$\begin{aligned} -\frac{\mathrm{i}}{2} \int \left(\Psi_2^* \frac{\partial \Psi_1}{\partial z} - \Psi_1 \frac{\partial \Psi_2^*}{\partial z} \right) d\boldsymbol{r} &= \frac{q_1 + q_1^*}{2|q_1|} \mathrm{e}^{\mathrm{i}(q_1 - q_1^*)z_0} \\ &\quad - \frac{q_2 - q_2^*}{2|q_2|} S(\boldsymbol{k}_2, \boldsymbol{k}_1) \mathrm{e}^{-\mathrm{i}(q_2 + q_2^*)z_0} \\ &\quad + \frac{q_1 - q_1^*}{2|q_1|} S^*(\boldsymbol{k}_1, \boldsymbol{k}_2) \mathrm{e}^{\mathrm{i}(q_1 + q_1^*)h} \\ &\quad - \int \frac{q + q'^*}{2|q'|} S(\boldsymbol{k}', \boldsymbol{k}_1) S^*(\boldsymbol{k}', \boldsymbol{k}_2) \mathrm{e}^{-\mathrm{i}(q' - q'^*)z_0} \, d\boldsymbol{k}' \\ &= 0 \ . \end{aligned}$$

It follows from this relation in the first that

$$\int_{|\boldsymbol{k}'| < K} S(\boldsymbol{k}', \boldsymbol{k}_1) S^*(\boldsymbol{k}', \boldsymbol{k}_2) \, d\boldsymbol{k}' = \delta(\boldsymbol{k}_1 - \boldsymbol{k}_2) \tag{2.3.10}$$

when $|\boldsymbol{k}_1| < K$, $|\boldsymbol{k}_2| < K$. Moreover,

$$\int_{|\boldsymbol{k}'| < K} S(\boldsymbol{k}', \boldsymbol{k}_1) \cdot S^*(\boldsymbol{k}', \boldsymbol{k}_2) \, d\boldsymbol{k}' = -\mathrm{i} S(\boldsymbol{k}_2, \boldsymbol{k}_1) \tag{2.3.11a}$$

when $|\boldsymbol{k}_1| < K$, $|\boldsymbol{k}_2| > K$,

$$\int_{|\boldsymbol{k}'| < K} S(\boldsymbol{k}', \boldsymbol{k}_1) \cdot S^*(\boldsymbol{k}', \boldsymbol{k}_2) \, d\boldsymbol{k} = -\mathrm{i} S(\boldsymbol{k}_2, \boldsymbol{k}_1) + \mathrm{i} S^*(\boldsymbol{k}_1, \boldsymbol{k}_2) \tag{2.3.11b}$$

when $|\boldsymbol{k}_1| > K$, $|\boldsymbol{k}_2| > K$. The total set of relations (2.3.10, 11) has been obtained in the work by *Waterman* [2.1]. Equation (2.3.10) is called the unitarity relation and represents the mathematical formulation of the law of conservation of energy.

Now let us formulate the unitarity relation for the case of multicomponental fields under the absence of energy flux through the boundary. We assume that the field incident onto the boundary is a superposition of homogeneous waves of the form (2.2.16) with arbitrary amplitude factors $a_{\sigma_0}(\boldsymbol{k}_0)$:

$$\Psi_i^{(\mathrm{in})} = \sum_{\sigma_0} \int_{|\boldsymbol{k}_0|<\omega/c_{\sigma_0}} d\boldsymbol{k}_0 \cdot {}^{\uparrow}A_i^{\sigma_0}(\boldsymbol{k}_0) \cdot q_{\sigma_0}^{-1/2}(k_0) \exp[\mathrm{i}\boldsymbol{k}_0\boldsymbol{r} + \mathrm{i}q_{\sigma_0}(k_0)z] \cdot a_{\sigma_0}(\boldsymbol{k}_0) \; . \tag{2.3.12}$$

Then the vertical component of energy flux averaged over horizontal coordinates $\boldsymbol{r}$ is

$$\begin{aligned} \bar{I}_z^{(\mathrm{in})} &= \sum_{\sigma_0} \int_{|\boldsymbol{k}_0|<\omega/c_{\sigma_0}} d\boldsymbol{k}_0 \, |a_{\sigma_0}(\boldsymbol{k}_0)|^2 \\ &= \sum_{\mu_0}^{(\mathrm{h})} |a_{\mu_0}|^2 \; . \end{aligned} \tag{2.3.13}$$

Determination of multi-index μ_0 is given in (2.2.31) and $\sum_{\mu_0}^{(\mathrm{h})}$ denotes the summation only over homogeneous waves:

$$\sum_{\mu_0}^{(\mathrm{h})} \stackrel{\mathrm{def}}{=} \sum_{\sigma_0} \int_{|\boldsymbol{k}_0|<\omega/c_{\sigma_0}} d\boldsymbol{k}_0 \; .$$

Here we used the fact that in the average energy fluxes for superposition of plane waves are just summed up. Quantity $\bar{I}_z$ for scattered waves averaged over $\boldsymbol{r}$ is represented quite similarly:

$$\begin{aligned} \bar{I}_z^{(\mathrm{sc})} &= \sum_{\sigma} \int_{|\boldsymbol{k}|<\omega/c_\sigma} d\boldsymbol{k} \left| \sum_{\sigma_0} \int_{|\boldsymbol{k}_0|<\omega/c_{\sigma_0}} d\boldsymbol{k}_0 \, S_{\sigma\sigma_0}(\boldsymbol{k}, \boldsymbol{k}_0) a_{\sigma_0}(\boldsymbol{k}_0) \right|^2 , \\ &= \sum_{\mu_0}^{(\mathrm{h})} \left| \sum_{\mu_0}^{(\mathrm{h})} S_{\mu\mu_0} a_{\mu_0} \right|^2 \; . \end{aligned} \tag{2.3.14}$$

It was taken into account that inhomogeneous waves do not transport energy in a vertical direction. In absence of energy losses

$$\bar{I}_z^{(\mathrm{in})} = \bar{I}_z^{(\mathrm{sc})} \tag{2.3.15}$$

or

$$\sum_{\mu_0}^{(\mathrm{h})} |a_{\mu_0}|^2 = \sum_{\mu_0, \mu_0'}^{(\mathrm{h})} \left(\sum_{\mu}^{(\mathrm{h})} S_{\mu\mu_0} \cdot S_{\mu\mu_0'}^{*} \right) a_{\mu_0} a_{\mu_0'}^{*} \; .$$

In virtue of the arbitrariness of amplitude a_{μ_0}, we obtain the unitarity relation

$$\sum_{\mu}^{(\mathrm{h})} S_{\mu\mu_0} \cdot S_{\mu\mu_0'}^{*} = \delta(\mu_0 - \mu_0') \; , \tag{2.3.16}$$

where

$$\delta(\mu_0 - \mu_0') = \delta_{\sigma_0\sigma_0'} \delta(\boldsymbol{k}_0 - \boldsymbol{k}_0')$$

and $\delta_{\sigma\sigma_0}$ is Kronecker's and $\delta(\boldsymbol{k}_0 - \boldsymbol{k}'_0)$ is Dirac's delta, $|\boldsymbol{k}_0| < \omega/c_{\sigma_0}$ and $|\boldsymbol{k}'_0| < \omega' c_{\sigma'_0}$.

The above reasoning is in fact valid for the boundary dividing two media, see Sect. 2.2. The difference is that in (2.3.15) in $\bar{I}_z^{(\mathrm{sc})}$ the energy fluxes leaving the boundary, both to the first and the second media, should be taken into account. As a result, instead of (2.3.16) we obtain the unitarity relation for the scattering matrix in the most general case

$$\sum_{\nu}^{(\mathrm{h})} S_{\nu\nu_0} \cdot S^*_{\nu\nu'_0} = \delta(\nu_0 - \nu'_0) \; , \tag{2.3.17}$$

where

$$\sum_{\nu}^{(\mathrm{h})} = \sum_{N=1,2} \sum_{\mu}{}^{(\mathrm{h})} = \sum_{N=1,2} \sum_{\sigma} \int_{|\boldsymbol{k}|<\omega/c_\sigma} d\boldsymbol{k}$$

and

$$\delta(\nu_0 - \nu'_0) = \delta_{N_0 N'_0} \cdot \delta(\mu_0 - \mu'_0) = \delta_{N_0 N'_0} \delta_{\sigma_0\sigma'_0} \delta(\boldsymbol{k}_0 - \boldsymbol{k}'_0) \; .$$

Let us assume

$$(S^+)_{\nu'\nu} \overset{\mathrm{def}}{=} S^*_{\nu\nu'} \; ,$$

where superscript "+" means the Hermitian conjugation of the corresponding matrix. In matrix notations (2.3.17) takes the form

$$\hat{S}^+ \hat{S} = \hat{1} \; .$$

Multiplying this relation from the right by $\hat{S}^+$ we get

$$\hat{S}^+ (\hat{S}\hat{S}^+ - \hat{1}) = 0 \; .$$

If equation $\hat{S}^+ x = 0$ has only a trivial solution, then it follows also that

$$\hat{S}\hat{S}^+ = \hat{1} \; .$$

In usual notations the unitary relation (2.3.17) takes the following form:

$$\sum_{N=1,2} \sum_{\sigma} \int_{|\boldsymbol{k}|<\omega/c_\sigma} d\boldsymbol{k} \cdot S^{NN_0}_{\sigma\sigma_0}(\boldsymbol{k}, \boldsymbol{k}_0) \cdot (S^{NN'_0}_{\sigma\sigma'_0}(\boldsymbol{k}, \boldsymbol{k}'_0))^*$$
$$= \delta_{N_0 N'_0} \delta_{\sigma_0\sigma'_0} \delta(\boldsymbol{k}_0 - \boldsymbol{k}'_0) \; ; \quad |\boldsymbol{k}_0| < \omega/c_{\sigma_0}, |\boldsymbol{k}'_0| < \omega/c_{\sigma'_0} \; . \tag{2.3.18}$$

It is complicated to make so simple and universal a formulation of the reciprocity theorem for the general case of multicomponental fields and the interface between two media. However, the concrete examples considered in the present section and Chap. 4 sufficiently show how to proceed in other cases.

The way is as follows: One should obtain from the initial motion equations for two solutions Ψ_1 and Ψ_2 the relation of the form

$$\operatorname{div} \boldsymbol{P} = 0 \ , \tag{2.3.19}$$

where vector $\boldsymbol{P}$ is a certain bilinear function with respect to wave fields: $\boldsymbol{P} = \boldsymbol{P}(\Psi_1, \Psi_2)$ (this function is usually anti-symmetric about Ψ_1 and Ψ_2). Moreover, vector $\boldsymbol{P}$ should be continuous while crossing the boundary between two media. Integrating (2.3.19) over a certain volume between two planes $z = z_0^+$, $z_0^+ > \max h(\boldsymbol{r})$ and $z = z_0^-$, $z_0^- < \min h(\boldsymbol{r})$ and taking into account the continuity of $\boldsymbol{P}$ at the boundary, we obtain

$$\int_{z=z_0^-} P_z(\Psi_1, \Psi_2)\, d\boldsymbol{r} = \int_{z=z_0^+} P_z(\Psi_1, \Psi_2)\, d\boldsymbol{r} \ . \tag{2.3.20}$$

Substituting into the latter solutions for the fields arising as a result of scattering of two different plane waves gives the sought relation of the kind of (2.3.6).

In the case of the half-space $z < h(\boldsymbol{r})$ the normal component of vector $\boldsymbol{P}$ should vanish at the boundary $z = h(\boldsymbol{r})$. Here the following relation will play the role of (2.3.20):

$$\int_{z=z_0^-} P_z(\Psi_1, \Psi_2)\, d\boldsymbol{r} = 0 \ .$$

2.4 Scattering Amplitude for Statistical Case

We start this section with formulating two simple transformation properties of scattering amplitude (SA).

Suppose $S(\boldsymbol{k}, \boldsymbol{k}_0)$ is known for a boundary of the form $z = h(\boldsymbol{r})$. Then the question arises what will be the formula for SA if the entire boundary is shifted in the horizontal direction by vector $\boldsymbol{d}$? It is clear that such a shift of the boundary does not in fact change the physical problem. In the new coordinates obtained after the same shift of primary coordinates by vector $\boldsymbol{d}$ both problems accurately coincide. However, in the new coordinates the incident wave amplitude acquires an additional phase factor $\exp(\mathrm{i}\boldsymbol{k}_0 \cdot \boldsymbol{d})$. In virtue of the linearity of the problem the scattered field will be multiplied by the same factor. Similarly, while returning to the primary coordinates, each scattered plane wave gets an additional change of phase $\exp(-\mathrm{i}\boldsymbol{k} \cdot \boldsymbol{d})$. Thus we reveal that when shifting the boundary horizontally as a whole, SA is transformed according to the law:

$$S_{h(\boldsymbol{r}-\boldsymbol{d})}(\boldsymbol{k}, \boldsymbol{k}_0) = \mathrm{e}^{-\mathrm{i}(\boldsymbol{k}-\boldsymbol{k}_0)\cdot \boldsymbol{d}} \cdot S_{h(\boldsymbol{r})}(\boldsymbol{k}, \boldsymbol{k}_0) \ . \tag{2.4.1}$$

It is apparent that the transformation law (2.4.1), having a pure geometrical origin, is fulfilled for each component of SA of multicomponental fields and for the boundary between two media, as well.

Now assume that we lift the boundary as a whole by value H: $h(\boldsymbol{r}) \to h(\boldsymbol{r}) + H$. Then each incident plane wave gets an additional change of phase with respect to the new location of the boundary equal to $\exp(\mathrm{i}q_0 H)$, where $q_0 =$

$q(\boldsymbol{k}_0)$. Analogous to the previous case, each scattered plane wave, while returning to the primary coordinates, acquires the phase factor $\exp(iqH)$ (since the scattered wave propagates in the direction of negative z: $\Psi_{\mathrm{sc}} \sim \exp(-iqz)$). As a result, the formula for transformation of SA at the vertical shift of the boundary becomes as follows:

$$S_{h(\boldsymbol{r})+H}(\boldsymbol{k}, \boldsymbol{k}_0) = \mathrm{e}^{\mathrm{i}(q+q_0)H} S_{h(\boldsymbol{r})}(\boldsymbol{k}, \boldsymbol{k}_0) \ . \tag{2.4.2}$$

Derivation of this formula shows that for multicomponental fields (2.4.2) is modified into

$$S_{\sigma\sigma_0}^{h(\boldsymbol{r})+H}(\boldsymbol{k}, \boldsymbol{k}_0) = \mathrm{e}^{\mathrm{i}(q_\sigma+q_{\sigma_0})H} S_{\sigma\sigma_0}^{h(\boldsymbol{r})}(\boldsymbol{k}, \boldsymbol{k}_0) \ . \tag{2.4.3}$$

Finally, for the interface between two media the given transformation property is formulated as

$$S_{h(\boldsymbol{r})+H}^{NN_0} = \begin{bmatrix} \exp(\mathrm{i}q_k^{(1)}H + \mathrm{i}q_0^{(1)}H)S^{11} & \exp(\mathrm{i}q_k^{(1)}H - \mathrm{i}q_0^{(2)}H)S^{12} \\ \exp(-\mathrm{i}q_k^{(2)}H + \mathrm{i}q_0^{(1)}H)S^{21} & \exp(-\mathrm{i}q_k^{(2)}H - \mathrm{i}q_0^{(2)}H)S^{22} \end{bmatrix} \tag{2.4.4}$$

(notation for $q^{(1,2)}$ used in this formula was introduced first in (2.2.21–23)).

Later in this section we will use relation (2.4.1). The transformation properties (2.4.2–4) will be required in the following sections.

Now suppose that irregularities of the boundary $z = h(\boldsymbol{r})$ form a statistical ensemble which is spatially homogeneous in the horizontal plane. The latter means that after a horizontal shift of every realization by an arbitrary vector $\boldsymbol{d}$, the statistical ensemble does not change. Therefore, the mean value of an arbitrary functional Φ of elevations should remain unchangeable for this shift:

$$\overline{\Phi[h(\boldsymbol{r} - \boldsymbol{d})]} = \overline{\Phi[h(\boldsymbol{r})]} \ . \tag{2.4.5}$$

Now we average relation (2.4.1) with respect to realizations $h(\boldsymbol{r})$. According to (2.4.5) we obtain for arbitrary $\boldsymbol{d}$

$$\bar{S}(\boldsymbol{k}, \boldsymbol{k}_0) = \exp[-\mathrm{i}(\boldsymbol{k} - \boldsymbol{k}_0)\cdot \boldsymbol{d}]\bar{S}(\boldsymbol{k}, \boldsymbol{k}_0) \ .$$

The last equality is possible only if

$$\bar{S}(\boldsymbol{k}, \boldsymbol{k}_0) = \bar{V}(\boldsymbol{k})\cdot \delta(\boldsymbol{k} - \boldsymbol{k}_0) \ . \tag{2.4.6}$$

Hence the averaged field consists only of the specular reflected wave. This fact follows directly from the horizontal homogeneity of the statistical ensemble of elevations. Coefficient $\bar{V}$ in formula (2.4.6) is referred to as the *mean reflection coefficient* [2.3]. For the case of multicomponental fields $\bar{V}$ will depend on the polarization of waves, $\bar{V} = \bar{V}_{\sigma\sigma_0}(\boldsymbol{k})$. When the statistical ensemble consists only of a single realization $h(\boldsymbol{r}) = 0$, i.e., the boundary is plane, in fact, then a set of mean reflection coefficients $\bar{V}_{\sigma\sigma_0}^{NN_0}(\boldsymbol{k})$ coincides with common reflection and transmission coefficients at the plane interface between two media.

Returning to the case of scalar fields and taking into account the reciprocity theorem (2.3.6), we have from (2.4.6)

$$\bar{V}(\boldsymbol{k}) = \bar{V}(-\boldsymbol{k}) \ . \tag{2.4.7}$$

The intensity of scattered fields is related to the second statistical moments of SA. We set

$$S(\boldsymbol{k},\boldsymbol{k}_0) = \bar{V}(\boldsymbol{k})\delta(\boldsymbol{k}-\boldsymbol{k}_0) + \Delta S(\boldsymbol{k},\boldsymbol{k}_0) \ , \tag{2.4.8}$$

where ΔS describes fluctuations of SA equal in average to zero, $\overline{\Delta S} = 0$. In the general case the second moments of SA are described with correlators of the form

$$\Delta S(\boldsymbol{k}_1,\boldsymbol{k}_2)\cdot \Delta S^*(\boldsymbol{k}_3,\boldsymbol{k}_4) \ .$$

It is clear that we can equally average the following quantity:

$$\Pi(\boldsymbol{k},\boldsymbol{k}_0;\boldsymbol{a},\boldsymbol{a}_0) = \Delta S\left(\boldsymbol{k}-\frac{\boldsymbol{a}}{2},\boldsymbol{k}_0-\frac{\boldsymbol{a}_0}{2}\right)\cdot \Delta S^*\left(\boldsymbol{k}+\frac{\boldsymbol{a}}{2},\boldsymbol{k}_0+\frac{\boldsymbol{a}_0}{2}\right) \ .$$

For horizontal shifts of boundary we have according to (2.4.1)

$$\Pi_{h(\boldsymbol{r}-\boldsymbol{d})} = \exp(\mathrm{i}\boldsymbol{a}\boldsymbol{d} - \mathrm{i}\boldsymbol{a}_0\boldsymbol{d})\Pi_{h(\boldsymbol{r})} \ .$$

Averaging this relationship similarly to the previous case yields $\bar{\Pi} \sim \delta(\boldsymbol{a}-\boldsymbol{a}_0)$. Thus,

$$\overline{\Pi(\boldsymbol{k},\boldsymbol{k}_0;\boldsymbol{a},\boldsymbol{a}_0)} = \Delta\mathscr{E}(\boldsymbol{k},\boldsymbol{k}_0;\boldsymbol{a})\delta(\boldsymbol{a}-\boldsymbol{a}_0) \ . \tag{2.4.9}$$

Relation (2.4.9) is a direct consequence of the spatial homogeneity of the statistical ensemble.

There is no difference, in a sense, in considering multicomponental fields and the scalar case. Now quantity $\Delta\mathscr{E}$ depends on four indices describing wave polarization

$$\overline{\Delta S_{\alpha\alpha_0}\cdot \Delta S^*_{\beta\beta_0}} = \Delta\mathscr{E}_{\alpha\alpha_0,\beta\beta_0}\cdot \delta(\boldsymbol{a}-\boldsymbol{a}_0) \ . \tag{2.4.10}$$

Suppose that there is no loss of energy and consequently the unitarity relation holds. Substituting (2.4.8) into (2.3.10) and averaging, we obtain

$$|\bar{V}_{\boldsymbol{k}}|^2 + \int\limits_{|\boldsymbol{k}'|<\omega/c} \sigma(\boldsymbol{k}',\boldsymbol{k})\,d\boldsymbol{k}' = 1 \ , \qquad |\boldsymbol{k}| < \omega/c \ , \tag{2.4.11}$$

where

$$\sigma(\boldsymbol{k}',\boldsymbol{k}) = \Delta\mathscr{E}(\boldsymbol{k}',\boldsymbol{k};0) \ . \tag{2.4.12}$$

Relation (2.4.11) represents the formulation of the law of conservation of energy for a statistical case. Quantity σ has dimension $[m^2]$ and it is naturally called the scattering cross-section. By virtue of (2.4.9) it is obvious that

$$\Delta\mathscr{E}(\boldsymbol{k},\boldsymbol{k}_0;\boldsymbol{a}) = \Delta\mathscr{E}^*(\boldsymbol{k},\boldsymbol{k}_0;-\boldsymbol{a})$$

so that the scattering cross-section is real.

In a similar way we obtain for multicomponental fields

$$\sum_{\alpha'}|\overline{V}_{\alpha'\alpha}(\boldsymbol{k})|^2 + \sum_{\alpha'}\int_{|\boldsymbol{k}'|<\omega/c'_\alpha}\sigma_{\alpha'\alpha}(\boldsymbol{k}',\boldsymbol{k})\,d\boldsymbol{k}' = 1 \ , \tag{2.4.13}$$

where

$$\sigma_{\alpha'\alpha}(\boldsymbol{k}',\boldsymbol{k}) = \Delta\mathscr{E}_{\alpha'\alpha,\alpha'\alpha}(\boldsymbol{k}',\boldsymbol{k};\boldsymbol{0}) \ . \tag{2.4.14}$$

Using the reciprocity theorem (2.3.6) in (2.4.9) we obtain

$$\Delta\mathscr{E}(\boldsymbol{k},\boldsymbol{k}_0;\boldsymbol{a}) = \Delta\mathscr{E}(-\boldsymbol{k}_0,-\boldsymbol{k};-\boldsymbol{a})$$

whence

$$\sigma(\boldsymbol{k},\boldsymbol{k}_0) = \sigma(-\boldsymbol{k}_0,-\boldsymbol{k}) \ . \tag{2.4.15}$$

We suppose that the statistical ensemble is invariant with respect to inversion and then obtain one more symmetry relation. This relation implies analogously to (2.4.5) that for an arbitrary functional Φ of elevations the following condition is fulfilled

$$\overline{\Phi[h(-\boldsymbol{r})]} = \overline{\Phi[h(\boldsymbol{r})]} \ .$$

At inversion $\boldsymbol{r}\to-\boldsymbol{r}$ the wave vector changes its sign as well. Therefore,

$$S_{h(-\boldsymbol{r})}(\boldsymbol{k},\boldsymbol{k}_0) = S_{h(\boldsymbol{r})}(-\boldsymbol{k},-\boldsymbol{k}_0) \ .$$

In virtue of the reciprocity theorem we have

$$S_{h(-\boldsymbol{r})}(\boldsymbol{k},\boldsymbol{k}_0) = S_{h(\boldsymbol{r})}(\boldsymbol{k}_0,\boldsymbol{k}) \ .$$

Using this relation in (2.4.9) we find

$$\sigma(\boldsymbol{k},\boldsymbol{k}_0) = \sigma(\boldsymbol{k}_0,\boldsymbol{k}) \ . \tag{2.4.16}$$

In order to establish a relation between the cross-section and experimentally measured values, we examine calculating the correlation function of the scattered field. We again start with the scalar case.

Analogous to (2.4.8) the field could be represented as a sum of averaged and fluctuation components, i.e.,

$$\Psi_{\rm sc} = \overline{\Psi}_{\rm sc} + \Delta\Psi_{\rm sc} \ , \tag{2.4.17}$$

where $\overline{\Delta\Psi_{sc}} = 0$. Existence of δ-function in (2.4.6) simplifies calculating $\overline{\Psi}_{sc}$ for the spatially homogeneous statistical ensembles and reduces the problem to calculating the field reflected from the layered half-space with the given reflection coefficient $\overline{V}_{\boldsymbol{k}}$.

Let us examine the calculation of the correlation function for the fluctuation field component. Assume that the point omnidirectional source is in the coordinates $(\boldsymbol{r}_0, z_0)$, and the scattered field is measured at two points A and B with coordinates $(\boldsymbol{r} + \Delta\boldsymbol{r}/2, z + \Delta z/2)$ and $(\boldsymbol{r} - \Delta\boldsymbol{r}/2, z - \Delta z/2)$, respectively. Here $(\Delta\boldsymbol{r}, \Delta z)$ is the vector connecting the points at which the field is measured, and $(\boldsymbol{r}, z)$ is the point lying halfway between A and B. Suppose that the source and the receivers are so far from the surface that the inhomogeneous waves' effect can be neglected. Then, according to (2.2.12)

$$\Delta\Psi_{sc}(A) = -\frac{\mathrm{i}}{8\pi^2} \int\limits_{|\boldsymbol{k}'|,|\boldsymbol{k}_0'|<\omega/c} q_{k'}^{-1/2} \exp\left[\mathrm{i}\boldsymbol{k}'\cdot\left(\boldsymbol{r} + \frac{\Delta\boldsymbol{r}}{2}\right) - \mathrm{i}q_{k'}\left(z + \frac{\Delta z}{2}\right)\right]$$

$$\times\ \Delta S(\boldsymbol{k}', \boldsymbol{k}_0') q_{k_0'}^{-1/2} \exp(-\mathrm{i}\boldsymbol{k}_0'\cdot\boldsymbol{r}_0 - \mathrm{i}q_0' z_0)\, d\boldsymbol{k}'\, d\boldsymbol{k}_0' \ ,$$

$$\Delta\Psi_{sc}^*(B) = \frac{\mathrm{i}}{8\pi^2} \int\limits_{|\boldsymbol{k}''|,|\boldsymbol{k}_0''|<\omega/c} q_{k''}^{-1/2} \exp\left[-\mathrm{i}\boldsymbol{k}''\cdot\left(\boldsymbol{r} - \frac{\Delta\boldsymbol{r}}{2}\right) + \mathrm{i}q_{k''}\left(z - \frac{\Delta z}{2}\right)\right]$$

$$\times\ \Delta S^*(\boldsymbol{k}'', \boldsymbol{k}_0'') q_{k_0''}^{-1/2} \exp(\mathrm{i}\boldsymbol{k}_0''\cdot\boldsymbol{r}_0 + \mathrm{i}q_0'' z_0)\, d\boldsymbol{k}''\, d\boldsymbol{k}_0'' \ .$$

Multiply these relations and average the result. Changing variables $\boldsymbol{k}' = \boldsymbol{k} + \boldsymbol{a}/2$, $\boldsymbol{k}'' = \boldsymbol{k} - \boldsymbol{a}/2$, $\boldsymbol{k}_0' = \boldsymbol{k}_0 + \boldsymbol{a}_0/2$, and $\boldsymbol{k}_0'' = \boldsymbol{k}_0 - \boldsymbol{a}_0/2$ and using formula (2.4.9) yields

$$\begin{aligned}\overline{\Delta\Psi_{sc}(A)\cdot\Delta\Psi_{sc}^*(B)} = \frac{1}{(8\pi^2)^2} \int \Delta\mathscr{E}(\boldsymbol{k}, \boldsymbol{k}_0; \boldsymbol{a}) \exp\{&\mathrm{i}\boldsymbol{a}\cdot(\boldsymbol{r} - \boldsymbol{r}_0) \\ &+ \mathrm{i}[q(\boldsymbol{k}_0 - \boldsymbol{a}/2) - q(\boldsymbol{k}_0 + \boldsymbol{a}/2)]z_0 \\ &+ \mathrm{i}[q(\boldsymbol{k} - \boldsymbol{a}/2) - q(\boldsymbol{k} + \boldsymbol{a}/2)]z + \mathrm{i}\boldsymbol{k}\cdot\Delta\boldsymbol{r} \\ &- (\mathrm{i}/2)[q(\boldsymbol{k} + \boldsymbol{a}/2) + q(\boldsymbol{k} - \boldsymbol{a}/2)]\cdot\Delta z\} \\ &\times [q(\boldsymbol{k} + \boldsymbol{a}/2)\cdot q(\boldsymbol{k} - \boldsymbol{a}/2)\cdot q(\boldsymbol{k}_0 + \boldsymbol{a}/2) \\ &\times q(\boldsymbol{k}_0 - \boldsymbol{a}/2)]^{-1/2}\, d\boldsymbol{k}\, d\boldsymbol{k}_0\, d\boldsymbol{a} \ . \end{aligned} \tag{2.4.18}$$

At first consider the integral over $\boldsymbol{a}$. Suppose that the source point and receiving points tend to infinity from the surface in the fixed directions. In other words, we introduce the formal small parameter, attaching it to the coordinates as a factor:

$$(\boldsymbol{r}, z) \to \varepsilon^{-1}\cdot(\boldsymbol{r}, z) \ , \qquad (\boldsymbol{r}_0, z_0) \to \varepsilon^{-1}\cdot(\boldsymbol{r}_0, z_0) \ .$$

It can be easily seen that in calculations in the lower order in ε it is sufficient to retain in exponent only terms linear with respect to $\boldsymbol{a}$, and set $\boldsymbol{a} = \boldsymbol{0}$ in the rest of the factors. Then the integral over $\boldsymbol{a}$ is computed using formula (2.1.20). Taking into account that

$$\frac{dq}{d\boldsymbol{k}} = -\frac{\boldsymbol{k}}{q} ,$$

we readily reduce formula (2.4.18) to the form

$$\overline{\Delta\Psi_{\mathrm{sc}}(A)\cdot\Delta\Psi^*_{\mathrm{sc}}(B)} = (4\pi)^{-2}\int\limits_{|\boldsymbol{k}|<K}\exp(\mathrm{i}\boldsymbol{k}\cdot\Delta\boldsymbol{r}-\mathrm{i}q_k\Delta z)\frac{d\boldsymbol{k}}{(Kq_k)}$$
$$\times\int\limits_{|\boldsymbol{k}_0|<K}\sigma(\boldsymbol{k},\boldsymbol{k}_0)K^2\delta(\boldsymbol{r}-\boldsymbol{r}_0+\boldsymbol{k}\cdot z/q_k+\boldsymbol{k}_0 z_0/q_0)\frac{d\boldsymbol{k}_0}{(Kq_0)} . \tag{2.4.19}$$

In the latter instead of variables k_x, k_y, the polar angles θ, φ of the vector $(\boldsymbol{k}, q_k)$ can be introduced by formulas $k_x = K\sin\theta\cdot\cos\varphi$, $k_y = K\sin\theta\cdot\sin\varphi$, and $K = \omega/c$ (polar axis is directed downward). Simple calculations yield

$$\left|\frac{\partial(k_x, k_y)}{\partial(\theta, \varphi)}\right| = K^2\sin\theta\cos\theta .$$

Therefore,

$$d\boldsymbol{k}/Kq_k = d\boldsymbol{n} , \tag{2.4.20}$$

where $d\boldsymbol{n} = \sin\theta\, d\theta\, d\varphi$ is the solid angle element in the wave vector $(\boldsymbol{k}, q_k)$ direction.

Note that expression (2.4.19) for the correlation function has the same form as for the random field consisting of the ensemble of plane waves δ-correlated in directions. Namely, if

$$\Psi(\boldsymbol{R}) = \int A(\boldsymbol{n})\exp(\mathrm{i}K\cdot\boldsymbol{n}\cdot\boldsymbol{R})\,d\boldsymbol{n} ,$$

where $\boldsymbol{R} = (\boldsymbol{r}, z)$ is the three-dimensional vector, then

$$\overline{\Psi(\boldsymbol{R})\cdot\Psi^*(\boldsymbol{R}')} = \int \mathscr{I}(\boldsymbol{n})\exp[\mathrm{i}K\boldsymbol{n}\cdot(\boldsymbol{R}-\boldsymbol{R}')]\,d\boldsymbol{n} , \tag{2.4.21}$$

where

$$\overline{A(\boldsymbol{n})\cdot A^*(\boldsymbol{n}')} = \mathscr{I}(\boldsymbol{n})\delta(\boldsymbol{n}-\boldsymbol{n}') .$$

The concept of such a field is in the base of the radiative transport theory and we use it in Chap. 7.

Now assume that the source has some directivity diagram which is given as the function of the horizontal projection of wave vector $\mathscr{D} = \mathscr{D}(\boldsymbol{k}_0)$. This implies that far away from the source in the direction $\boldsymbol{n}_0 = (\boldsymbol{k}_0, q_0)/K$ the incident field has the form

$$\Psi_{\text{in}} = -\mathscr{D}(\boldsymbol{k}_0)\cdot\frac{\exp(\mathrm{i}KR)}{4\pi R} . \tag{2.4.22}$$

For the directed source, formula (2.4.19) is modified in the following obvious manner:

$$\begin{aligned}\overline{\Delta\Psi_{\text{sc}}(A)\cdot\Delta\Psi^*_{\text{sc}}(B)} &= \frac{1}{(4\pi)^2}\int\limits_{|\boldsymbol{k}|<K}\exp(\mathrm{i}\boldsymbol{k}\cdot\Delta\boldsymbol{r} - \mathrm{i}q_k\Delta z)\,d\boldsymbol{k}/(Kq_k)\\ &\times\int\limits_{|\boldsymbol{k}_0|<K}\sigma(\boldsymbol{k},\boldsymbol{k}_0)K^2\delta(\boldsymbol{r} - \boldsymbol{r}_0 + \boldsymbol{k}\cdot z/q + \boldsymbol{k}_0 z_0/q_0)\\ &\times|\mathscr{D}(\boldsymbol{k}_0)|^2\,d\boldsymbol{k}_0/(Kq_0) .\end{aligned} \tag{2.4.23}$$

Now, we transform formula (2.4.33) from variables $\boldsymbol{k}_0$ to some new integration variables $\boldsymbol{\varrho}$ by the formula

$$\boldsymbol{\varrho} = \boldsymbol{r}_0 - \boldsymbol{k}_0\cdot z_0/q_0 = \boldsymbol{r}_0 + \boldsymbol{k}_0\cdot|z_0|/q_0 ,$$

where $\boldsymbol{\varrho}$ are the horizontal coordinates of the point at which the ray launched from the source $(\boldsymbol{r}_0, z_0)$ in direction $(\boldsymbol{k}_0, q_0)$ arrives at the plane $z = 0$ (Fig. 2.6). Calculation of the Jacobian gives

$$\left|\frac{\partial(\varrho_x, \varrho_y)}{\partial(k_0^x, k_0^y)}\right| = \frac{K^2}{q_0^4}|z_0| ,$$

whence

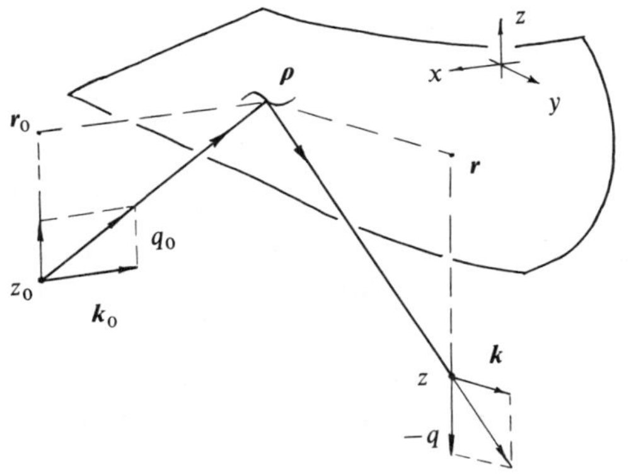

Fig. 2.6. Physical meaning of integration variables $\boldsymbol{\varrho}$

$$(4\pi)^{-2}|\mathscr{D}(\boldsymbol{k}_0)|^2\,d\boldsymbol{k}_0/Kq_0 = |\Psi_{\rm in}(\boldsymbol{\varrho},0)|^2(q_0/K)\,d\boldsymbol{\varrho}\ ,$$

where

$$|\Psi_{\rm in}(\boldsymbol{\varrho},0)|^2 = (4\pi)^{-2}|\mathscr{D}(\boldsymbol{k}_0)|^2[(\boldsymbol{\varrho}-\boldsymbol{r}_0)^2+z_0^2]^{-1}$$

is the intensity of the incident field at the point $(\boldsymbol{\varrho},0)$. Formula (2.4.23) in these new variables takes the form

$$\overline{\Delta\Psi_{\rm sc}(A)\cdot\Delta\Psi^*_{\rm sc}(B)} = \int \exp(\mathrm{i}\boldsymbol{k}\cdot\Delta\boldsymbol{r} - \mathrm{i}q_k\Delta z)\cdot d\boldsymbol{k}/(Kq_k)\cdot\int\sigma(\boldsymbol{k},\boldsymbol{k}_0)$$
$$\times\, Kq_0\cdot\delta(\boldsymbol{r}+\boldsymbol{k}z/q_k-\boldsymbol{\varrho})\cdot|\Psi_{\rm in}(\boldsymbol{\varrho},0)|^2\,d\boldsymbol{\varrho}\ . \qquad (2.4.24)$$

Integrating (2.4.24) over $\boldsymbol{\varrho}$ and comparing the result with (2.4.21) we obtain that in the reference case

$$\mathscr{I}(\boldsymbol{n}) = Kq_0\cdot\sigma(\boldsymbol{k},\boldsymbol{k}_0)\cdot|\Psi_{\rm in}(\boldsymbol{\varrho},0)|^2\ ,$$

where $\boldsymbol{n}=(\boldsymbol{k},-q_k)/K$ and $\boldsymbol{\varrho}$ and $\boldsymbol{k}_0$ for given $\boldsymbol{k}$ are calculated by formulas

$$\boldsymbol{\varrho} = \boldsymbol{r}+\boldsymbol{k}\cdot z/q_k = \boldsymbol{r}-\boldsymbol{k}\cdot|z|/q_k\ ,$$

$$\boldsymbol{k}_0 = K\cdot(\boldsymbol{\varrho}-\boldsymbol{r}_0)\cdot[(\boldsymbol{\varrho}-\boldsymbol{r}_0)^2+Z_0^2]^{-1/2}\ .$$

Now consider the average fluctuation intensity. For this we set in (2.4.24) $\Delta\boldsymbol{r} = \Delta z = 0$ and integrate over $\boldsymbol{k}$, using the δ-function. Calculating similarly to the previous Jacobian, we easily find

$$\overline{|\Delta\Psi_{\rm sc}(\boldsymbol{r},z)|^2} = \int q_k q_{k_0}\sigma(\boldsymbol{k},\boldsymbol{k}_0)\cdot|\Psi_{\rm in}(\boldsymbol{\varrho},0)|^2[(\boldsymbol{r}-\boldsymbol{\varrho})^2+z^2]^{-1}\,d\boldsymbol{\varrho}\ , \qquad (2.4.25)$$

where for given $\boldsymbol{\varrho}$ the arguments $\boldsymbol{k}$ and $\boldsymbol{k}_0$ are calculated in virtue of geometrical relations shown in Fig. 2.6:

$$\boldsymbol{k} = \frac{K\cdot(\boldsymbol{r}-\boldsymbol{\varrho})}{[(\boldsymbol{r}-\boldsymbol{\varrho})^2+z^2]^{1/2}}\ ,$$

$$\boldsymbol{k}_0 = \frac{K\cdot(\boldsymbol{\varrho}-\boldsymbol{r}_0)}{[(\boldsymbol{\varrho}-\boldsymbol{r}_0)^2+z_0^2]^{1/2}}\ .$$

The dimensionless value,

$$m_{\rm s}(\boldsymbol{k},\boldsymbol{k}_0) = q_k q_{k_0}\sigma(\boldsymbol{k},\boldsymbol{k}_0)\ , \qquad (2.4.26)$$

characterizing scattering in formula (2.4.25) is called the *scattering coefficient*. Being measured in dB it is called the *scattering strength*:

$$I_{\rm s} = 10\log_{10} m_{\rm s}\ . \qquad (2.4.27)$$

The law of conservation of energy (2.4.11) in terms of the scattering coefficient is written as

$$\int m_s(\boldsymbol{k}', \boldsymbol{k})\, d\boldsymbol{k}'/Kq' = (1 - |\bar{V}_{\boldsymbol{k}}|^2)\cos\theta_k \ ,$$

where

$$\cos\theta_k = q_k/K$$

and θ_k is the angle of incidence of the wave.

In the case of a two-dimensional problem there is no scattering in transverse directions and one can set

$$m_s = m_s^{(1)}\delta(k_y) = K^{-1}\cdot m_s^{(1)}\cdot\delta(\theta_y) \ .$$

Here θ_y is the scattering angle in the transverse direction. Hence, in the two-dimensional problem the scattering strength is usually defined as

$$I_s = 10\log_{10}(m_s^{(1)}/K) \ . \tag{2.4.28}$$

Finally, we consider the case of multicomponental fields. In this case the scattered field of the point source is the superposition of waves of various polarizations, according to (2.2.20). This formula differs from the scalar case only in the existence of factors ${}^{\downarrow}A_i^{\sigma}(\boldsymbol{k})$ and $\alpha^{\sigma_0}(\boldsymbol{k}_0)$. The above calculations for the field correlation function remain without changing. Assume for simplicity that the source launches the wave of only one polarization γ_0 (i.e., $\alpha^{\sigma_0}(\boldsymbol{k}_0)$ is not equal to zero only at $\sigma_0 = \gamma_0$). Then instead of (2.2.25) for the correlation between ith and jth field components at the point $(\boldsymbol{r}, z)$ we obtain

$$\overline{\Delta\Psi_i^{(sc)}\cdot(\Delta\Psi_j^{(sc)})^*} = \sum_{\alpha_1,\alpha_2}\int {}^{\downarrow}A_i^{\alpha_1}(\boldsymbol{k})\cdot[{}^{\downarrow}A_j^{\alpha_2}(\boldsymbol{k})]^*\cdot[q_k q_{k_0}\Delta\mathscr{E}_{\alpha_1\gamma_0,\alpha_2\gamma_0}(\boldsymbol{k},\boldsymbol{k}_0;0)]$$

$$\times\ |a_{\gamma_0}^{(\text{in})}(\boldsymbol{\varrho},0)|^2\cdot[(\boldsymbol{r}-\varrho)^2 + z^2]^{-1}\, d\boldsymbol{\varrho} \ , \tag{2.4.29}$$

where the values ${}^{\downarrow}A_i^{\alpha}(\boldsymbol{k})$ are determined in (2.2.16) and $a_{\gamma_0}^{(\text{in})}$ is the amplitude of a wave of γ_0th polarization at the point $(\boldsymbol{\varrho}, 0)$

$$|a_{\gamma_0}^{(\text{in})}(\boldsymbol{\varrho},0)|^2 = (4\pi)^{-2}\cdot|\alpha^{\gamma_0}(\boldsymbol{k}_0)|^2\cdot|\mathscr{D}(\boldsymbol{k}_0)|^2\cdot[(\boldsymbol{\varrho}-\boldsymbol{r}_0)^2 + z_0^2]^{-1} \ .$$

Thus for multicomponental fields the values,

$$\sigma_{\alpha\beta,\alpha_0\beta_0}(\boldsymbol{k},\boldsymbol{k}_0) = \Delta\mathscr{E}_{\alpha\beta,\alpha_0\beta_0}(\boldsymbol{k},\boldsymbol{k}_0;\boldsymbol{0}) \ ,$$

play role of scattering cross-sections.

Hence the scattering cross-section $\sigma(\boldsymbol{k},\boldsymbol{k}_0)$ is closely related to calculations of the field correlation functions which are usually measured in experiment. On the other hand, it is directly calculated from the formula for the second moments of SA.

In practice the rough surface sometimes adjoins not to the homogeneous medium as was supposed before, but to some refractive medium which is able to return the scattered field back to the rough surface. It turns out (Chap. 7) that in this case values of the mean reflection coefficient and the scattering cross-section also participate in the description of wave propagation.

In conclusion we can say that calculation of $\overline{V}_{\boldsymbol{k}}$ and $\sigma(\boldsymbol{k}, \boldsymbol{k}_0)$ is the main task in the theory of wave scattering from the statistically rough surfaces.

3. Helmholtz Formula and Rayleigh Hypothesis

This chapter concerns two problems. First, we consider the Helmholtz formula which enables us to express the field at any point of some region through boundary values of this field and its normal derivative. Section 3.1 presents a simple derivation of the Helmholtz formula when it directly follows from the differentiation rule for discontinuous functions. The advantage of this derivation is a clear demonstration of the Helmholtz formula ambiguity which is used in the following chapters. Section 3.1 also discusses some consequences of the formula and derives integral equations for surface density at the boundary.

The next three sections consider an important and interesting problem which is repeatedly discussed in the literature – the validity of the Rayleigh hypothesis. Its essence is the following. The scattered field at the points below the boundary is presented as a superposition of outgoing plane waves generated by the surface sources. However, *Rayleigh* [3.1] applied a representation of field in the form of outgoing waves for the points at the boundary as well.

Lippman [3.2] noted in 1953 that, generally speaking, this is incorrect, since the waves reradiated upward by other surface sources arrive at these points also. Therefore, the structure of the scattered field near the boundary must differ from that used by Rayleigh and include upgoing waves also. In 1966 *Petit* and *Cadilhac* [3.3] proved that for the Dirichlet problem and a sinusoidal boundary with a slope exceeding 0.448, the Rayleigh hypothesis is wrong – a solution in the form of outgoing waves cannot take required values at all boundary points. However, *Millar* [3.4] showed that for a sinusoidal boundary with a rather small slope the Rayleigh hypothesis is fulfilled, and in 1971 he proved [3.5] that the Rayleigh hypothesis is fulfilled for all slopes less 0.448. Sections 3.2, 3 present the main ideas and results of Millar's works and Sect. 3.4 clears up the contradiction between Lippman's argument and Millar's results. Lippman's argument appeared to be incorrect.

The problem of validity of the Rayleigh hypothesis is of significant practical importance. Its application allows us to write down immediately an equation for SA and to avoid using integral equations for surface sources density. This circumstance simplifies many of the calculations while analyzing scattering in the framework of perturbation theory and reduces the problem to a simple calculation. Discussion of the Rayleigh hypothesis in Sect. 3.4 shows that such a procedure is *completely true*. Proof of this fact follows from Millar's results and the theorem proved by *Van den Berg* and *Fokkema* [3.6] in 1979, see also Sect. 3.2.

3.1 Helmholtz Formula

The Helmholtz formula allows us to express the wave field at an arbitrary point inside some region through the field and its normal derivative at the boundary of this region. This formula is widely used in various wave problems. In the present section we give a simple derivation of the Helmholtz formula and discuss some applications of it. For clearness the derivation will be shown first for a one-dimensional case. Consider

$$\frac{d^2y}{dx^2} + K^2y = Q(x) , \tag{3.1.1}$$

where function Q describing the field sources vanishes at large x. To find a unique solution to (3.1.1), two additional conditions should be imposed. First, the radiation condition for $x \to +\infty$ should be taken. It implies that this solution represents the right-going wave:

$$y \to A \cdot \exp(\mathrm{i}Kx) , \qquad x \to +\infty , \tag{3.1.2}$$

where A is some coefficient. The similar condition for $x \to -\infty$ takes the form

$$y \to B \cdot \exp(-\mathrm{i}Kx) , \qquad x \to -\infty \tag{3.1.3}$$

any (3.1.3) represents the left-going wave.

When verifying the radiation conditions (3.1.2, 3), one can avoid calculating the asymptotic of the solution by using the limiting absorption principle. The latter implies that with introducing into the medium some absorption and attaining a small positive imaginary portion to the wave number K, solution $y(x)$ for $|x| \to \infty$ should decrease:

$$K^2 \to K^2 + \mathrm{i}\alpha , \qquad \alpha > 0 \quad \Rightarrow \quad y(x) \to 0 , \qquad |x| \to \infty . \tag{3.1.4}$$

Then solution of (3.1.1) can be written as

$$y(x) = \int g_0(x - x')Q(x')\,dx' , \tag{3.1.5}$$

where

$$g_0(x) = (2\mathrm{i}K)^{-1} \exp(\mathrm{i}K|x|) \tag{3.1.6}$$

is the Green's function of (3.1.1), i.e.,

$$\left(\frac{d^2}{dx^2} + K^2\right) g_0 = \delta(x) .$$

It is obvious that both the Green's function g_0 and solution (3.1.5) satisfy conditions (3.1.2–4). By virtue of evenness of the Green's function formula, (3.1.5) can also be represented in the form

$$y(x) = \int g_0(x' - x)Q(x')\,dx' \ . \tag{3.1.5a}$$

Now consider on the interval $(-\infty, x_0)$ some boundary problem determined by homogeneous equation of (3.1.1):

$$\left(\frac{d^2}{dx^2} + K^2\right)y = 0 \ . \tag{3.1.7}$$

For the boundary conditions we take the radiation condition (3.1.3) at $x \to -\infty$, and at $x = x_0$ impose condition

$$y(x_0) = y_0 \ . \tag{3.1.8}$$

Solution of this boundary problem certainly is

$$y(x) = y_0 \cdot \exp[-iK(x - x_0)] \ , \qquad x \in (-\infty, x_0) \ . \tag{3.1.9}$$

We obtain this solution using the method which can be generalized for the multidimensional case also. For this we need the formula for differentiating the piecewise smooth functions with a final jump at some point. Hence, let

$$f(x) = \begin{cases} f_1(x), & x > x_0 \ , \\ f_2(x), & x < x_0 \ , \end{cases}$$

where f_1 and f_2 are arbitrary smooth functions. Then,

$$\frac{df}{dx} = \left\{\frac{df}{dx}\right\} + [f] \cdot \delta(x - x_0) \ . \tag{3.1.10}$$

Here

$$\left\{\frac{df}{dx}\right\} = \begin{cases} f_1'(x), & x > x_0 \ , \\ f_2'(x), & x < x_0 \ , \end{cases}$$

is an ordinary "classical" derivative of function f, and $[f]$ is the jump of function f at the point $x = x_0$:

$$[f] = \lim_{\varepsilon \to +0} (f_1(x + \varepsilon) - f_2(x - \varepsilon)) \ .$$

Differentiating formula (3.1.10) with respect to x according to the same rule, we obtain

$$\frac{d^2 f}{dx^2} = \left\{\frac{d^2 f}{dx^2}\right\} + \left[\frac{df}{dx}\right]\delta(x - x_0) + [f]\delta'(x - x_0) \ ,$$

where

$$\left[\frac{df}{dx}\right] = \lim_{\varepsilon \to +0} [f_1'(x + \varepsilon) - f_2'(x - \varepsilon)]$$

is the jump of the derivative at $x = x_0$. Adding function $K^2 f$ to both sides of the last equality we eventually obtain for arbitrary piecewise smooth function f with discontinuity at $x = x_0$:

$$\left(\frac{d^2}{dx^2} + K^2\right)f = \left\{\left(\frac{d^2}{dx^2} + K^2\right)f\right\} + \left[\frac{df}{dx}\right]\delta(x - x_0) + [f]\delta'(x - x_0) \; . \tag{3.1.11}$$

If functions f_1 and f_2 are solutions of (3.1.7) for $x > x_0$ and $x < x_0$, respectively, then the first term in the right-hand side of (3.1.11) vanishes[1]. With the given right-hand side relation (3.1.11) can be considered as an equation for function f. Solution of this equation can be obtained according to formula (3.1.5a):

$$f(x) = \left[\frac{df}{dx}\right]_{x_0} \cdot g_0(x_0 - x) - [f]_{x_0} g_0'(x_0 - x) \; . \tag{3.1.12}$$

We emphasize that when solving (3.1.11) the Green's function (3.1.6) was used satisfying radiation conditions (3.1.2, 3). Hence the transition from (3.1.11) to (3.1.12) is valid only when f_1 satisfies the boundary condition (3.1.2) and f_2 satisfies the boundary condition (3.1.3).

Now we take for $f(x)$ the function

$$\tilde{y}(x) = \begin{cases} 0, & x > x_0 \; , \\ y(x), & x < x_0 \; , \end{cases} \tag{3.1.13}$$

where $y(x)$ is the unknown solution of the boundary problem (3.1.7), (3.1.3), and (3.1.8). Both $f_1 = 0$ and $f_2 = y(x)$ satisfy (3.1.7, 2) and (3.1.3), respectively, and therefore, formula (3.1.12) can be applied to $\tilde{y}$:

$$\tilde{y}(x) = -y_0' g_0(x_0 - x) + y_0 g_0'(x_0 - x) = \begin{cases} 0, & x > x_0 \; , & (3.1.14a) \\ y(x), & x < x_0 \; . & (3.1.14b) \end{cases}$$

Quantity y_0 is known from the boundary condition (3.1.8), however, $y_0' = dy/dx|_{x \to x_0 - 0}$ is unknown. It is true that this quantity cannot be given arbitrarily as an additional boundary condition, since two conditions were already imposed on our equation of the second order: the boundary (3.1.8) and radiation (3.1.3) condition at $x \to -\infty$. Consequently, quantity y_0' is determined completely in terms of y_0. It is easy to express y_0' through y_0 in a one-dimensional case. Consider equality (3.1.14a) for $x \to x_0 + 0$. Since at $x_0 - x \to -0$ $g_0'(x_0 - x) \to -1/2$, then

$$y_0' = -\mathrm{i}K y_0 \; . \tag{3.1.15}$$

---

[1] In particular, applying formula (3.1.11) for function g_0 from (3.1.6) we are convinced that it is really the Green's function of (3.1.1).

Unfortunately, we fail to obtain such an explicit formula for the multidimensional case. Hence, in diffraction problems, various approximated relations should be used instead of the exact relation (3.1.15).

Substituting (3.1.15) into (3.1.14), we readily obtain

$$\tilde{y}(x) = \mathrm{i}K y_0 \cdot g_0(x_0 - x) + y_0 \cdot g_0'(x_0 - x)$$

$$= \begin{cases} 0, & x > x_0 \ , \\ y_0 \cdot \exp[-\mathrm{i}K(x - x_0)], & x < x_0 \ . \end{cases}$$

The upper line in this equality coincides with (3.1.14a) and confirms the correctness of the calculations. The lower line gives the sought solution of the boundary problem coinciding with (3.1.9).

Equation (3.1.11) shows that the wave propagating at $x < x_0$ is generated by the sum of monopole and dipole sources with the strengths $[df/dx] = -y_0'$ and $[f] = -y_0$, respectively, situated at the point $x = x_0$, i.e., at the boundary of the area. However, the strength of the sources generating this wave at $x < x_0$ are not defined unambiguously and depend on how we define the solution for $x > x_0$. For example, instead of (3.1.13) we could set

$$\tilde{y}(x) = \begin{cases} C \cdot \exp[\mathrm{i}K(x - x_0)], & x > x_0 \ , \quad (3.1.16a) \\ y(x), & x < x_0 \ , \quad (3.1.16b) \end{cases}$$

and then the strength of the monopole and dipole at $x = x_0$ would change:

$$\left[\frac{df}{dx}\right]_{x_0} = -y_0' + \mathrm{i}KC \ , \qquad [f]_{x_0} = -y_0 + C \ .$$

However, these additional sources do not generate the field at $x < x_0$, and change the field only at $x > x_0$ – which one can easily check with (3.1.14b).

All above considerations are directly generalized for the multidimensional case. Consider the Helmholtz equation

$$(\Delta + K^2)\Psi = Q(\boldsymbol{R}) \ , \tag{3.1.17}$$

where $K = \omega/c = \text{const}$. Solution of this equation takes the form

$$\Psi(\boldsymbol{R}) = \int G_0(\boldsymbol{R} - \boldsymbol{R}')Q(\boldsymbol{R}')\,d\boldsymbol{R}' = \int G_0(\boldsymbol{R}' - \boldsymbol{R})Q(\boldsymbol{R}')\,d\boldsymbol{R}' \ , \tag{3.1.18}$$

where G_0 is the Green's function for (3.1.17)

$$(\Delta + K^2)G_0 = \delta(\boldsymbol{R})$$

satisfying the radition condition for $|\boldsymbol{R}| \to \infty$. This function, according to (2.1.21), has the form

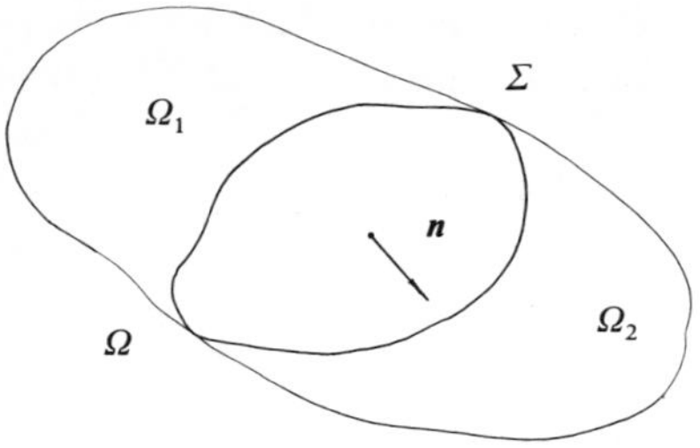

Fig. 3.1. On derivation of differentiation formula for functions with the discontinuity surface

$$G_0(R) = -\exp(\mathrm{i}KR)/4\pi R$$

(here and below $R = |\boldsymbol{R}| = (x^2 + y^2 + z^2)^{1/2}$).

Now we need to obtain the analog of formula (3.1.11). Let $\varphi(\boldsymbol{R})$ be an arbitrary smooth function unequal to zero only inside some region Ω, see Fig. 3.1, so that at the boundary of the region function φ and all its derivatives vanish. Let the region Ω be divided by some smooth surface Σ into two parts Ω_1 and Ω_2. Some function $f(\boldsymbol{R})$ is assumed to be piecewise smooth in the region Ω:

$$f(\boldsymbol{R}) = \begin{cases} f_1(\boldsymbol{R}), & \boldsymbol{R} \in \Omega_1 \ , \\ f_2(\boldsymbol{R}); & \boldsymbol{R} \in \Omega_2 \ , \end{cases} \tag{3.1.19}$$

so that $f(\boldsymbol{R})$ and its derivatives at the surface can undergo jumps. Consider the following relation:

$$\int_{\Omega} f(\boldsymbol{R}) \cdot \Delta\varphi(\boldsymbol{R})\, d\boldsymbol{R} = \left(\int_{\Omega_1} + \int_{\Omega_2} \right) f(\boldsymbol{R}) \Delta\varphi(\boldsymbol{R})\, d\boldsymbol{R} \tag{3.1.20}$$

and try to get rid of operator Δ acting on φ in the right-hand side of this equality. This can be done using the Green's formula which is obtained as follows: We apply to the obvious identity

$$u \cdot \Delta v = v \cdot \Delta u - \operatorname{div}(v\nabla u - u\nabla v)$$

in which u and v are some smooth functions, the Gauss theorem

$$\int_{\omega} \operatorname{div} \boldsymbol{A}\, d\boldsymbol{R} = \int_{\partial\omega} \boldsymbol{A} \cdot d\boldsymbol{\sigma} = \int_{\partial\omega} \boldsymbol{A} \cdot \boldsymbol{n}\, d\sigma \ .$$

Here ω is some region, $\partial\omega$ its boundary and $\boldsymbol{n}$ the external normal to the boundary. As a result, we obtain

$$\int_{\omega} u \Delta v\, d\boldsymbol{R} = \int_{\omega} v \Delta u\, d\boldsymbol{R} - \int_{\partial\omega} v \frac{\partial u}{\partial \boldsymbol{n}} d\sigma + \int_{\partial\omega} u \frac{\partial v}{\partial \boldsymbol{n}} d\sigma \ .$$

Apply this formula to the first and second terms in (3.1.20):

$$\int_{\Omega_1} f\Delta\varphi\, d\boldsymbol{R} = \int_{\Omega_1} \Delta f_1 \varphi\, d\boldsymbol{R} - \int_{\Sigma} \frac{\partial f_1}{\partial \boldsymbol{n}} \varphi\, d\Sigma + \int_{\Sigma} f_1 \frac{\partial \varphi}{\partial \boldsymbol{n}} d\Sigma \ ,$$

$$\int_{\Omega_2} f\Delta\varphi\, d\boldsymbol{R} = \int_{\Omega_2} \Delta f_2 \varphi\, d\boldsymbol{R} + \int_{\Sigma} \frac{\partial f_2}{\partial \boldsymbol{n}} \varphi\, d\Sigma - \int_{\Sigma} f_2 \frac{\partial \varphi}{\partial \boldsymbol{n}} d\Sigma \ .$$

(the last formula takes into account that normal $\boldsymbol{n}$ is internal with respect to region Ω_2; $\partial/\partial\boldsymbol{n}$ designates derivative in the $\boldsymbol{n}$-direction, i.e., $\partial/\partial\boldsymbol{n} = \boldsymbol{n}\cdot\nabla$). As a result, we have

$$\int_{\Omega} f\Delta\varphi\, d\boldsymbol{R} = \int_{\Omega} \{\Delta f\}\varphi\, d\boldsymbol{R} + \int_{\Sigma} \left[\frac{\partial f}{\partial \boldsymbol{n}}\right] \varphi\, d\Sigma - \int_{\Sigma} [f] \frac{\partial \varphi}{\partial \boldsymbol{n}} d\Sigma \ , \tag{3.1.21}$$

where

$$\{\Delta f\} = \begin{cases} \Delta f_1, & \boldsymbol{R} \in \Omega_1 \ , \\ \Delta f_2, & \boldsymbol{R} \in \Omega_2 \ , \end{cases}$$

is the "classical" Laplacian of function f and symbol $[\cdot]$ denotes, as previously, an appropriate jump at the boundary Σ:

$$[g] = \lim_{\varepsilon\to+0} [g(\boldsymbol{R} + \varepsilon\cdot\boldsymbol{n}) - g(\boldsymbol{R} - \varepsilon\boldsymbol{n})] \ , \qquad \boldsymbol{R} \in \Sigma \ .$$

Relation (3.1.21) in terms of "distributions" (generalized functions) is written as

$$\Delta f = \{\Delta f\} + \left[\frac{\partial f}{\partial \boldsymbol{n}}\right]\delta_\Sigma + \frac{\partial}{\partial \boldsymbol{n}}([f]\delta_\Sigma) \ , \tag{3.1.22}$$

where distributions $\mu\cdot\delta_\Sigma$ and $\partial/\partial\boldsymbol{n}(\nu\delta_\Sigma)$ act on the smooth functions φ according to the rules

$$\int \mu\delta_\Sigma\cdot\varphi\, d\boldsymbol{R} = \int_\Sigma \mu\cdot\varphi\, d\Sigma \ , \tag{3.1.23a}$$

$$\int \frac{\partial}{\partial \boldsymbol{n}}(\nu\delta_\Sigma)\varphi\, d\boldsymbol{R} = -\int \nu\cdot\frac{\partial\varphi}{\partial \boldsymbol{n}} d\Sigma \tag{3.1.23b}$$

(to avoid misunderstanding we stress that notation $\mu\delta_\Sigma$ and $(\partial/\partial\boldsymbol{n})(\nu\delta_\Sigma)$ should be considered as uniform symbols whose meaning is given by equalities (3.1.23). In particular, here $\partial/\partial\boldsymbol{n}$ does not imply any real differentiation of function ν, but helps us to remember formula (3.1.23b) which resembles integration by parts). Adding the term $K^2 f$ to both parts of (3.1.22) we finally obtain

$$(\Delta + K^2)f = \{(\Delta + K^2)f\} + \left[\frac{\partial f}{\partial \boldsymbol{n}}\right]\delta_\Sigma + \frac{\partial}{\partial \boldsymbol{n}}([f]\delta_\Sigma) \ . \tag{3.1.24}$$

Relation (3.1.24) is quite similar to (3.1.11) and the former can be considered as an equation for function f at the given right-hand side.

Now suppose that for function f_1 in (3.1.19) a certain solution Ψ of the inhomogeneous Helmholtz equation is taken

$$(\Delta + K^2)\Psi = Q(\boldsymbol{R}) \;, \qquad \boldsymbol{R} \in \Omega_1 \tag{3.1.25}$$

and function f_2 in the region Ω_2 satisfies the homogeneous Helmholtz equation

$$(\Delta + K^2) f_2 = 0 \;, \qquad \boldsymbol{R} \in \Omega_2 \tag{3.1.26}$$

Moreover, assume that $f_1 = \Psi$ and f_2 at $|\boldsymbol{R}| \to \infty$ satisfies the radiation condition (if points exist with infinitely large R in the regions Ω_1 and Ω_2). Under these assumptions (3.1.24) takes the following form:

$$f(R_*) = \Psi_{\text{in}}(R_*) + \int_\Sigma \left[\frac{\partial f}{\partial \boldsymbol{n}}\right] G_0(\boldsymbol{R} - \boldsymbol{R}_*)\, d\Sigma - \int_\Sigma [f] \frac{\partial}{\partial \boldsymbol{n}} G_0(\boldsymbol{R} - \boldsymbol{R}_*)\, d\Sigma \;, \tag{3.1.27}$$

where

$$\Psi_{\text{in}}(\boldsymbol{R}_*) = \int G_0(\boldsymbol{R}_* - \boldsymbol{R}')Q(\boldsymbol{R}')\, d\boldsymbol{R}' \tag{3.1.28}$$

represents the primary "incident" field, i.e., the field which is generated by the sources Q in an infinite medium. Formula (3.1.27) for $\Psi_{\text{in}} = 0$ is quite analogous to (3.1.12). Similar to the one-dimensional case, the requirement for the fields $f_1 = \Psi$ and f_2 to satisfy the radiation condition is related to the choice of the Green's function G_0 consisting of outgoing waves. In particular, function $f_2 = 0$, can be taken as a solution of (3.1.26). In this case (3.1.27) takes the following form:

$$\Psi_{\text{in}}(\boldsymbol{R}_*) + \int \Psi \frac{\partial}{\partial \boldsymbol{n}} G_0(\boldsymbol{R} - \boldsymbol{R}_*)\, d\Sigma - \int \frac{\partial \Psi}{\partial \boldsymbol{n}} G_0(\boldsymbol{R} - \boldsymbol{R}_*)\, d\Sigma$$

$$= \begin{cases} \Psi(\boldsymbol{R}_*), & R_* \in \Omega_1 \;, \qquad (3.1.29a) \\ 0, & R_* \in \Omega_2 \;. \qquad (3.1.29b) \end{cases}$$

Relation (3.1.29a) is the *Helmholtz formula* in its most often applied formulation, and (3.1.29b) is usually called the *extinction theorem.*

Formula (3.1.29) can serve as a base for solving diffraction problems. In addition to (3.1.25) and the radiation condition for $R \to \infty$, let the boundary condition be imposed on function Ψ:

$$\Psi|_\Sigma = 0 \tag{3.1.30}$$

(the Dirichlet problem) or

$$\left.\frac{\partial \Psi}{\partial \boldsymbol{n}}\right|_{\Sigma} = 0 \qquad (3.1.31)$$

(the Neumann problem). Then one of the terms in the left-hand side of (3.1.29) vanishes. For example, in the Dirichlet problem the scattered field is represented in the form of monopoles distributed over the surface Σ:

$$\Psi(\boldsymbol{R}_*) = \Psi_{\text{in}}(\boldsymbol{R}_*) - \int_{\Sigma} \frac{\partial \Psi}{\partial \boldsymbol{n}} \cdot G_0(\boldsymbol{R} - \boldsymbol{R}_*)\, d\Sigma\ , \qquad \boldsymbol{R}_* \in \Omega_1\ . \qquad (3.1.32)$$

The unknown surface density of these monopoles equal to $-\partial\psi/\partial\boldsymbol{n}$ can, in principal, be found from the extinction theorem (3.1.29b). In particular, if the point of observation tends to the surface $\boldsymbol{R} \to \boldsymbol{r} \in \Sigma$:

$$\int_{\Sigma} \frac{\partial \Psi}{\partial \boldsymbol{n}} G_0(\boldsymbol{R} - \boldsymbol{r})\, d\Sigma = \Psi_{\text{in}}(\boldsymbol{r})\ ; \qquad \boldsymbol{r}, \boldsymbol{R} \in \Sigma\ . \qquad (3.1.33)$$

Here we used the fact that the second term in formula (3.1.32) is a continuous function of $\boldsymbol{R}$ in the whole space and does not undergo a jump while passing through Σ, see (3.1.27). It is obvious that the field determined by formula (3.1.32) satisfies the boundary condition (3.1.30), if $\partial\Psi/\partial\boldsymbol{n}$ is a solution of (3.1.33). Hence (3.1.33) can be obtained immediately without applying the extinction theorem.

The situation changes somewhat for the Neumann problem. In this case

$$\Psi(\boldsymbol{R}_*) = \Psi_{\text{in}}(\boldsymbol{R}_*) + \int_{\Sigma} \Psi \frac{\partial}{\partial \boldsymbol{n}} G_0(\boldsymbol{R} - \boldsymbol{R}_*)\, d\Sigma\ , \qquad R_* \in \Omega_1\ . \qquad (3.1.34)$$

Now the scattered field is generated by dipole sources distributed over the surface Σ with unknown density Ψ. To obtain an equation for this density we can also write down the extinction theorem (3.1.29b). Passing to the limit $\boldsymbol{R}_* \to \boldsymbol{r} \in \Sigma$ should be performed thoroughly, since function $|\nabla G_0|$ has a rather strong singularity: $|\nabla G_0(\varrho)| \sim \varrho^{-2}$, $\varrho \to 0$. First we prove the useful formula

$$\lim_{\varepsilon \to \pm 0} \int_{\Sigma} v(\boldsymbol{R}) \frac{\partial}{\partial \boldsymbol{n}_{\boldsymbol{R}}} G_0(\boldsymbol{R} - \boldsymbol{r} \mp \varepsilon \boldsymbol{n}_{\boldsymbol{r}})\, d\boldsymbol{\Sigma}_{\boldsymbol{R}}$$

$$= \mp\frac{1}{2} v(\boldsymbol{r}) + \fint_{\Sigma} v(\boldsymbol{R}) \frac{\partial}{\partial \boldsymbol{n}_{\boldsymbol{R}}} G_0(\boldsymbol{R} - \boldsymbol{r})\, d\boldsymbol{\Sigma}_{\boldsymbol{R}}\ , \qquad (3.1.35)$$

where $v(\boldsymbol{R})$ is an arbitrary smooth function at the surface Σ, and $\fint$ denotes the principal value integral. When $\varepsilon \to +0$ the observation point tends to the surface from the outer side, and when $\varepsilon \to -0$, from the inner side. To prove (3.1.35) we cut out the circle $C(\boldsymbol{r}, \delta)$ of small radius δ around the point $\boldsymbol{r}$ at the surface Σ, see Fig. 3.2, and calculate the integrals separately over Σ without the circle

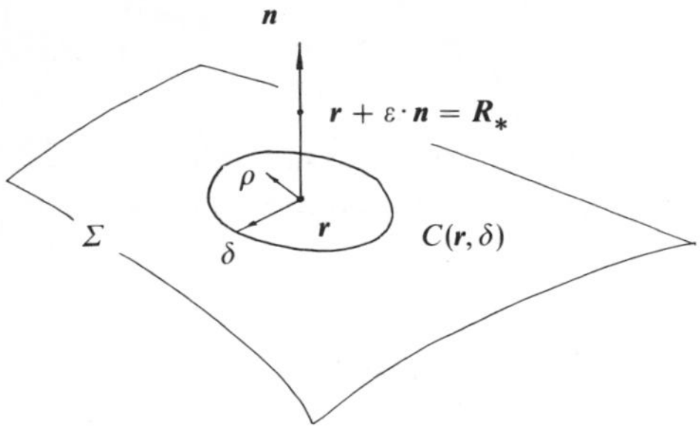

Fig. 3.2. On derivation of limiting value for the field of surface dipole sources when the point of observation $\boldsymbol{R}_*$ approaches the boundary

$C(\boldsymbol{r}, \delta)$ and over the circle C. In the limit $\delta \to 0$ the value $v(\boldsymbol{R})$ at C can be replaced by $v(\boldsymbol{r})$, and the surface shape inside the circle can be approximated by a plane. In other words we can write

$$\int_{\Sigma} v(\boldsymbol{R}) \frac{\partial}{\partial \boldsymbol{n_R}} G_0(\boldsymbol{R} - \boldsymbol{r} - \varepsilon \boldsymbol{n})\, d\Sigma = \int_{\Sigma / C(\boldsymbol{r}, \delta)} v(\boldsymbol{R}) \frac{\partial}{\partial \boldsymbol{n_R}} G_0(\boldsymbol{R} - \boldsymbol{r} - \varepsilon \boldsymbol{n})\, d\Sigma$$
$$+ v(\boldsymbol{r}) \int_0^{\delta} \left\{ -\frac{1}{4\pi} \frac{\partial}{\partial z} [\varrho^2 + (\varepsilon - z)^2]^{-1/2} \right\}_{z=0}$$
$$\times 2\pi \varrho\, d\varrho + \mathrm{O}(\delta) \ . \tag{3.1.36}$$

In the limit $\delta \to 0$ the first term transforms into the principal value integral. With respect to the second term, we have

$$\int_0^{\delta} \frac{\partial}{\partial z} [\varrho^2 + (\varepsilon - z)^2]^{-1/2} \bigg|_{z=0} \varrho\, d\varrho = \int_0^{\delta/\varepsilon} \frac{x\, dx}{(1 + x^2)^{3/2}} \to \pm 1 \quad \text{when} \quad \varepsilon \to \pm 0 \ .$$

Thus passing in (3.1.36) to the limit $\varepsilon \to \pm 0$ and then shrinking the circle $C(\boldsymbol{r}, \delta)$ to the point: $\delta \to 0$, we obtain formula (3.1.35).

Then tending to the limit $\boldsymbol{R}_* \to \boldsymbol{r}$ in formula (3.1.29b) for the Neumann problem and using (3.1.35) for $\varepsilon \to +0$ we find

$$\Psi(\boldsymbol{r})/2 - \int_{\Sigma} \Psi(\boldsymbol{R}) \frac{\partial}{\partial \boldsymbol{n_R}} G_0(\boldsymbol{R} - \boldsymbol{r})\, d\Sigma = \Psi_{\text{in}}(\boldsymbol{r}) \ ; \qquad \boldsymbol{r}, \boldsymbol{R} \in \Sigma \ . \tag{3.1.37}$$

It can be easily shown that the same equation is obtained in tending to the limit $\boldsymbol{R}_* \to \boldsymbol{r}$ in (3.1.34) from inside the region Ω_1. Hence for the Neumann problem the density of dipole sources at Σ equal to the total field $\Psi|_{\Sigma}$ can be determined from (3.1.37). Then the field to be sought can be calculated at an arbitrary point in the region Ω_1 by (3.1.34).

The described procedures of solving the Dirichlet and Neumann boundary problems which reduce to integral equations (3.1.33) and (3.1.37) are most often encountered in applications. These procedures are based on the classical Helmholtz formula resulting from (3.1.24) for differentiating the piecewise smooth function, if $f_2 = 0$ is chosen in (3.1.24). However, quite similarly to the one-dimensional case, the densities of monopole and dipole sources at the surface, which could generate the given fixed field at Ω_1, are determined ambiguously. These densities depend on the way of determining the field for Ω_2. For function f_2 we can take an arbitrary solution of the homogeneous Helmholtz equation $F(\boldsymbol{R})$, satisfying the radiation condition at infinity. For example, such a solution will be the function

$$F(\boldsymbol{R}) = \int_{\Omega_1} G_0(\boldsymbol{R} - \boldsymbol{R}')Q_1(\boldsymbol{R}')\,d\boldsymbol{R}'$$

with arbitrary density of sources Q_1 situated at Ω_1. When chosing $f_2 = F$ in (3.1.19) the densities of surface sources in (3.1.29) transform as follows:

$$\frac{\partial \Psi}{\partial \boldsymbol{n}} \to \frac{\partial \Psi}{\partial \boldsymbol{n}} - \frac{\partial F}{\partial \boldsymbol{n}} , \qquad \Psi \to \Psi - F . \tag{3.1.38}$$

Introducing these additional sources does not affect the field in Ω_1, but changes it in Ω_2. Since the solution of the boundary problem in region Ω_1 is of interest, we can chose F arbitrarily. For example, independent of the form of boundary condition which is imposed on function Ψ, we can take for $F(\boldsymbol{R})$, $\boldsymbol{R} \in \Omega_2$ the solution of the following boundary problem:

$$\begin{aligned} &(\Delta + K^2)F = 0 , \\ &\left.\frac{\partial F}{\partial \boldsymbol{n}}\right|_{\Sigma} = \left.\frac{\partial \Psi}{\partial \boldsymbol{n}}\right|_{\Sigma} . \end{aligned} \tag{3.1.39}$$

As usual, F satisfies the radiation condition at infinity. Then for the field Ψ we obtain the following representation:

$$\Psi(R_*) = \Psi_{\text{in}}(R_*) + \int_{\Sigma} \nu(\boldsymbol{R}) \frac{\partial}{\partial \boldsymbol{n}_{\boldsymbol{R}}} G_0(\boldsymbol{R} - \boldsymbol{R}_*)\,d\Sigma , \qquad R_* \in \Omega_1 , \tag{3.1.40}$$

where $\nu = \Psi - F$. Now the unknown density of surface dipole sources is determined from the boundary condition imposed on function Ψ.

For example, consider the boundary Dirichlet problem:

$$\begin{cases} (\Delta + K^2)\Psi = Q_1(\boldsymbol{R}) , & \boldsymbol{R} \in \Omega_1 , \\ \Psi|_{\Sigma} = 0 . \end{cases} \tag{3.1.41}$$

We shall seek the solution of this problem in the form (3.1.40). To obtain the equation for ν we pass in (3.1.40) to the limit $\boldsymbol{R}_* \to \boldsymbol{r} \in \Sigma$, $\boldsymbol{R}_* \in \Omega_1$. Using formula (3.1.35) for $\varepsilon \to -0$ we find

$$v(\boldsymbol{r})/2 + \int_{\Sigma} v(\boldsymbol{R})\frac{\partial}{\partial \boldsymbol{n}} G_0(\boldsymbol{R}-\boldsymbol{r})\, d\Sigma = -\Psi_{\text{in}}(\boldsymbol{r}) \; ; \qquad \boldsymbol{r}, \boldsymbol{R} \in \Sigma \; . \tag{3.1.42}$$

Solution of (3.1.42) is a linear functional with respect to Ψ_{in}, and can be represented in the following form:

$$v(\boldsymbol{R}) = -2\Psi_{\text{in}}(\boldsymbol{R}) + 2\int d\Sigma' \, \Psi_{\text{in}}(\boldsymbol{R}')L(\boldsymbol{R}', \boldsymbol{R}) \; ; \qquad \boldsymbol{R}, \boldsymbol{R}' \in \Sigma \; . \tag{3.1.43}$$

The kernel $L(\boldsymbol{R}', \boldsymbol{R})$ can be written in the form of the following iteration series:

$$\begin{aligned} L(\boldsymbol{R}', \boldsymbol{R}) = {} & 2\frac{\partial}{\partial \boldsymbol{n}_{\boldsymbol{R}'}} G_0(\boldsymbol{R}', \boldsymbol{R}) - 4\int \frac{\partial}{\partial \boldsymbol{n}_{\boldsymbol{R}'}} G_0(\boldsymbol{R}' - \boldsymbol{R}_1)\, d\Sigma_1 \frac{\partial}{\partial \boldsymbol{n}_{\boldsymbol{R}_1}} G_0(\boldsymbol{R}_1 - \boldsymbol{R}) \\ & + 8\iint \frac{\partial}{\partial \boldsymbol{n}_{\boldsymbol{R}'}} G_0(\boldsymbol{R}' - \boldsymbol{R}_1)\, d\Sigma_1 \frac{\partial}{\partial \boldsymbol{n}_{\boldsymbol{R}_1}} G_0(\boldsymbol{R}_1 - \boldsymbol{R}_2)\, d\Sigma_2 \\ & \times \frac{\partial}{\partial \boldsymbol{n}_{\boldsymbol{R}_2}} G_0(\boldsymbol{R}_2 - \boldsymbol{R}) - \cdots \; . \end{aligned} \tag{3.1.44}$$

Indices at the normal $\boldsymbol{n}_{\boldsymbol{R}_i}$ in (3.1.44) indicate the point of the surface where the appropriate derivative $\partial/\partial \boldsymbol{n}$ is calculated.

It is clear that (3.1.42, 37) for the Dirichlet and the Neumann problems transform into each other at replacements $\Psi_{\text{in}} \to -\Psi_{\text{in}}$ and $\boldsymbol{n} \to -\boldsymbol{n}$. Reversal of the sign of normal $\boldsymbol{n}$ means interchanging the first and the second media. Therefore, solution of the Neumann boundary problem for region Ω_2 reduces, in fact, to the same integral equation (3.1.42).

3.2 Rayleigh Hypothesis

Solution of the boundary problem

$$\begin{cases} (\Delta + K^2)\Psi = Q \; , \qquad \boldsymbol{R} \in \Omega_1 \; , & (3.2.1\text{a}) \\ \hat{B}\Psi|_{\Sigma} = 0 \; , & (3.2.1\text{b}) \end{cases}$$

($\hat{B}$ is some linear operator) can be represented in the form

$$\Psi = \Psi_{\text{in}} + \Psi_{\text{sc}} \; , \tag{3.2.2}$$

where Ψ_{in} is the field incident onto the boundary Σ and given by formula (3.1.28); Ψ_{sc} is the scattered field which can be calculated by (3.1.27), for example,

$$\Psi_{\text{sc}}(\boldsymbol{R}_*) = -\int_{\Sigma} \left\{ [\Psi]_{z=h(\boldsymbol{r})} \cdot \frac{\partial}{\partial \boldsymbol{n}} G_0(\boldsymbol{R} - \boldsymbol{R}_*) - \left[\frac{\partial \Psi}{\partial \boldsymbol{n}}\right]_{z=h(\boldsymbol{r})} \cdot G_0(\boldsymbol{R} - \boldsymbol{R}_*) \right\} d\Sigma \; . \tag{3.2.3}$$

For the Green's function in (3.2.3) we can use representation (2.1.24)

$$G_0(\boldsymbol{R} - \boldsymbol{R}_*) = -\frac{\mathrm{i}}{8\pi^2} \int \exp[-\mathrm{i}\boldsymbol{k} \cdot (\boldsymbol{r} - \boldsymbol{r}_*) + \mathrm{i}q_k |z - z_*|] \, d\boldsymbol{k}/q_k \ . \tag{3.2.4}$$

Suppose that the observation point $\boldsymbol{R}_* = (\boldsymbol{r}_*, z_*)$ is below the rough boundary: $z_* < \min h(\boldsymbol{r})$ (general geometry of the problem is shown in Fig. 2.1). Here one should take into account that when substituting $z = h(\boldsymbol{r})$ into formula (3.2.4) $|z - z_*| = h(\boldsymbol{r}) - z_*$. As a result for the scattered field at $z_* < \min h(\boldsymbol{r})$ we obtain representation (2.2.14) used above

$$\Psi_{\mathrm{sc}}(\boldsymbol{r}_*, z_*) = \int S(\boldsymbol{k}) \cdot q_k^{-1/2} \exp(\mathrm{i}\boldsymbol{k} \cdot \boldsymbol{r}_* - \mathrm{i}q_k z_*) \, d\boldsymbol{k} \ , \tag{3.2.5}$$

where

$$S(\boldsymbol{k}) = -\frac{1}{2} \int \Bigg\{ (q_k + \boldsymbol{k} \nabla k) \cdot [\Psi]_{z=h(\boldsymbol{r})} + \mathrm{i}[1 + (\nabla h)^2]^{1/2} \left[\frac{\partial \Psi}{\partial \boldsymbol{n}}\right]_{z=h(\boldsymbol{r})} \Bigg\} q_k^{-1/2} \exp[-\mathrm{i}\boldsymbol{k} \cdot \boldsymbol{r} + \mathrm{i}q_k h(\boldsymbol{r})] \frac{d\boldsymbol{r}}{(2\pi)^2} \ . \tag{3.2.6}$$

Recall that representation (3.2.3) and, consequently, representation (3.2.6) are not unique. Namely, values $[\partial \Psi / \partial \boldsymbol{n}]$ and $[\Psi]$ can be changed according to transformation (3.1.38).

Quantity $S(\boldsymbol{k})$ depends on the incident field. In particular, if the latter is a plane wave

$$\Psi_{\mathrm{in}} = q_0^{-1/2} \exp(\mathrm{i}\boldsymbol{k}_0 \cdot \boldsymbol{r} + \mathrm{i}q_0 z) \ , \qquad q_0 = q(k_0) \tag{3.2.7}$$

then $S(k)$ coincides with the scattering amplitude $S(\boldsymbol{k}, \boldsymbol{k}_0)$ considered in Sect. 2.2.

Let us examine the problem of convergence of integral (3.2.5). Since

$$q_k \to \mathrm{i}|\boldsymbol{k}| \ , \qquad |\boldsymbol{k}| \to \infty \tag{3.2.8}$$

then we easily obtain from (3.2.6)

$$|S(\boldsymbol{k})| < \mathrm{const} \cdot |\boldsymbol{k}|^{1/2} \exp[-|\boldsymbol{k}| \cdot \min h(\boldsymbol{r})] \ , \qquad |\boldsymbol{k}| \to \infty \ . \tag{3.2.9}$$

For const in (3.2.9) to be finite, one should, strictly speaking, assume that the wave beam incident on the rough surface has a finite aperture. Using (3.2.9) in (3.2.5) shows that the integral converges if

$$z_* < \min h(\boldsymbol{r}) \tag{3.2.10}$$

and this proves the existence of representation (3.2.5) used in Sect. 2.2. Since the wave beam aperture can be arbitrarily large, then we can in fact apply it for a plane wave also.

Hence, formula (3.2.5) is obtained under condition (3.2.10). However, Rayleigh, who was probably the first to consider scattering waves at rough surfaces [3.1], also applied representation (3.2.5) for the points not satisfying this condition, namely for $\boldsymbol{R}_* = (\boldsymbol{r}_*, h(\boldsymbol{r}_*))$. If it is possible, we can substitute (3.2.5) into the boundary condition (3.2.1b) and obtain equation for S. This is the way Rayleigh proceeded. This assumption is referred to as the Rayleigh hypothesis.

But in 1953 *Lippman* [3.2] observed that, generally speaking, this is incorrect. In fact, when substituting (3.2.4) into (3.2.3) we obtain the following result:

$$\begin{aligned}\Psi_{\mathrm{sc}}(\boldsymbol{r}_*, z_*) &= \int S_-(\boldsymbol{k}) q_k^{-1/2} \exp(\mathrm{i}\boldsymbol{k}\cdot\boldsymbol{r}_* - \mathrm{i}q_k z_*)\, d\boldsymbol{k} \\ &\quad + \int S_+(\boldsymbol{k}) q_k^{-1/2} \exp(\mathrm{i}\boldsymbol{k}\cdot\boldsymbol{r}_* + \mathrm{i}q_k z_*)\, d\boldsymbol{k} \ . \end{aligned} \tag{3.2.11}$$

Here

$$\begin{aligned} S_-(\boldsymbol{k}) = -q_k^{-1/2}/2 \int\limits_{\boldsymbol{r}:h(\boldsymbol{r})>z_*} &\Big\{(q_k + \boldsymbol{k}\nabla h)[\Psi]_{z=h(\boldsymbol{r})} + \mathrm{i}[1+(\nabla h)^2]^{1/2} \\ &\times \left[\frac{\partial\Psi}{\partial\boldsymbol{n}}\right]_{z=h(\boldsymbol{r})}\Big\}\cdot \exp[-\mathrm{i}\boldsymbol{k}\cdot\boldsymbol{r} + \mathrm{i}q_k h(\boldsymbol{r})]\, d\boldsymbol{r}/(2\pi)^2 \ , \end{aligned} \tag{3.2.12a}$$

$$\begin{aligned} S_+(\boldsymbol{k}) = -q_k^{-1/2}/2 \int\limits_{\boldsymbol{r}:h(\boldsymbol{r})<z_*} &\Big\{(-q_k + \boldsymbol{k}\nabla h)[\Psi]_{z=h(\boldsymbol{r})} + \mathrm{i}[1+(\nabla h)^2]^{1/2} \\ &\times \left[\frac{\partial\Psi}{\partial\boldsymbol{n}}\right]_{z=h(\boldsymbol{r})}\Big\}\cdot \exp[-\mathrm{i}\boldsymbol{k}\cdot\boldsymbol{r} - \mathrm{i}q_k h(\boldsymbol{r})]\, d\boldsymbol{r}/(2\pi)^2 \ . \end{aligned} \tag{3.2.12b}$$

Integration in (3.2.12a) is only over those $\boldsymbol{r}$ for which the boundary is above the observation point, see Fig. 3.3, and in (3.2.12b), below the latter. Thus, the field in the region immediately adjoining the boundary is represented as a sum of upgoing and downgoing waves. The first and second terms in formula (3.2.11)

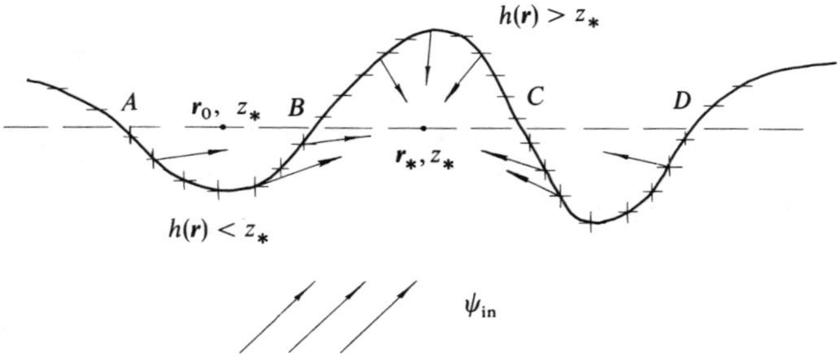

Fig. 3.3. The surface sources and the waves they launch. Upgoing and downgoing waves at the transects AB, CD accurately compensate each other

correspond to the latter. The downgoing and upgoing waves are generated by the sources situated above and below the observation point, respectively. Therefore, existence of these upgoing waves should be taken into account for the points at the boundary, which is in contradiction with (3.2.5) at first glance. This was Lippman's argument.

In 1966 *Petit* and *Cadilhac* [3.3] demonstrated that representation for the field of the kind (3.2.5) for all points of region Ω: $(\boldsymbol{r}_*, z_*)$, $z_* < h(\boldsymbol{r}_*)$ cannot exist for the case of a sinusoidal surface $h = a\cos(px)$ with rather large slopes $ap > 0.448$. Hence, in such a case the Rayleigh hypothesis is not fulfilled.

On the other hand, *Marsh* [3.7] emphasized that if the integral in (3.2.5) converges for all $z_* < h(\boldsymbol{r}_*)$ and the boundary condition (3.2.1b) is satisfied, then (3.2.5) is undoubtedly the solution of (3.2.1a) satisfying the radiation condition. Therefore, (3.2.5) is the sought solution of the boundary problem.

Convergence of integral (3.2.5) has been rigorously and in detail examined in the works by *Millar* [3.4, 5]. It turned out that under certain conditions the above integral was convergent and, consequently, the Rayleigh hypothesis was true in this case.

Consider the reference problem using Millar's approach, see also [3.8].

We restrict ourselves to the two-dimensional problem, assuming that $h = h(x)$ and the incident field $\Psi_{\rm in}$ are y independent. In this case, both $[\Psi]_{z=h(x)}$ and $[\partial\Psi/\partial\boldsymbol{n}]_{z=h(x)}$ are the functions of x only, and integrating over y can be performed in (3.2.6) which leads to the factor $2\pi\delta(k_y)$:

$$S(\boldsymbol{k}) = S(k_x, k_y) = \tilde{S}(k_x)\cdot\delta(k_y) \ ,$$

where

$$\tilde{S}(k) = -q_k^{-1/2}/(4\pi)\int\limits_{-\infty}^{+\infty}\left\{\left(q_k + k\frac{dh}{dx}\right)[\Psi]_{z=h(x)} + \mathrm{i}\left[1+\left(\frac{dh}{dx}\right)^2\right]^{1/2}\right.$$
$$\left.\times\left[\frac{\partial\Psi}{\partial\boldsymbol{n}}\right]_{z=h(x)}\right\}\exp[-\mathrm{i}kx + \mathrm{i}q_k h(x)]\,dx \ . \quad (3.2.13a)$$

When considering the Dirichlet problem and using representation (3.1.40) for the field, then for $\tilde{S}$ the following formula can be written down:

$$\tilde{S}(k) = q_k^{-1/2}/(4\pi)\int\limits_{-\infty}^{+\infty} \nu(x)\left(q_k + k\frac{dh}{dx}\right)\exp[-\mathrm{i}kx + \mathrm{i}q_k h(x)]\,dx \ . \quad (3.2.13b)$$

The scattered field can be expressed in terms of $\tilde{S}(k)$ as

$$\Psi_{\rm sc}(x_*, z_*) = \int\limits_{-\infty}^{+\infty}\tilde{S}(k)q_k^{-1/2}\exp(\mathrm{i}kx_* - \mathrm{i}q_k z_*)\,dk \ . \quad (3.2.14)$$

Convergence of the integral in (3.2.14) depends on the behaviour of function $\tilde{S}(k)$ at $|k|\to\infty$. In virtue of estimation (3.2.8)

$$\exp[-ikx + iq_k h(x)] \to \exp\{k[-ix \mp h(x)]\} ,$$

where the upper sign is taken for $k \to +\infty$ and the lower one – for $k \to -\infty$. Thus, for $k \to \pm\infty$ the amplitude $\tilde{S}$ can be given in the form

$$\tilde{S}(k) = -(4\pi)^{-1} \int_{-\infty}^{+\infty} \left\{ \left(q_k + k\frac{dh}{dx} \right) [\Psi]_{z=h(x)} + i\left[1 + \left(\frac{dh}{dx}\right)^2\right]^{1/2} \right.$$
$$\left. \times \left[\frac{d\Psi}{d\boldsymbol{n}}\right]_{z=h(x)} \right\} (i|k|)^{-1/2} \exp\{k[-ix \mp h(x)]\}\, dx . \qquad (3.2.15)$$

Note that values $[\Psi]$ and $[\partial\Psi/\partial\boldsymbol{n}]$ are k independent. The simplest estimation of this integral has the form (3.2.9) and leads to the convergence criterion (3.2.10). However, for roughness described by analytic functions $h(x)$, the asymptotic of function $\tilde{S}(k)$ at $|k| \to \infty$ can be specified using the steepest descent method. We briefly consider the essence of this method.

Let $f(z)$ be an analytic function. Consider at the complex plane lines

$$C_I\colon \mathrm{Im}\{f(z)\} = \mathrm{const} .$$

It is easy to check that along these lines

$$\frac{d}{dl}\mathrm{Re}\{f(z)\} = |f'(z)| ,$$

where l is the arclength. Thus, the quantity $\mathrm{Re}\{f(z)\}$ changes monotonously along the lines C_I when $f'(z) \neq 0$. The points z_s for which $f'(z_s) = 0$ (saddle points) are crossed at the straight angle by two lines which are determined by one and the same equation

$$\mathrm{Im}\{f(z)\} = \mathrm{Im}\{f(z_s)\} = \mathrm{const} .$$

While going away from the saddle point $z = z_s$ the value $\mathrm{Re}\{f(z)\}$ increases monotonously along one line and decreases along another (the saddle point is assumed as undegenerated: $f''(z_s) \neq 0$). The last line is called the steepest descent contour. Now consider the integral of the form

$$\mathcal{I} = \int_C \varphi(z) \exp[A \cdot f(z)]\, dz ,$$

where φ and f are analytic functions and $A > 0$ is a real constant. If the primary integration path C can be deformed into the steepest descent path C_I, then one can write

$$\mathcal{I} = \exp(Af(z_s)) \cdot \int_{C_I} \varphi(z) \cdot \exp[A(f(z) - f(z_s)]\, dz .$$

It is obvious that the real value with maximum at $z = z_s$ is in the exponent. Therefore, for $A \to +\infty$ outside the small vicinity of the point $z = z_s$, the integrand very rapidly tends to zero. For $z \approx z_s$ one can set $\varphi(z) \approx \varphi(z_s)$ and $f(z) - f(z_s) \approx f''(z_s)(z - z_s)^2/2$, $z - z_s \approx \varrho \cdot e^{i\theta_s}$ where θ_s is the angle of the steepest descent contour with respect to the x-axis. After all mentioned approximations, the obtained (Gaussian) integral is easily calculated:

$$\mathscr{I} \simeq \varphi(z_s) \cdot \exp[Af(z_s) + i\theta_s] \cdot [2\pi/(|f''(z_s)|A)]^{1/2} , \qquad A \to +\infty . \tag{3.2.16}$$

Before estimating the integral (3.2.15) one should deform the integration path running along the real axis into a path running through the complex saddle point x_s being the root of equation

$$\frac{d}{dx}[-ix \mp h(x)]\bigg|_{x=x_s} = 0 \tag{3.2.17}$$

or

$$h'(x_s) = \mp i . \tag{3.2.18}$$

Here we need to assume that $h(x)$ is an analytic function of x, and functions $[\Psi]_{z=h(x)}$ and $[\partial\Psi/\partial\boldsymbol{n}]_{z=h(x)}$ can be analytically continued into the complex plane. The steepest descent contour is determined by equation

$$\operatorname{Im}\{-ix \mp h(x)\} = \operatorname{Im}\{-ix_s \mp h(x_s)\} = \text{const} \tag{3.2.19}$$

with the requirement that $\operatorname{Re}\{-ix \mp h(x)\}$ decreases when going away from the saddle point x_s.

Since $h(x)$ becomes real for real x, then according to the symmetry principle for complex x $h(x^*) = [h(x)]^*$ and $h'(x^*) = [h'(x)]^*$ (the asterisk denotes the complex conjugation). Hence, saddle points and steepest descent contours for both signs in (3.2.18, 19) are symmetric in the complex plane with respect to the real axis. First, we consider the case $k \to +\infty$. Using the standard formula of the saddle point method (3.2.16), we find

$$\tilde{S}(k) \to M_+(k) \cdot \exp[-ikx_s - kh(x_s)] , \qquad k \to +\infty , \tag{3.2.20}$$

where $M_+(k)$ is limited and even vanishes as k^{-1}. Correspondingly, for $k \to +\infty$ the integrand in (3.2.14) behaves as

$$M_+(k) \cdot k^{-1/2} \cdot \exp(ikx_*) \cdot \exp\{k[z_* - h(x_s) - ix_s]\} .$$

For the integral convergence the following condition is necessary and sufficient:

$$z_* < \operatorname{Re}\{h(x_s) + ix_s\} . \tag{3.2.21}$$

Now examine the case $k \to -\infty$. Since x_s goes over to x_s^*, we then have

$$\tilde{S}(k) \to M_-(k) \cdot \exp\{-ikx_s^* + k[h(x_s)]^*\} , \qquad k \to -\infty$$

and the integrand behaves as

$$M_-(k)\cdot|k|^{-1/2}\exp(ikx_*)\cdot\exp\{k(-z_* - \mathrm{i}x_s^* + [h(x_s)]^*)\}\ , \qquad k \to -\infty\ .$$

It is obvious that integral (3.2.14) converges at $k \to -\infty$ also when condition (3.2.21) is fulfiled. Since convergence must exist at any $z_* < h(x_*)$, then we obtain the following criterion for the Rayleigh hypothesis validity

$$\max h(x) < \mathrm{Re}\{h(x_s) + \mathrm{i}x_s\}\ , \tag{3.2.22}$$

where x_s is the root of equation

$$h'(x_s) = -\mathrm{i} \tag{3.2.23}$$

Let us shift up the integration contour in (3.2.15) running along the real axis by iε. As

$$\mathrm{Re}\{-\mathrm{i}(x + \mathrm{i}\varepsilon) - h(x + \mathrm{i}\varepsilon)\} - \mathrm{Re}\{-\mathrm{i}x - h(x)\} = \varepsilon + o(\varepsilon)$$

then the integrand in (3.2.15) decreases at $\varepsilon < 0$, i.e., in the lower half-plane. Consider the saddle points lying in the latter. In the general case, the steepest descent contours corresponding to these points and determined by (3.2.19) cannot intersect. Therefore, the primary integration contour along the real axis can be continuously deformed into the nearest to real axis saddle contour in the lower half-plane. This requirement controls the choice of the concrete root among the solutions of equation (3.2.23) which should be set up into (3.2.22).

Note that the criterion (3.2.22) is frequency independent.

The best known example of the application of the above theory is the case of a sinusoidal surface [3.3–6]:

$$h(x) = a\cos(px)\ .$$

Equation (3.2.23) takes the form

$$-ap\cdot\sin(px_s) = -\mathrm{i}\ .$$

Hence two series of solutions arise

$$px_s = \mathrm{i}t_s + 2\pi n \tag{3.2.24a}$$

and

$$px_s = -\mathrm{i}t_s + \pi\cdot(2m+1)\ , \tag{3.2.24b}$$

where

$$t_s = \sinh^{-1}(1/ap) = \ln\{1/ap + [(1/ap)^2 + 1]^{1/2}\}$$

and $n, m = 0, \pm 1, \ldots$. According to the above consideration we should use the saddle points in the lower half-plane, or solutions (3.2.24b). Assuming that $px = u + \mathrm{i}v$, we obtain from (3.2.19) the following equation for the steepest descent contour

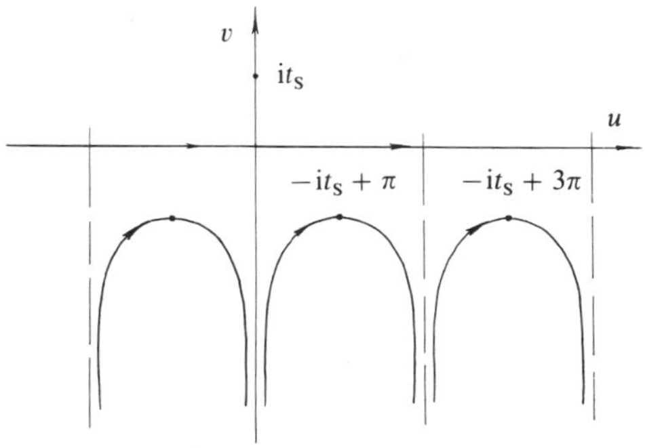

Fig. 3.4. The saddle contour for sinusoidal roughness

$$\sinh v = \frac{1}{ap}\,\frac{u - \pi\cdot(1 + 2m)}{\sin u} \; .$$

The latter is schematically shown in Fig. 3.4. It is clear that the primary integration contour (real axis) can be continuously deformed into the steepest descent contour. On the other hand, the steepest descent contours running through the points (3.2.24a) are the straight lines parallel to imaginary axis: $u = 2\pi n$. Obviously, the real axis cannot be continuously deformed into these straight lines, and the roots of (3.2.24a) cannot be applied. Substituting (3.2.24b) into (3.2.22) yields

$$a < -a\cdot\cosh t_s + t_s/p$$

or

$$ap + [1 + (ap)^2]^{1/2} < \ln\{1/ap + [(1/ap)^2 + 1]^{1/2}\} \; .$$

Whence

$$ap < (ap)_{cr} = (\xi^2 - 1)/2\xi = 0.448 \; ,$$

where $\xi = 1.544$ is the root of equation

$$\exp(\xi) = (\xi + 1)/(\xi - 1) \; .$$

Hence the Rayleigh hypothesis holds if the maximum slope of sinusoidal roughness is bounded,

$$ap < 0.448 \; .$$

Another example gives roughness of the following kind:

$$h(x) = a\,\mathrm{tg}^{-1}(px) \; .$$

We find from (3.2.23) that

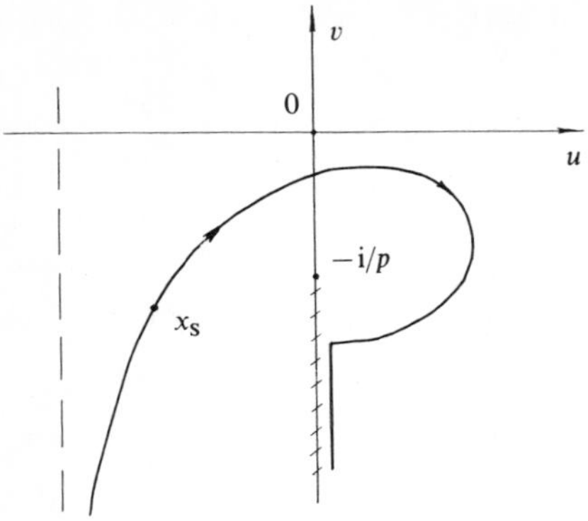

Fig. 3.5. The saddle contour for roughness of the type $h(x) = a \cdot \tan^{-1} px$

$$px_s = (-1 + \mathrm{i}ap)^{1/2} \ .$$

Thus, this is the only saddle point in the lower half-plane:

$$px_s = -\sinh\alpha - \mathrm{i}\cosh\alpha \ ,$$

where

$$ap = \sinh(2\alpha) \ .$$

The steepest descent contour is shown in Fig. 3.5. Function $\mathrm{tg}^{-1}(px)$ has the branching points at $px = \pm\mathrm{i}$ from which the cuts should be drawn. The integration contour fracture in approaching the cut is unimportant, since only a small vicinity of the point x_s of the order of $k^{-1/2}$ contributes to the integral. The condition (3.2.22) acquires the form

$$\sinh(2\alpha) < \sinh(2\alpha)[-\pi/4 - \mathrm{tg}^{-1}(\tanh(\alpha/2)] + \cosh\alpha \ .$$

From which it follows that

$$ap < (ap)_{\mathrm{cr}} = \sinh(2\xi) = 0.540$$

where $\xi = 0.2585$ is the root of equation

$$1 + \pi/4 = -\mathrm{tg}^{-1}[(\tanh(\xi/2)] + 1/2\sinh\xi \ .$$

Now consider the set of rough surface of the form

$$h(x) = aH(x) \ ,$$

where $H(x)$ is an analytic function, $\max|H(x)| = 1$ and $a > 0$ is some amplitude parameter. In this case (3.2.23) takes the form

$$H'(x_s) = -\mathrm{i}/a \ .$$

Since function $H(x)$ is regular at finite x, then obviously $\mathrm{Im}\{x_s\} \to -\infty$ at $a \to 0$. Let function $H(x)$ have the property that if $|H'(x)| \to \infty$, then $|H(x)/H'(x)| <$ const. This condition is satisfied by functions consisting of any finite number of harmonics:

$$H(x) = \sum_{n=1}^{N} A_n \cos(\xi_n x + \varphi_n) \ .$$

It is clear that functions of this type can accurately approximate any roughness profile of physical interest. In this case condition (3.2.22) is always fulfilled for sufficiently small a (at the expense of $\mathrm{i}x_s$). Assume that for some $a = a_0$ (3.2.22) reduces to the exact equality

$$a_0 = \mathrm{Re}[a_0 \cdot H(x_s^{(0)})] + \mathrm{i}x_s^{(0)} \ ,$$

where

$$H'(x_s^{(0)}) = -\mathrm{i}/a_0 \ .$$

Then we have

$$\frac{d}{da}\{\max h(x) - \mathrm{Re}[h(x_s) + \mathrm{i}x_s]\}_{a=a_0}$$

$$= 1 - \mathrm{Re}\{H(x_s^{(0)})\} - \mathrm{Re}\left\{[h'(x_s) + \mathrm{i}]\frac{dx_s}{da}\right\}_{a=a_0}$$

$$= -a_0^{-1}\,\mathrm{Im}\{x_s^{(0)}\} > 0$$

(as $x_s^{(0)}$ belongs to the lower half-plane). Whence it follows that inequality (3.2.22) is fulfilled for all $a < a_0$ and it reverses sign for all $a > a_0$. Thus the Rayleigh hypothesis holds for roughness of the form $h(x) = aH(x)$ for all $a\colon 0 < a < a_0$.

The conclusion that the Rayleigh hypothesis is fulfilled for a wide class of profiles $h(x)$ with sufficiently small slopes is of great practical significance and will often be applied further.

This result has been obtained by *Van den Berg* and *Fokkema* [3.6] in 1979.

3.3 Analytical Properties of Surface Sources Density Function

In estimating the integrals (3.2.13) by the steepest descent method analyticity of the pre-exponential factor in the integrand was significant. We will consider this question in more detail, taking for example the Dirichlet problem.

Now we shall use representation (3.2.13b). Function $v(x)$ is a solution of (3.1.42). Since both the incident field and density v are y independent, then we can integrate in (3.1.42) over y, which leads to replacement of the three-dimensional Green's function by the two-dimensional one, according to (2.1.23a). Taking into account that

$$d\Sigma \frac{\partial}{\partial \boldsymbol{n}} = dx\left(\frac{\partial}{\partial z} - \frac{dh}{dx}\frac{\partial}{\partial x}\right)$$

and

$$\nabla\left[-\frac{\mathrm{i}}{4} H_0^{(1)}(Kr)\right] = \frac{\mathrm{i}K}{4} H_1^{(1)}(Kr)\cdot\frac{\boldsymbol{r}}{r}$$

we obtain the following equation:

$$v(x)/2 = -\Psi_{\mathrm{in}}(x) + (\mathrm{i}K/4)\int_{-\infty}^{+\infty} v(x')\left[h(x) - h(x') - \frac{dh}{dx'}(x - x')\right] \times H_1^{(1)}[K\varrho(x,x')]\,dx'/\varrho(x,x')\ , \tag{3.3.1}$$

where

$$\varrho(x,x') = \{(x - x')^2 + [h(x) - h(x')]^2\}^{1/2}\ . \tag{3.3.2}$$

At $\varrho \to 0$ $H_1^{(1)}(K\varrho) \approx -2\mathrm{i}/\pi K\varrho$, but, as can be easily seen, at $x \to x'$

$$h(x) - h(x') - \frac{dh}{dx'}(x - x') = \mathrm{O}((x - x')^2)\ . \tag{3.3.3}$$

Therefore, the kernel of integral equation (3.3.1) becomes continuous at $x = x'$ and the integral principal value reduces to an ordinary one.

Suppose that the incident field is a superposition of a finite number of plane waves

$$\Psi_{\mathrm{in}} = \sum_{n=1}^{N} A_n \exp[\mathrm{i}k_0^{(n)}\cdot x + \mathrm{i}q_0^{(n)}h(x)]\ .$$

In this case function $\Psi_{\mathrm{in}}(x)$ is analytic in the plane of the complex variable x, because $h(x)$ is assumed such. We integrate in (3.3.1) over x' along the real axis, but complex values can be substituted in this integral for parameter x as well. In this case the right-hand side of formula (3.3.1) gives an analytic continuation of function $v(x)$ from real to complex x. However, in an analytic continuation of an integral one should keep in mind that the integrand singularities should not cross the integration path when varying x.

We explain the arising problem by giving a simple example. Consider expression

$$I(a) = \int_{-\infty}^{+\infty} (a^2 + x^2)^{-1}\,dx\ .$$

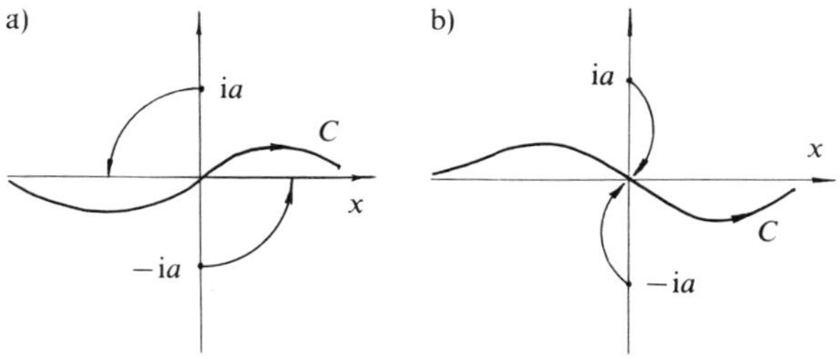

Fig. 3.6a,b. Continuously changing parameter a and distortion of integration contour; pinch takes place for $a = 0$

Setting $a = \mathrm{i}$ we reveal that nonintegrable singularities originate at the integration contour for $x = \pm 1$ and it could be supposed that function $I(a)$ has a singularity at $a = \mathrm{i}$. However, direct calculation yields

$$I(a) = \pi/a \tag{3.3.4}$$

so that the point $a = \mathrm{i}$ turns out to be regular. To solve the appeared contradiction we refer to Fig. 3.6, showing the plane of complex variable x and the integrand singularities at $x = \pm \mathrm{i}a$. It is easily seen that for continuous changing parameter a: $a \to \mathrm{i}$ the integrand poles really go over to the real axis. However, the primary integration contour running along the real axis can be deformed into contour C for which the point $a = \mathrm{i}$ is not singular. Such deformation of the integration contour becomes impossible only when both the singularities of integrand $x = \mathrm{i}a$ and $x = -\mathrm{i}a$ "squeeze" the initial contour between them. It is apparent that this phenomenon ("pinch") originates at $a = 0$. The point $a = 0$, is really the only singularity of function $I(a)$, according to (3.3.4).

Apply the above consideration to formula (3.3.1). The integrand singularities arise at $\rho(x, x') = 0$ and that is possible if

$$h(x') + \mathrm{i}x' = h(x) + \mathrm{i}x \tag{3.3.5a}$$

or

$$h(x') - \mathrm{i}x' = h(x) - \mathrm{i}x \ . \tag{3.3.5b}$$

Relations (3.3.5) should be considered as equations for x' with x being a parameter (note that (3.3.5a) goes over into (3.3.5b) when applying the complex conjugation).

For each fixed x (3.3.5a) generally has an infinite number of roots $x' = \xi_n(x)$, $n = 1, 2, \ldots$. One of the solutions is obvious: $\xi_1(x) = x$. It is clear also that for real x the only real solution of (3.3.5a) is $x' = x$. Further, if the pinch is performed by two points, which are solutions of (3.3.5a) and (3.3.5b), respectively, then again $x' = x$ (this is confirmed while subtracting (3.3.5b) from (3.3.5a)). It was emphasized above that the point $x' = x$ of the integrand is regular, hence singularity does not arise in these cases. Therefore, it is sufficient to consider one of the equations (3.3.5), (3.3.5a), for example. Condition of pinch, that is existence of double roots for (3.3.5a), takes the form

$$h'(x') + \mathrm{i} = 0 \ .$$

This equation turns out to be coincident with (3.2.23). Its solutions are $x' = x_s(m)$, $m = 1, 2, \ldots$. Thus pinch can occur only at these points.

Therefore running along the real axis the integration contour could be shifted down to the level of saddle points

$$\mathrm{Im}\{x_s\} = -t_s = -\sinh^{-1}(1/ap) \ .$$

Pinch, which could occur at this moment, is related to solution $x' = x$ and does not produce singularity. Thus, function $v(x)$ could be analytically continued into the strip $-t_s < \mathrm{Im}\{x\} < 0$. As the main contribution to the integral (3.2.15) at $|k| \to \infty$ comes from an arbitrarily small vicinity of saddle point, application of formula (3.2.16) is thus justified.

3.4 Rayleigh Hypothesis and Lippman Argument: Completeness of Set of the Plane Waves

At the end of Sect. 3.2 we showed that for roughness of the type $h(x) = aH(x)$ where $H(x)$ is an analytic function, at rather small a: $0 < a < a_0$ the Rayleigh hypothesis is fulfilled. On the other hand, at the beginning of Sect. 3.2 Lippman's argument was presented implying that in the vicinity of the boundary the upgoing scattered waves should be considered, which are generated by boundary sources situated below the point of observation at $\min h(x) < z < z_*$. However, it is clear that these waves exist at any slope and when the Lippman argument is valid the Rayleigh hypothesis would never be fulfilled.

How does one explain this contradiction? The point is that in the region not being an infinite horizontal homogeneous layer, dividing the waves into upgoing and downgoing is ambigious. We shall show this. The straight line $z =$ const, see Fig. 3.3, surely contains the portions (for instance AB, CD and so on) lying *outside* the region Ω. Consider some point O, $(\boldsymbol{r}_0, z_*)$ lying on the straight line $z = z_* =$ const *outside* the region Ω, see Fig. 3.3, and use representation for the field of the form (3.1.32) when the scattered field is generated by monopole surface sources with density $\mu(\boldsymbol{r}) = -\partial\Psi/\partial\boldsymbol{n}$. Let the incident field be the plane wave of the form (3.2.7). Operating as in obtaining formulas (3.2.11, 12) we can write

$$\begin{aligned}\Psi(\boldsymbol{r}_0, z_0) = q_0^{-1/2} \exp(\mathrm{i}\boldsymbol{k}_0 \cdot \boldsymbol{r}_0 + \mathrm{i}q_0 z_0) + \int S_+(\boldsymbol{k}, \boldsymbol{k}_0) \cdot q_k^{-1/2} \\ \times \exp(\mathrm{i}\boldsymbol{k} \cdot \boldsymbol{r}_0 + \mathrm{i}q_k z_0)\, d\boldsymbol{k} \\ + \int S_-(\boldsymbol{k}, \boldsymbol{k}_0) q_k^{-1/2} \exp(\mathrm{i}\boldsymbol{k} \cdot \boldsymbol{r}_0 - \mathrm{i}q_k z_0)\, d\boldsymbol{k} \ ,\end{aligned} \tag{3.4.1}$$

where

$$S_+(\boldsymbol{k}, \boldsymbol{k}_0) = -\frac{\mathrm{i}}{2q_k^{1/2}} \int\limits_{\boldsymbol{r}: h(\boldsymbol{r}) < z_0} \mu(\boldsymbol{r}) \exp[-\mathrm{i}\boldsymbol{k} \cdot \boldsymbol{r} - \mathrm{i}q_k h(\boldsymbol{r})] \times [1 + (\nabla h)^2]^{1/2} \, d\boldsymbol{r}/(2\pi)^2$$

and

$$S_-(\boldsymbol{k}, \boldsymbol{k}_0) = -\frac{\mathrm{i}}{2q_k^{1/2}} \int\limits_{\boldsymbol{r}: h(\boldsymbol{r}) > z_0} \mu(\boldsymbol{r}) \exp[-\mathrm{i}\boldsymbol{k} \cdot \boldsymbol{r} + \mathrm{i}q_k h(\boldsymbol{r})] \times [1 + (\nabla h)^2]^{1/2} \, d\boldsymbol{r}/(2\pi)^2 \; .$$

However, the extinction theorem (3.1.29b) gives

$$\Psi(\boldsymbol{r}_0, z_0) = 0 \; . \tag{3.4.2}$$

Consequently for the points $(\boldsymbol{r}_0 \cdot z_0)$ lying outside the region Ω it can be written, according to (3.4.1, 2)

$$\int S_-(\boldsymbol{k}, \boldsymbol{k}_0) q_k^{-1/2} \exp(\mathrm{i}\boldsymbol{k} \cdot \boldsymbol{r}_0 - \mathrm{i}q_k z_0) \, d\boldsymbol{k} = -q_0^{-1/2} \exp(\mathrm{i}\boldsymbol{k}_0 \cdot \boldsymbol{r} + \mathrm{i}q_0 z) - \int S_+(\boldsymbol{k}, \boldsymbol{k}_0) q_k^{-1/2} \exp(\mathrm{i}\boldsymbol{k} \cdot \boldsymbol{r}_0 + \mathrm{i}q_k z_0) \, d\boldsymbol{k} \; . \tag{3.4.3}$$

There is superposition of *downgoing* plane waves in the left-hand side of this equality, however, it is seen from (3.4.3) that the field can be represented as well as superposition of *upgoing* waves only. The same can occur with respect to the points belonging to the region Ω.

Hence in some regions the field of sources situated below the point of observation can be given as superposition of *downgoing* plane waves. This seems strange but is possible according to (3.4.3). Therefore, Lippman's argument is generally wrong.

Independent of whether or not the Rayleigh hypothesis is applicable, to know SA is sufficient for calculating the field at any point of region Ω to the boundary. First, suppose that point (x_0, z_0) and near to it point (x, z): $x = x_0 + \varrho \cos\varphi$, $z = z_0 + \varrho \sin\varphi$ are both below roughness: $z, z_0 < \min h(x)$, see Fig. 3.7. Then the following transformations are justified:

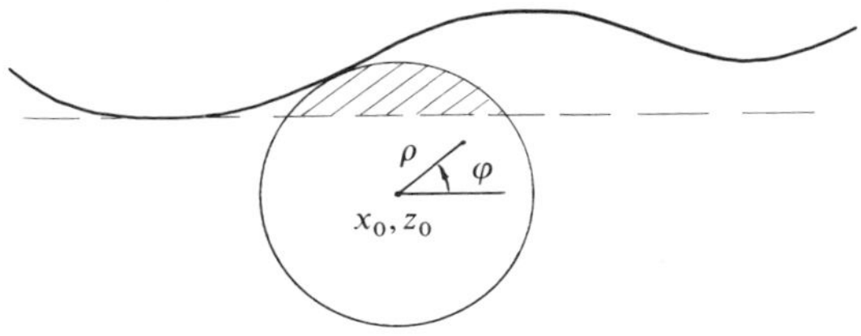

Fig. 3.7. On calculation of the field near the boundary via SA

$$\Psi_{sc}(x,z) = \int_{-\infty}^{+\infty} S(k) q_k^{-1/2} \exp(ikx - iq_k z)\, dk$$

$$= \int_{-\infty}^{+\infty} S(k) q_k^{-1/2} \exp(ikx_0 - iq_k z_0) \sum_{n=0}^{\infty} \frac{(i\varrho)^n}{n!} (k\cos\varphi - q_k \sin\varphi)^n\, dk$$

$$= \sum_{n=0}^{\infty} C_n \varrho^n \ , \tag{3.4.4}$$

where

$$C_n = \frac{i^n}{n!} \int_{-\infty}^{+\infty} S(k) q_k^{-1/2} \exp(ikx_0 - iq_k z_0)(k\cos\varphi - q_k \sin\varphi)^n\, dk \ . \tag{3.4.5}$$

The integral in (3.4.5) exists at any n by estimation (3.2.9). In fact series (3.4.4) converges in virtue of analyticity of the solution of the Helmholtz equation for all points inside the circle (Fig. 3.7) including those at $z > \min h(x)$. Note that while passing into region $z > \min h(x)$ we did not consider additional waves radiating from the parts of the boundary situated at $h(x) < z$.

Validity of the Rayleigh hypothesis is an issue which regards, in fact, the way of calculating the integral (3.2.5). The standard approach is the following:

$$\Psi_{sc}(x,z) = \lim_{M\to\infty} \Psi_{sc}^{(M)}(x,z) \ , \tag{3.4.6a}$$

where

$$\Psi_{sc}^{(M)}(x,z) = \int_{-M}^{M} S(k) q_k^{-1/2} \exp(ikx - iq_k z)\, dk \ . \tag{3.4.6b}$$

However the limit (3.4.6a) does not exist for some points (x, z), $z > \min h(x)$. This means that in the general case the Rayleigh hypothesis is wrong. Then the appropriate integral should be calculated in some other way, for instance, according to (3.4.4, 5). Other techniques of calculating Ψ_{sc} via SA are also available [3.9].

Regarding the Rayleigh hypothesis, it is worth noting that the system of functions $\{\Psi_{\boldsymbol{k}}\}$:

$$\Psi_{\boldsymbol{k}}(\boldsymbol{R}) = q_k^{-1/2} \exp(i\boldsymbol{k}\cdot\boldsymbol{r} - iq_k z) \tag{3.4.7}$$

is complete in Ω. This implies that any scattered field can be approximated with any accuracy by linear combinations of functions of the form (3.4.7):

$$\left| \Psi_{sc}(\boldsymbol{R}) - \int_{|\boldsymbol{k}|<M} S_M(\boldsymbol{k}) q_k^{-1/2} \exp(i\boldsymbol{k}\cdot\boldsymbol{r} - iq_k z)\, d\boldsymbol{k} \right| < \varepsilon \ , \tag{3.4.8}$$

where $M = M(\varepsilon)$. It should be emphasized that in contrast to (3.4.6b) quantity $S_M(\boldsymbol{k})$ in (3.4.8) depends on M. Moreover,

$$S(\boldsymbol{k}) = \lim_{M \to \infty} S_M(\boldsymbol{k}) \; . \tag{3.4.9}$$

This statement does not contradict the fact that the Rayleigh hypothesis does not generally stand. For example, nearly any continuous function can be approximated with any accuracy by polynomials, although this function cannot be expanded in a power series at all. This problem is discussed in detail by *Millar* [3.10]. We point out only key ideas of proving the completeness of a system of functions (3.4.7).

Let the function $[\mu(\boldsymbol{R})]^*$ given at the boundary Σ be orthogonal to functions of the form (3.4.7) in L_2:

$$\int \mu(\boldsymbol{R}')q_k^{-1/2} \exp[\mathrm{i}\boldsymbol{k}\cdot\boldsymbol{r}' - \mathrm{i}q_k h(\boldsymbol{r}')]\, d\Sigma_{\boldsymbol{R}'} = 0 \tag{3.4.10}$$

for all $\boldsymbol{k}$. We arrange the monopole sources with density $\mu(\boldsymbol{R})$ at the surface Σ and consider the function

$$\Psi(\boldsymbol{R}) = \int \mu(\boldsymbol{R}')G_0(\boldsymbol{R}' - \boldsymbol{R})\, d\Sigma_{\boldsymbol{R}'} \; . \tag{3.4.11}$$

Take an arbitrary point $(\boldsymbol{r}, z)$ with $z > \max h(\boldsymbol{r})$, multiply (3.4.10) by value

$$-\mathrm{i}q_k^{-1/2} \exp(-\mathrm{i}\boldsymbol{k}\cdot\boldsymbol{r} + \mathrm{i}q_k z)/8\pi^2$$

and integrate the result over all $\boldsymbol{k}$. In virtue of (2.1.24) we find

$$\Psi(\boldsymbol{R}) = 0 \; , \qquad \boldsymbol{R} = (\boldsymbol{r}, z) \; , \qquad z > \max h(\boldsymbol{r}) \; . \tag{3.4.12}$$

Function $\Psi(\boldsymbol{R})$ determined for any $\boldsymbol{R}$ by equality (3.4.11) is a radiative solution of the Helmholtz equation for both half-spaces separated by interface Σ. Solutions of the Helmholtz equation are analytic functions of coordinates. Therefore, it follows from (3.4.11) that this function is identically equal to zero not only at $z > \max h(\boldsymbol{r})$, but completely over all the upper half-space:

$$\Psi(\boldsymbol{R}) = 0 \; , \qquad z > h(\boldsymbol{r}) \; .$$

When setting in (3.1.27) $\Psi_{\text{in}} = 0$ and comparing the result to (3.4.11), one can see that function Ψ does not undergo a jump in the boundary: $[\Psi]_\Sigma = 0$. Therefore, function Ψ at the boundary should vanish:

$$\Psi(\boldsymbol{R})|_{\boldsymbol{R}\in\Sigma} = 0 \; .$$

Consider now the lower half-space. The radiative solution of the Helmholtz equation equal to zero at the boundary, should be zero identically in region $z < h(\boldsymbol{r})$, since the energy flux into the lower medium is zero, see (2.1.5). It follows that

$$\mu(\boldsymbol{R}) = 0 \; , \qquad \boldsymbol{R} \in \Sigma \; .$$

Turning the function μ into zero means that the set of functions (3.4.7) is complete at Σ. Therefore any function Ψ can be approximated with an arbitrary accuracy by linear combinations of these functions

$$\left\| \Psi[\boldsymbol{r}, h(\boldsymbol{r})] - \int\limits_{|\boldsymbol{k}|<M} S_M(\boldsymbol{k}) q_k^{-1/2} \exp[\mathrm{i}\boldsymbol{k}\cdot\boldsymbol{r} - \mathrm{i}q_k h(\boldsymbol{r})]\, d\boldsymbol{k} \right\|_{L_2} < \varepsilon \ , \tag{3.4.13}$$

where

$$\|f\|_{L_2} = (\textstyle\int |f|^2\, d\Sigma)^{1/2} \ .$$

Two radiative solutions of the Helmholtz equation whose difference tends to zero at the boundary, also tend to each other at any point of space. This can be proved by formula (3.4.16) which is of interest itself.

Let φ be an arbitrary radiative solution of the following boundary problem:

$$\begin{cases} \Delta\varphi + K^2\varphi = 0 \ , \qquad \boldsymbol{R} \in \Omega \ , & (3.4.14\text{a}) \\ \varphi|_\Sigma = f \ . & (3.4.14\text{b}) \end{cases}$$

We express solution φ at arbitrary point $\boldsymbol{R}_* \in \Omega$ in terms of the boundary value f. Let $G(\boldsymbol{R}, \boldsymbol{R}_*)$ be the Green's function of the Dirichlet boundary problem:

$$\begin{cases} (\Delta + K^2)G = \delta(\boldsymbol{R} - \boldsymbol{R}_*) \ , & (3.4.15\text{a}) \\ G|_\Sigma = 0 \ . & (3.4.15\text{b}) \end{cases}$$

Multiplying (3.4.14a) and (3.4.15a) by G and φ, respectively, substracting the results and integrating over Ω yields

$$\varphi(\boldsymbol{R}_*) = \int\limits_\Omega \operatorname{div}(\varphi\nabla G - G\nabla\varphi)\, d\boldsymbol{R} = \int\limits_\Sigma f(\boldsymbol{R}) \frac{\partial}{\partial\boldsymbol{n}} G(\boldsymbol{R}, \boldsymbol{R}_*)\, d\Sigma \ . \tag{3.4.16}$$

Relation (3.4.16) is obtained using the Gauss formula (2.3.2) and the boundary conditions (3.4.14b, 15b). Besides that, we took into account that the surface integral at an infinite part of the boundary vanishes when an arbitrarily small attenuation is introduced into the medium. Applying the Schwarz inequality to (3.4.16) we find

$$|\varphi(R_*)| \leqslant \|f\|_{L_2} \cdot \left(\int \left| \frac{\partial G(\boldsymbol{R}, \boldsymbol{R}_*)}{\partial \boldsymbol{n}} \right|^2 d\Sigma \right)^{1/2} \ .$$

By virtue of (3.4.13) and from this last inequality (3.4.8) follows.

Completeness of the set of functions (3.4.7) at Σ can be directly used for the numerical solution of boundary problems. For example, for the Dirichlet problem the field Ψ_{sc} at Σ is known, according to the boundary condition (2.1.15):

$$\Psi_{sc}[\boldsymbol{r}, h(\boldsymbol{r})] = -\Psi_{in}[\boldsymbol{r}, h(\boldsymbol{r})] \ .$$

Minimizing the form (3.4.13) with respect to S_M for rather large M by virtue of (3.4.9), $S(\boldsymbol{k})$ can be found. In particular, this approach was used in the work [3.11].

4. Small Peturbation Method

The problems of wave scattering at roughness of more or less general form have no exact analytic solutions. To solve them, various approximative methods were developed. Of the most significance among these methods are the small perturbation method (SPM) and the quasi-classical approximation (or geometrical optics approximation). These two approaches can be referred to as "classical". In the present chapter we consider the former.

It is based on the assumption that scattering is weak, so that the scattered field is small and does not distort significantly the original field in the absence of roughness. Mathematically this implies that the case of small undulations, $h \to 0$, is considered. The true small parameters of the present theory will be obtained after calculating corrections to the principal formulas of the method.

It will be shown further that SPM yields a relation for SA (scattering amplitude) in the form of the integral-power series in elevations h, which takes the following form:

$$S(\boldsymbol{k}, \boldsymbol{k}_0) = V_0(\boldsymbol{k})\delta(\boldsymbol{k} - \boldsymbol{k}_0) + \int C_1(k, k_0; \boldsymbol{r}_1)h(\boldsymbol{r}_1)d\boldsymbol{r}_1 + \int C_2(\boldsymbol{k}, \boldsymbol{k}_0; \boldsymbol{r}_1, \boldsymbol{r}_2)h(\boldsymbol{r}_1)h(\boldsymbol{r}_2)d\boldsymbol{r}_1 d\boldsymbol{r}_2 + \cdots \ . \tag{4.0.1}$$

In the absence of roughness, $h = 0$, SA reduces to the first term of this series describing plane wave reflection in specular direction with the reflection coefficient $V_0(\boldsymbol{k})$. In the general case of multicomponental fields and the interface between two media, for S considered as a matrix, the coefficient functions of the series (4.0.1) V_0, C_1, $C_2, \ldots$ are the corresponding matrices, as well. It is obvious that functions C_2, $C_3, \ldots$ can be considered from the very beginning as symmetric functions of $\boldsymbol{r}_1$, $\boldsymbol{r}_2, \ldots$.

The concrete form of the boundary conditions is related to the coefficient functions only. The simplest way to calculate them is as follows: Postulate that an asymptotic series of the form (4.0.1) exists for SA. This means that when considering the set of irregularities of the form $\varepsilon h(\boldsymbol{r})$ and setting $\varepsilon \to 0$, then the first $(n + 1)$ terms of series (4.0.1) yield a value of SA accurate within ε^n. In other words, the difference between the genuine SA and its approximation by the first $(n + 1)$ terms of series (4.0.1) has an order of $o(\varepsilon^n)$, i.e., vanishes quicker than ε^n.

In many cases, see for example, Sect. 4.2, the integral-power series of the kind (4.0.1) can be analytically obtained directly from the explicit equations for the surface sources of the form (3.1.42). Although the asymptotic character of this series (and only with this do they make sense) was probably not proved rigorously, physically, it is rather likely, and is confirmed by a great number of experiments.

For asymptotic series (4.0.1) the coefficient functions C_1, C_2, ... are determined unambiguously. For example, let two asymptotic series of the kind (4.0.1) with noncoincident coefficients C_1 and C_1' exist. Then

$$|\int C_1 h(\boldsymbol{r}_1)\,d\boldsymbol{r}_1 - \int C_1' h(\boldsymbol{r}_1)\,d\boldsymbol{r}_1|$$

$$\leqslant \varepsilon^{-1}|S(\boldsymbol{k},\boldsymbol{k}_0) - \int C_1\cdot\varepsilon h(\boldsymbol{r}_1)\,d\boldsymbol{r}_1| + \varepsilon^{-1}|S(\boldsymbol{k},\boldsymbol{k}_0) - \int C_1' h(\boldsymbol{r}_1)\,d\boldsymbol{r}_1|$$

$$= \varepsilon^{-1}\cdot \mathrm{o}(\varepsilon) \to 0 \ , \qquad \varepsilon \to 0 \ .$$

Consequently for an arbitrary function $h(\boldsymbol{r})$,

$$\int (C_1 - C_1')h(\boldsymbol{r}_1)\,d\boldsymbol{r}_1 = 0 \ ,$$

whence $C_1 = C_1'$. The coefficients $C_2, C_3, \ldots$ are considered in the same manner.

Since the coefficient functions are determined unambiguously, we may proceed as follows: Suppose that roughness of the form $\varepsilon h(\boldsymbol{r})$ is given by analytic functions of $\boldsymbol{r} = (x, y)$, and parameter ε is chosen so small that the Rayleigh hypothesis discussed in Chap. 3 is fulfilled. Then the field in the region adjoining the rough boundary can be calculated directly from representation of the form (2.2.10). Substituting this expression into the boundary conditions immediately gives an equation for SA. The method of calculating SA based on Rayleigh's equation was applied afterwards by *Fano* [4.1] and *Rice* [4.2], see also [4.3].

Since the arguments permitting us to apply the Rayleigh hypothesis when calculating expansion (4.0.1) are of abstract character, it is desirable to compare immediately two expansions: the first – obtained with the help of the Rayleigh hypothesis, and the second – obtained in a "fair" way. "Fair" way implies application of the perturbation theory to the exact integral equation for surface sources. Section 4.2 demonstrates that for the Dirichlet problem both expansions coincide in all orders.

The rest of the sections of this chapter are devoted to some concrete problems which cover all typical situations – scalar waves in a half-space (Sect. 4.1–3); vector waves in a half-space (Sect. 4.4); scalar waves at the interface between two half-spaces (Sect. 4.5); and finally, multicomponent (vector) fields at the boundary between two half-spaces (Sect. 4.6). All of them are of practical importance and are considered in a similar manner. In particular, for each case some antisymmetric bilinear form of two solutions of the boundary problem is introduced which allows us to obtain the reciprocity theorem and unitarity condition in a standard way.

Finally, we point out that it is convenient to use in the integral-power expansion (4.0.1), instead of elevations $h(\boldsymbol{r})$, their Fourier transforms $h(\boldsymbol{k})$, see formulas (4.1.7, 8) below. In this case expansion (4.0.1) takes the form

$$S(\boldsymbol{k},\boldsymbol{k}_0) = V_0(\boldsymbol{k})\delta(\boldsymbol{k}-\boldsymbol{k}_0) + \int \tilde{C}_1(\boldsymbol{k},\boldsymbol{k}_0;\boldsymbol{k}_1)h(\boldsymbol{k}_1)\delta(\boldsymbol{k}-\boldsymbol{k}_0-\boldsymbol{k}_1)\,d\boldsymbol{k}_1$$
$$+ \int \tilde{C}_2(\boldsymbol{k},\boldsymbol{k}_0;\boldsymbol{k}_1,\boldsymbol{k}_2)h(\boldsymbol{k}_1)h(\boldsymbol{k}_2)\delta(\boldsymbol{k}-\boldsymbol{k}_0-\boldsymbol{k}_1-\boldsymbol{k}_2)\,d\boldsymbol{k}_1\,d\boldsymbol{k}_2 + \cdots \,. \tag{4.0.2}$$

Functions $\tilde{C}_n(\boldsymbol{k},\boldsymbol{k}_0;\boldsymbol{k}_1,\ldots,\boldsymbol{k}_n)$ can obviously be assumed to be symmetric with respect to the arguments $\boldsymbol{k}_1$, ..., $\boldsymbol{k}_n$. Presence of δ-functions in this formula is due to transformation property (2.4.1), since at the shift of boundary $h(\boldsymbol{r}) \to h(\boldsymbol{r}-\boldsymbol{d})$ we have $h(\boldsymbol{k}) \to \exp(-\mathrm{i}\boldsymbol{k}\cdot\boldsymbol{d})h(\boldsymbol{k})$. Integrating the δ-functions in (4.0.2) we obtain

$$S(\boldsymbol{k},\boldsymbol{k}_0) = V_0(\boldsymbol{k})\delta(\boldsymbol{k}-\boldsymbol{k}_0) + 2\mathrm{i}(q_k q_0)^{1/2}B(\boldsymbol{k},\boldsymbol{k}_0)h(\boldsymbol{k}-\boldsymbol{k}_0)$$
$$+ (q_k q_0)^{1/2}\cdot\int B_2(\boldsymbol{k},\boldsymbol{k}_0;\boldsymbol{k}')h(\boldsymbol{k}-\boldsymbol{k}')h(\boldsymbol{k}'-\boldsymbol{k}_0)\,d\boldsymbol{k}' + \cdots \,. \tag{4.0.3}$$

Extracting the factor $(q_k q_0)^{1/2}$ in the coefficients in (4.0.3) simplifies the following formulas. This term describes the Bragg's wave scattering at the spectral components of roughness, see Sect. 4.1 below, and dimensionless function $B(\boldsymbol{k},\boldsymbol{k}_0)$ is independent of elevations and is totally related to boundary conditions; it is determined for all the problems considered in the present chapter.

4.1 Dirichlet Problem

The present section deals with the scalar problem with boundary condition

$$\Psi|_{z=h(\boldsymbol{r})} = 0 \,. \tag{4.1.1}$$

In virtue of the above procedure we obtain the following (Rayleigh's) equation:

$$q_0^{-1/2}\exp(\mathrm{i}\boldsymbol{k}_0\cdot\boldsymbol{r} + \mathrm{i}q_0 h(\boldsymbol{r})] + \int q_k^{-1/2}\exp[\mathrm{i}\boldsymbol{k}\cdot\boldsymbol{r} - \mathrm{i}q_k h(\boldsymbol{r})]S(\boldsymbol{k},\boldsymbol{k}_0)\,d\boldsymbol{k} = 0 \tag{4.1.2}$$

or

$$\int q_{k'}^{-1/2}S(\boldsymbol{k}',\boldsymbol{k}_0)\exp(\mathrm{i}\boldsymbol{k}'\cdot\boldsymbol{r})\,d\boldsymbol{k}' + \int q_{k'}^{-1/2}S(\boldsymbol{k}',\boldsymbol{k}_0)\exp(\mathrm{i}\boldsymbol{k}'\cdot\boldsymbol{r})$$
$$\times\{\exp(-\mathrm{i}q_{k'}h(\boldsymbol{r})] - 1)\}\,d\boldsymbol{k}' = -q_0^{-1/2}\exp[\mathrm{i}\boldsymbol{k}_0\cdot\boldsymbol{r} + \mathrm{i}q_0 h(\boldsymbol{r})] \,.$$

Multiplying the last equation by $\exp(-\mathrm{i}\boldsymbol{k}\cdot\boldsymbol{r})/(2\pi)^2$ and integrating over $\boldsymbol{r}$ according to (2.1.20) we have

$$S(\boldsymbol{k},\boldsymbol{k}_0) = -\int (q_k/q_{k'})^{1/2}S(\boldsymbol{k}',\boldsymbol{k}_0)\exp[-\mathrm{i}(\boldsymbol{k}-\boldsymbol{k}')\cdot\boldsymbol{r}]\{\exp[-\mathrm{i}q_{k'}h(\boldsymbol{r})] - 1\}$$
$$\times\frac{d\boldsymbol{r}\,d\boldsymbol{k}'}{(2\pi)^2} - \int (q_k/q_0)^{1/2}\exp[-\mathrm{i}(\boldsymbol{k}-\boldsymbol{k}_0)\cdot\boldsymbol{r} + \mathrm{i}q_0 h(\boldsymbol{r})]\,d\boldsymbol{r}/(2\pi)^2 \,. \tag{4.1.3}$$

For small $h \to 0$ it is convenient to solve this equation by iterations. For this we set

$$S = S^{(0)} + S^{(1)} + S^{(2)} + \cdots \ , \tag{4.1.4}$$

where $S^{(h)}$ has an order h^n and substitute (4.1.4) into (4.1.3), separating the terms of the same order. Assuming in (4.1.3) $h = 0$ we have

$$S^{(0)}(\boldsymbol{k}, \boldsymbol{k}_0) = -\delta(\boldsymbol{k} - \boldsymbol{k}_0) \ . \tag{4.1.5}$$

Further,

$$S^{(1)}(\boldsymbol{k}, \boldsymbol{k}_0) = -\int (q_k/q_{k'})^{1/2} S^{(0)}(\boldsymbol{k}', \boldsymbol{k}) \exp[-\mathrm{i}(\boldsymbol{k} - \boldsymbol{k}')\cdot\boldsymbol{r}]\cdot[-\mathrm{i}q_{k'}h(\boldsymbol{r})]$$

$$\times \frac{d\boldsymbol{r}\, d\boldsymbol{k}'}{(2\pi)^2} - \int (q_k/q_0)^{1/2} \exp[-\mathrm{i}(\boldsymbol{k} - \boldsymbol{k}_0)\cdot\boldsymbol{r}]\cdot \mathrm{i}q_0 h(\boldsymbol{r})\, d\boldsymbol{r}/(2\pi)^2$$

or

$$S^{(1)}(\boldsymbol{k}, \boldsymbol{k}_0) = -2\mathrm{i}(q_k q_{k_0})^{1/2} h(\boldsymbol{k} - \boldsymbol{k}_0) \ , \tag{4.1.6}$$

where

$$h(\boldsymbol{k}) = \int \exp(-\mathrm{i}\boldsymbol{k}\cdot\boldsymbol{r}) h(\boldsymbol{r})\, d\boldsymbol{r}/(2\pi)^2 \ . \tag{4.1.7}$$

We shall not introduce new notation for the Fourier transform of the roughness shape, since the argument ($h(\boldsymbol{r})$, $h(\boldsymbol{r}')$, $h(\boldsymbol{r}_n)$ etc., or $h(\boldsymbol{k})$, $h(\boldsymbol{k}')$, $h(\boldsymbol{k}_n)$ etc.) and the structure of corresponding formulas clearly show what value is considered. The inverse Fourier transform yields

$$h(\boldsymbol{r}) = \int \exp(\mathrm{i}\boldsymbol{k}\cdot\boldsymbol{r}) h(\boldsymbol{k})\, d\boldsymbol{k} \ . \tag{4.1.8}$$

It is seen from (4.1.6) that in the first order of perturbation theory we have an ordinary process of Bragg's scattering at the roughness Fourier-components. The most important feature of this process is due to the fact that $S(\boldsymbol{k}, \boldsymbol{k}_0) \sim h(\boldsymbol{k} - \boldsymbol{k}_0)$. In particular, the angle distribution of the scattered field is related to the width of the roughness spectrum $\Delta k \sim l^{-1}$ where l is the characteristic horizontal length scale. Horizontal components of the wave vector of the most part of the scattered field will be bounded within the limits $k_0 \pm l^{-1}$. In particular, for smooth undulations with a narrow spectrum: $k_0 l \gg 1$, scattering exists in a narrow range of angles in the vicinity of specular direction. In the reverse limiting case: $k_0 l \ll 1$ with the characteristic horizontal scale of roughness much less than the incident wave length, scattering occurs in all directions.

One more iteration of (4.1.3) allows us to calculate the second order corrections. Simple calculations give

$$S^{(2)}(\boldsymbol{k}, \boldsymbol{k}_0) = 2(q_k q_{k_0})^{1/2} \int q_{k_1} h(\boldsymbol{k} - \boldsymbol{k}_1) h(\boldsymbol{k}_1 - \boldsymbol{k}_0)\, d\boldsymbol{k}_1 \ . \tag{4.1.9a}$$

Combining formulas (4.1.5, 6, 9), we write within an accuracy to the values of $O(h^2)$ order

$$S(\boldsymbol{k},\boldsymbol{k}_0) = -\delta(\boldsymbol{k}-\boldsymbol{k}_0) - 2\mathrm{i}(q_k q_{k_0})^{1/2} h(\boldsymbol{k}-\boldsymbol{k}_0) + 2(q_k q_{k_0})^{1/2} \int q_{k_1} h(\boldsymbol{k}-\boldsymbol{k}_1) h(\boldsymbol{k}_1-\boldsymbol{k}_0)\, d\boldsymbol{k}_1 \ . \tag{4.1.10a}$$

One more iteration of (4.1.3) gives

$$S^{(3)}(\boldsymbol{k},\boldsymbol{k}_0) = \mathrm{i}(q_k q_{k_0})^{1/2} \int [2q_{k_1}q_{k_2} - \tfrac{1}{2}(q_{k_1}^2 + q_{k_2}^2) + \tfrac{1}{6}(q_k^2 + q_{k_0}^2)] h(\boldsymbol{k}-\boldsymbol{k}_1) h(\boldsymbol{k}_1-\boldsymbol{k}_2) h(\boldsymbol{k}_2-\boldsymbol{k}_0)\, d\boldsymbol{k}_1\, d\boldsymbol{k}_2 \ . \tag{4.1.9b}$$

Hence applying the Rayleigh hypothesis for the Dirichlet problem enables us to obtain a representation in the form of an integral-power of (4.0.2) type, i.e.,

$$S(\boldsymbol{k},\boldsymbol{k}_0) = -\delta(\boldsymbol{k}-\boldsymbol{k}_0) - 2\mathrm{i}(q_k q_{k_0})^{1/2} \int h(\boldsymbol{k}_1)\delta(\boldsymbol{k}-\boldsymbol{k}_0-\boldsymbol{k}_1)\, d\boldsymbol{k}_1 + (q_k q_{k_0})^{1/2} \times \int h(\boldsymbol{k}_1) h(\boldsymbol{k}_2)(q_{\boldsymbol{k}-\boldsymbol{k}_1} + q_{\boldsymbol{k}-\boldsymbol{k}_2})\delta(\boldsymbol{k}-\boldsymbol{k}_0-\boldsymbol{k}_1-\boldsymbol{k}_2)\, d\boldsymbol{k}_1\, d\boldsymbol{k}_2 + \cdots \ . \tag{4.1.10b}$$

Let us calculate SA for undulations of the some concrete form. The simplest and most convenient is the sin-function,

$$h(\boldsymbol{r}) = a\cos(px) \ , \tag{4.1.11}$$

where a is an amplitude and

$$p = 2\pi/L$$

with L being a period of irregularities. Respectively, in this case

$$h(\boldsymbol{k}) = \frac{a}{2}[\delta(k_x - p) + \delta(k_x + p)]\delta(k_y) \ .$$

If the incident wave propagates in the plane (x, z), i.e., $k_y^{(0)} = 0$, then for the scattered wave $k_y = 0$ also. This implies that in all formulas of the type (4.1.6–10) $k_y = 0$ should automatically be assumed and integrating over k_y is excluded. In the two-dimensional problem index x at the wave vector k_x will be omitted: $k_x = k$, etc. Hence we write

$$h(k) = \frac{a}{2}\cdot\delta(k-p) + \frac{a}{2}\cdot\delta(k+p) \tag{4.1.12}$$

and substitute this equation into (4.1.6, 9). Summing up the results, we obtain, to an accuracy of a^2

$$S(k,k_0) = [-1 + (q_k q_{k_0})^{1/2} \tfrac{1}{2} a^2 \cdot (q_{k-p} + q_{k+p})]\delta(k-k_0)$$
$$- \mathrm{i}(q_k q_{k_0})^{1/2} \cdot a \cdot [\delta(k-k_0-p) + \delta(k-k_0+p)]$$
$$+ \tfrac{1}{2}(q_k q_{k_0})^{1/2} a^2 \cdot [q_{k-p}\delta(k-k_0-2p) + q_{k+p}\delta(k-k_0+2p)] \ . \tag{4.1.13}$$

It is obvious that SA is proportional to δ-functions of the arguments of the kind $(k - k_0 + pn)$. In virtue of (2.2.10) this means that the scattered wave field consists of a discrete set of plane waves with the horizontal components of the wave vector equal to

$$k = k_0 + pn \ , \qquad n = 0, \pm 1, \dots \ . \tag{4.1.14}$$

This fact is not related to SPM but is valid in the general case for any periodical undulations and any boundary conditions. To demonstrate this we apply the translation property of SA (2.4.1) and shift the irregularities horizontally by a value equal to their period. According to (2.4.1)

$$S_{h(x-L)}(k,k_0) = \exp[-\mathrm{i}(k-k_0)L] \cdot S_{h(x)}(k,k_0)$$

However, shifting by the period preserves the shape of the boundary and therefore, SA must not change what is possible only under the condition

$$(k - k_0)L = 2\pi n \ , \qquad n = 0, \pm 1, \dots$$

which coincides with (4.1.14). Relation (4.1.14) could be interpreted in the following way, using the quantum-mechanical language. Irregularities with the period L contain states with momenta $pn = 2\pi n/L$, $n = 0, \pm 1, \dots$ and a phonon with a momentum equal to (k_0, q_{k_0}) as a result of interaction with roughness, acquires a horizontal momentum pn. Hence scattered phonons have momenta $(k_0 + pn, q_{k_0+pn})$. Thus, in the general case of periodical roughness SA has the form

$$S(k,k_0) = \sum_{-\infty}^{+\infty} S_n \delta(k - k_0 - pn) \tag{4.1.15}$$

and the corresponding expression for the field according to (2.2.10) is

$$\Psi = q_0^{-1/2} \exp(\mathrm{i}k_0 x + \mathrm{i}q_0 z) + \sum_n q_n^{-1/2} \exp(\mathrm{i}k_n x - \mathrm{i}q_n z) \cdot S_n \ , \tag{4.1.16}$$

where

$$k_n = k_0 + pn \ , \qquad q_n = q(k_n) \ .$$

The plane waves in (4.1.16) are often referred to as spectra (of the nth order).

The law of conservation of energy (2.3.10) for the case of periodical roughness reads

$$\sum_n^{(h)} |S_n|^2 = 1 \ , \tag{4.1.17}$$

where $\sum^{(h)}$ denotes summing over n corresponding to homogeneous waves: $|k_n| < K$.

Now consider the problem of small parameter of the theory. Suppose that roughness has some characteristic horizontal scale l. The typical example of such roughness is the above sinusoid. Write down expressions for the amplitudes of some first spectra which account for third order effects, according to (4.1.9b):

$$S_0 = -1 + a^2 q_0 \cdot (q_1 + q_{-1})/2 \ ,$$

$$S_1 = -\mathrm{i}a(q_0 q_1)^{1/2}\left\{1 - \frac{a^2}{4}\cdot[q_0 q_1 + q_0 q_{-1} + q_1 q_2 - (q_{-1}^2 + q_0^2 + q_1^2 + q_2^2)/4]\right\} \ ,$$

$$S_{\pm 2} = a^2 (q_0 q_{\pm 2})^{1/2} q_{\pm 1} \ , \qquad \text{etc.}$$

For successive approximations to have a sense, the higher order corrections in these formulas should be relatively small. In other words the following conditions:

$$a q_0 \ll 1 \ , \tag{4.1.18a}$$

$$a q_{k_0 \pm p} \ll 1 \ , \qquad a q_{k_0 \pm 2p} \ll 1 \tag{4.1.18b}$$

should be fulfilled. The product of characteristic value of elevation and the vertical component of wave number appeared in (4.1.18) is called the Rayleigh parameter. Generalizing conditions (4.1.18) shows that the applicability of SPM implies the small Rayleigh parameter with respect to the incident wave and to all spectra (including the virtual inhomogeneous ones) arising in the calculations as well. The last statement follows from (4.1.2) or (4.1.3), since an iterative solution of this equation is related to a series expansion of appropriate exponentials, whose arguments are proportional to the Rayleigh parameter.

Criterion (4.1.18a) yields

$$aK \sin \chi_0 \ll 1 \ . \tag{4.1.19}$$

It is reasonable to consider criterion (2.3.18b) in various limiting cases:

1. Let $p/K \ll 1$. In addition we suppose that
   1a: $k_0 + p \ll K$ or

$$p/K \ll 1 - \cos \chi_0 = 2 \sin^2(\chi_0/2) \ ,$$

where χ_0 is the grazing angle of the incident wave. In this case the spectra excited by interaction with roughness are far from inhomogeneous ones. In this case we can set $p = 0$ in (4.1.18b) that reduces to (4.1.19).

Now assume that the grazing angle χ_0 is small and 1b:

$$2\sin^2(\chi_0/2) \approx \chi_0^2/2 \ll p/K \ll 1$$

Then $|q_{k_0 \pm p}| \approx (2Kp)^{1/2}$ and (4.1.18b) gives

$$a(Kp)^{1/2} \ll 1 \tag{4.1.20}$$

Note that $(Kp)^{-1/2}$ is the size of the Fresnel zone $(\lambda R)^{1/2}$, for distances of the horizontal scale p^{-1} order. Therefore condition (4.1.20) means that for every grazing incidence the geometrical shadowing is insignificant due to strong diffraction effects.

2. Now examine the inverse limit case: $p/K \gg 1$. Hence $q_{k_0+p} \approx ip$ and (4.1.18) gives

$$ap \ll 1 \ . \tag{4.1.21}$$

The latter condition implies a small slope of roughness.

It is necessary to emphasize that a small slope is always needed, in fact. So for the case "1a" we represent the slope in the form

$$ap = aK \sin\chi_0 \{p/[K(1 - \cos\chi_0)]\} \operatorname{tg}(\chi_0/2) \ .$$

Here the first two factors are small as compared to unity (the latter is simply less then unity). For the case "1b" we write

$$ap = a(Kp)^{1/2}(p/K)^{1/2}$$

which is also much smaller than unity.

Now we consider the statistical case. It is clear that direct averaging (4.1.10) leads to an expression of the form (2.4.6), the mean reflection coefficient being given by formula

$$\bar{V}(\boldsymbol{k}) = -1 + 2q_k \int q_{k_1} W(\boldsymbol{k}_1 - \boldsymbol{k})\, d\boldsymbol{k}_1 \ . \tag{4.1.22}$$

Here $W(\boldsymbol{k})$ is the spectrum of roughness:

$$\overline{h(k_1)h(k_2)} = W(\boldsymbol{k}_1)\delta(\boldsymbol{k}_1 + \boldsymbol{k}_2) \ . \tag{4.1.23}$$

Function $W(\boldsymbol{k})$ is related to the correlation function of elevations

$$\tilde{W}(\boldsymbol{\varrho}) = \overline{h(\boldsymbol{r} + \boldsymbol{\varrho})h(\boldsymbol{r})}$$

by common relation

$$\tilde{W}(\boldsymbol{\varrho}) = \int \exp(i\boldsymbol{k}\cdot\boldsymbol{\varrho}) W(\boldsymbol{k})\, d\boldsymbol{k} \ . \tag{4.1.24}$$

Obviously both $\tilde{W}(\boldsymbol{\varrho})$ and $W(\boldsymbol{k})$ are even and real functions.

Now proceed to the second moments. We have

$$\Delta S(\boldsymbol{k}, \boldsymbol{k}_0) = -2\mathrm{i}(q_k q_{k_0})^{1/2} h(\boldsymbol{k} - \boldsymbol{k}_0) + \mathrm{O}(h^2)$$

and in virtue of (2.4.9) find

$$\Delta \mathscr{E}(\boldsymbol{k}, \boldsymbol{k}_0; \boldsymbol{a}) = 4(q_{\boldsymbol{k}-\boldsymbol{a}/2} q_{\boldsymbol{k}_0-\boldsymbol{a}/2} q^*_{\boldsymbol{k}+\boldsymbol{a}/2} q^*_{\boldsymbol{k}_0+\boldsymbol{a}/2})^{1/2} W(\boldsymbol{k} - \boldsymbol{k}_0) + \mathrm{O}(W^2) \ .$$

Respectively, we have for the cross-section of scattering within an accuracy to the W-order terms

$$\sigma(\boldsymbol{k}, \boldsymbol{k}_0) = 4 q_k q_{k_0} W(\boldsymbol{k} - \boldsymbol{k}_0) \ . \qquad (4.1.25)$$

It is obvious that the reciprocity theorem (2.4.15) holds in this case. Moreover, the symmetry relation (2.4.16) is automatically fulfilled. Let us verify whether the law of conservation of energy (2.4.11) holds. Simple calculation yields

$$1 - |\overline{V}_k|^2 = 2 q_k \int (q_{k_1} + q^*_{k_1}) W(\boldsymbol{k} - \boldsymbol{k}_1)\, d\boldsymbol{k}_1 + \mathrm{O}(W^2)$$

$$= 4 q_k \int_{|\boldsymbol{k}_1| < K} q_{k_1} W(\boldsymbol{k} - \boldsymbol{k}_1)\, d\boldsymbol{k}_1 + \mathrm{O}(W^2) \ .$$

Consequently, we, in fact, have with an accuracy to W:

$$1 - |\overline{V}_{\boldsymbol{k}}|^2 = \int_{|\boldsymbol{k}'| < K} \sigma(\boldsymbol{k}', \boldsymbol{k})\, d\boldsymbol{k}' \ .$$

Formulas (4.1.10, 22, 25) are the main results of application of SPM to the Dirichlet problem.

4.2 Small Perturbation Method and the Rayleigh Hypothesis

From Sect. 4.1 and other sections of the present chapter it follows that the application of the Rayleigh hypothesis for determining SA in the form of integral-power expansion reduces the problem to simple calculation. Reasoning in Sect. 4.1 showed that the coefficient functions of series (4.0.1) found in such way will be determined correctly. However, discussions concerning the Rayleigh hypothesis make it desirable to verify this fact by direct calculation. Attention to this problem was put in literature (in particular, see the works by *Burrows* [4.4], *Toigo et al.* [4.5], *Kazandjian* [4.6]). Concrete calculations of the first series terms confirmed this thesis, see, e.g., *Jackson et al.* [4.7]. For the terms of any order in the case of the Dirichlet problem, this was shown in the work by *Voronovich* [4.8]. In the present section we outline the proof from this work.

In the general case the scattering problem solution for the Dirichlet boundary condition reduces to a solution of some integral equation for surface sources density. We shall use (3.1.42) which for the plane incident wave has the form

$$v(\boldsymbol{r})/2 + \int v(\boldsymbol{r}')\frac{\partial}{\partial \boldsymbol{n}_{\boldsymbol{R}'}}G_0(\boldsymbol{R}' - \boldsymbol{R})\,d\Sigma' = -q_0^{-1/2}\exp[\mathrm{i}\boldsymbol{k}_0\cdot\boldsymbol{r} + \mathrm{i}q_0 h(\boldsymbol{r})] \; . \tag{4.2.1}$$

Here density of dipoles $v(\boldsymbol{r})$ at the rough boundary Σ is given as a function of horizontal coordinates and the points $\boldsymbol{R}' = [\boldsymbol{r}', h(\boldsymbol{r}')]$ and $\boldsymbol{R} = [\boldsymbol{r}, h(\boldsymbol{r})]$ belong to Σ. Symbol $\partial/\partial \boldsymbol{n}_{\boldsymbol{R}'}$ denotes the derivative with respect to external normal to the boundary Σ at the point $\boldsymbol{R}'$, and the principal value integral in (4.2.1) is supposed. With (4.2.1) we can determine the field at any point of space $\boldsymbol{R}_*$ by the formula

$$\Psi_{\mathrm{sc}}(\boldsymbol{R}_*) = \int v(\boldsymbol{r}')\frac{\partial}{\partial \boldsymbol{n}_{\boldsymbol{R}'}}G_0(\boldsymbol{R}' - \boldsymbol{R}_*)\,d\Sigma' \; .$$

Assuming that the point $\boldsymbol{R}_*$ is below the boundary, $z_* < \min h(\boldsymbol{r})$, and using for G_0 expansion in plane waves (2.1.24), after simple calculations by virtue of (2.2.10) we obtain

$$S(\boldsymbol{k}, \boldsymbol{k}_0) = \frac{q_k^{1/2}}{8\pi^2}\int v(\boldsymbol{r})\left(1 + \frac{\boldsymbol{k}\nabla h}{q_k}\right)\exp[-\mathrm{i}\boldsymbol{k}\cdot\boldsymbol{r} + \mathrm{i}q_k h(\boldsymbol{r})]\,d\boldsymbol{r} \; . \tag{4.2.2}$$

In these calculations formulas (2.1.10, 11) were used by which

$$d\Sigma\frac{\partial}{\partial \boldsymbol{n}} = d\boldsymbol{r}\left(\frac{\partial}{\partial z} - \nabla h\cdot\nabla\right) .$$

It can be easily proved that at $h \to 0$ the kernel of integral equation $(\partial/\partial \boldsymbol{n}_{\boldsymbol{R}'})G_0(\boldsymbol{R}' - \boldsymbol{R})$ vanishes also. Consequently, for small h we can apply the iterations to solve (4.2.1).

Iterating (4.2.1), we write its formal solution as the following series:

$$q_0^{1/2}v = -2\cdot\exp[\mathrm{i}\boldsymbol{k}_0\cdot\boldsymbol{r} + \mathrm{i}q_0 h(\boldsymbol{r})] + \sum_{m=2}^{\infty}(-2)^m\int\exp[\mathrm{i}\boldsymbol{k}_0\cdot\boldsymbol{r}_1 + \mathrm{i}q_0 h(\boldsymbol{r}_1)] \times\prod_{j=1}^{m-1}\int d\Sigma_j\frac{\partial G_0(\boldsymbol{R}_j - \boldsymbol{R}_{j+1})}{\partial \boldsymbol{n}_{\boldsymbol{R}_j}} \; . \tag{4.2.3}$$

Here $\boldsymbol{R}_j = [\boldsymbol{r}_j, h(\boldsymbol{r}_j)]$ and for $j = m - 1$, $\boldsymbol{R}_{j+1} = [\boldsymbol{r}, h(\boldsymbol{r})]$ by definition. We first obtain the power expansion of the kernel participating in iterations in (4.2.3). Write down the obvious relation

$$\frac{dG_0}{dz} = \frac{z}{R}\frac{dG_0}{dR} \; ,$$

where $G_0 = G_0(R)$, $R = (r^2 + z^2)^{1/2}$. As a result we find

$$d\Sigma'(\partial/\partial \boldsymbol{n}_{\boldsymbol{R}'})G_0(\boldsymbol{R}' - \boldsymbol{R}) = d\boldsymbol{r}'(h_{\boldsymbol{r}'} - h_{\boldsymbol{r}} - \nabla h_{\boldsymbol{r}'}\cdot(\boldsymbol{r}' - \boldsymbol{r})]\frac{1}{z}\frac{\partial G_0}{\partial z}\bigg|_{\boldsymbol{R}=(\boldsymbol{r}'-\boldsymbol{r},h_{\boldsymbol{r}'}-h_{\boldsymbol{r}})} . \tag{4.2.4}$$

Now expand G_0 in powers of z:

$$G_0 = -\frac{\exp[iK(r^2 + z^2)^{1/2}]}{4\pi(r^2 + z^2)^{1/2}} = -\sum_{n=0}^{\infty}\frac{z^{2n}}{(2n)!}\frac{\partial^{2n}}{\partial z^{2n}}\frac{\exp[iK(r^2 + z^2)^{1/2}]}{4\pi(r^2 + z^2)^{1/2}}\bigg|_{z=0} . \tag{4.2.5}$$

We emphasize that the principal value integrals in (4.2.3) are calculated. As at $R \neq 0$ G_0 satisfies the homogeneous Helmholtz equation

$$\frac{\partial^2}{\partial z^2}G_0 = -(K^2 + \nabla_{\boldsymbol{r}}^2)G_0$$

we can write down

$$\begin{aligned} d\Sigma'\frac{\partial}{\partial \boldsymbol{n}_{\boldsymbol{R}'}}G_0(\boldsymbol{R}' - \boldsymbol{R}) &= d\boldsymbol{r}'[h_{\boldsymbol{r}'} - h_{\boldsymbol{r}} - \nabla h_{\boldsymbol{r}'}\cdot(\boldsymbol{r}' - \boldsymbol{r})] \\ &\quad\times \sum_{n=0}^{\infty}(-1)^n\frac{(h_{\boldsymbol{r}'} - h_{\boldsymbol{r}})^{2n}}{(2n+1)!}(K^2 + \nabla_{\boldsymbol{r}'}^2)^{n+1} \\ &\quad\times \frac{\exp(iK|\boldsymbol{r}' - \boldsymbol{r}|)}{4\pi|\boldsymbol{r}' - \boldsymbol{r}|} . \end{aligned} \tag{4.2.6}$$

Series (4.2.5) converges for $z^2 < r^2$ and (4.2.6) under condition

$$(h_{\boldsymbol{r}'} - h_{\boldsymbol{r}})^2 < (\boldsymbol{r}' - \boldsymbol{r})^2 .$$

The latter will be fulfilled if the slope of roughness does not exceed unity. Substitution of series (4.2.6) into (4.2.3) and then into (4.2.2), and expansion in powers of h two exponentials in (4.2.3) give the sought integral-power expansion over elevations for SA.

We point out that expansion of the kernel (4.2.6) begins from the first order terms in h. Therefore the mth term in series (4.2.3) has an order $O(h^{m-1})$, and to obtain the sought expansion with an accuracy to arbitrary finite order in (4.2.3), the finite number of terms should be considered.

Substitutions of (4.2.6) into (4.2.3) and the result into (4.2.2) yield the expansion of SA which is the perturbation theory applied to the exact integral equation (4.2.1). It does not use any reasonings related to the Rayleigh hypothesis and in this sense is correct. The question is whether this expansion coincides with that obtained in solving the Rayleigh equation (4.1.2)? We will show that this is the case.

For this substitute into the Rayleigh equation (4.1.2) the expansion obtained above. According to (4.2.2), we have

$$\int q_k^{-1/2} \exp[\mathrm{i}\boldsymbol{k}\cdot\boldsymbol{r} - \mathrm{i}q_k h(\boldsymbol{r})] S(\boldsymbol{k},\boldsymbol{k}_0)\, d\boldsymbol{k}$$
$$= (8\pi^2)^{-1} \int v_{\boldsymbol{r}'}(1 + \boldsymbol{k}\cdot\nabla h_{\boldsymbol{r}'}/q_k)\exp[-\mathrm{i}\boldsymbol{k}\cdot(\boldsymbol{r}'-\boldsymbol{r}) + \mathrm{i}q_k(h_{\boldsymbol{r}'} - h_{\boldsymbol{r}})]\, d\boldsymbol{r}'\, d\boldsymbol{k}\ . \tag{4.2.7}$$

Write down the following identity which can be easily verified:

$$(1 + \boldsymbol{k}\cdot\nabla h_{\boldsymbol{r}'}/q_k)\exp[-\mathrm{i}\boldsymbol{k}\cdot(\boldsymbol{r}'-\boldsymbol{r}) + \mathrm{i}q_k(h_{\boldsymbol{r}'} - h_{\boldsymbol{r}})]$$
$$= \exp[-\mathrm{i}\boldsymbol{k}\cdot(\boldsymbol{r}'-\boldsymbol{r})] + \exp[-\mathrm{i}\boldsymbol{k}\cdot(\boldsymbol{r}'-\boldsymbol{r})][h_{\boldsymbol{r}'} - h_{\boldsymbol{r}} - (\boldsymbol{r}'-\boldsymbol{r})\cdot\nabla h_{\boldsymbol{r}'}]$$
$$\times \{\exp[\mathrm{i}q_k(h_{\boldsymbol{r}'} - h_{\boldsymbol{r}})] - 1\}(h_{\boldsymbol{r}'} - h_{\boldsymbol{r}})^{-1} - (\partial/\partial\boldsymbol{k})$$
$$\times \{\nabla h_{\boldsymbol{r}'}\exp[-\mathrm{i}\boldsymbol{k}\cdot(\boldsymbol{r}'-\boldsymbol{r})]\{\exp[\mathrm{i}q_k(h_{\boldsymbol{r}'} - h_{\boldsymbol{r}})] - 1\}$$
$$\times [\mathrm{i}(h_{\boldsymbol{r}'} - h_{\boldsymbol{r}})]^{-1}\}\ . \tag{4.2.8}$$

Substitute (4.2.8) into (4.2.7) and expand $\exp[\mathrm{i}q_k(h_{\boldsymbol{r}'} - h_{\boldsymbol{r}})]$ in the power series. In the calculation of (4.2.7) we are concern with a finite order in h, so only finite sums are integrated. Therefore, it is allowable to change in (4.2.7) the order of integration over $\boldsymbol{r}'$ and $\boldsymbol{k}$. If we assume function $h(\boldsymbol{r})$ to be smooth and rapidly decaying at $|\boldsymbol{r}| \to \infty$, then its Fourier transform will also decay at $|\boldsymbol{k}| \to \infty$ faster than any power of k. Hence the last term in (4.2.8) vanishes as a result of integration over $\boldsymbol{k}$. We select in the second term in (4.2.8) the terms corresponding to even and odd powers of $q_k = (K^2 - k^2)^{1/2}$. We have

$$(8\pi^2)^{-1}\int d\boldsymbol{k}\int v_{\boldsymbol{r}'}\, d\boldsymbol{r}'\exp[-\mathrm{i}\boldsymbol{K}\cdot(\boldsymbol{r}'-\boldsymbol{r})]\frac{\exp[\mathrm{i}q_k(h_{\boldsymbol{r}'} - h_{\boldsymbol{r}})] - 1}{(h_{\boldsymbol{r}'} - h_{\boldsymbol{r}})}$$
$$\times [h_{\boldsymbol{r}'} - h_{\boldsymbol{r}} - (\boldsymbol{r}'-\boldsymbol{r})\cdot\nabla h_{\boldsymbol{r}'}]$$
$$= (8\pi^2)^{-1}\int d\boldsymbol{k}\int v_{\boldsymbol{r}'}\, d\boldsymbol{r}'\exp[-\mathrm{i}\boldsymbol{k}\cdot(\boldsymbol{r}'-\boldsymbol{r})]$$
$$\times\left\{\sum_{n=1}^{\infty}\frac{(-1)^n}{(2n)!}(K^2 - k^2)^n(h_{\boldsymbol{r}'} - h_{\boldsymbol{r}})^{2n-1}\right.$$
$$\left.+\sum_{n=1}^{\infty}\frac{(-1)^{n+1}}{(2n+1)!}(h_{\boldsymbol{r}'} - h_{\boldsymbol{r}})^{2n}(K^2 - k^2)^{n+1}/q_k\right\}$$
$$\times [h_{\boldsymbol{r}'} - h_{\boldsymbol{r}} - (\boldsymbol{r}'-\boldsymbol{r})\cdot\nabla h_{\boldsymbol{r}'}]\ . \tag{4.2.9}$$

Setting

$$\exp[-\mathrm{i}\boldsymbol{k}\cdot(\boldsymbol{r}'-\boldsymbol{r})](K^2 - k^2)^n = (K^2 + \nabla_{\boldsymbol{r}'}^2)^n\exp[-\mathrm{i}\boldsymbol{k}\cdot(\boldsymbol{r}'-\boldsymbol{r})]$$

and integrating over $\boldsymbol{k}$ and $\boldsymbol{r}'$, we find

$$(8\pi^2)^{-1} \int d\boldsymbol{k} \int v_{\boldsymbol{r}'} d\boldsymbol{r}' \exp[-\mathrm{i}\boldsymbol{k}\cdot(\boldsymbol{r}'-\boldsymbol{r})] \sum_{n=1}^{\infty} \frac{(-1)^n}{(2n)!}(K^2-k^2)^n (h_{\boldsymbol{r}'}-h_{\boldsymbol{r}})^{2n-1}$$
$$= \frac{1}{2}\sum_{n=1}^{\infty} \frac{(-1)^n}{(2n)!}(K^2+\nabla_{\boldsymbol{r}'}^2)^n \Big|_{\boldsymbol{r}'=\boldsymbol{r}} v_{\boldsymbol{r}'}(h_{\boldsymbol{r}'}-h_{\boldsymbol{r}})^{2n-1}(h_{\boldsymbol{r}'}-h_{\boldsymbol{r}}-(\boldsymbol{r}'-\boldsymbol{r})\nabla h_{\boldsymbol{r}'}) \ . \tag{4.2.10}$$

It is obvious that at $\boldsymbol{r}' \to \boldsymbol{r}$ we have

$$(h_{\boldsymbol{r}'}-h_{\boldsymbol{r}})^{2n-1}[h_{\boldsymbol{r}'}-h_{\boldsymbol{r}}-(\boldsymbol{r}'-\boldsymbol{r})\cdot\nabla h_{\boldsymbol{r}'}] = \mathrm{O}(|\boldsymbol{r}'-\boldsymbol{r}|^{2n+1}) \ .$$

After the action of the operator $(K^2+\nabla_{\boldsymbol{r}'}^2)^n$ the power order of coordinates decrease by not more than $2n$, and therefore the expression written down in (4.2.10) becomes zero.

The rest of the second sum in (4.2.9) is transformed as follows:

$$(8\pi^2)^{-1} \int d\boldsymbol{k} \int v_{\boldsymbol{r}'} d\boldsymbol{r}' \exp[-\mathrm{i}\boldsymbol{k}\cdot(\boldsymbol{r}'-\boldsymbol{r})][h_{\boldsymbol{r}'}-h_{\boldsymbol{r}}-(\boldsymbol{r}'-\boldsymbol{r})\cdot\nabla h_{\boldsymbol{r}'}]$$
$$\times \sum_{n=1}^{\infty} \frac{(-1)^{n+1}}{(2n+1)!}(h_{\boldsymbol{r}'}-h_{\boldsymbol{r}})^{2n}(K^2-k^2)^{n+1}/\mathrm{i}q_k$$
$$= \int d\boldsymbol{r}' v_{\boldsymbol{r}'}[h_{\boldsymbol{r}'}-h_{\boldsymbol{r}}-(\boldsymbol{r}'-\boldsymbol{r})\cdot\nabla h_{\boldsymbol{r}'}]$$
$$\times \sum_{n=0}^{\infty} (h_{\boldsymbol{r}'}-h_{\boldsymbol{r}})^{2n} \frac{(-1)^n}{(2n+1)!}(K^2+\nabla_{\boldsymbol{r}'}^2)^{n+1} \frac{\mathrm{i}}{8\pi^2}\int \frac{\exp[\mathrm{i}\boldsymbol{k}\cdot(\boldsymbol{r}'-\boldsymbol{r})]}{q_k} d\boldsymbol{k}. \tag{4.2.11}$$

In virtue of (2.1.24) we have

$$\frac{\mathrm{i}}{8\pi^2}\int \frac{\exp[\mathrm{i}\boldsymbol{k}\cdot(\boldsymbol{r}'-\boldsymbol{r})]}{q_k} d\boldsymbol{k} = \frac{\exp[\mathrm{i}K|\boldsymbol{r}'-\boldsymbol{r}|]}{4\pi|\boldsymbol{r}'-\boldsymbol{r}|} \ .$$

Then comparing the expansion obtained on the right-hand side of (4.2.11) with (4.2.6) we find

$$(4.2.11) = \int v_{\boldsymbol{R}'} \frac{\partial}{\partial \boldsymbol{n}_{\boldsymbol{R}'}} G_0(\boldsymbol{R}'-\boldsymbol{R})\, d\Sigma' \ .$$

Taking into account the first term in (4.2.8) yields

$$\int q_k^{-1/2} \exp[\mathrm{i}\boldsymbol{k}\cdot\boldsymbol{r} - \mathrm{i}q_k h(\boldsymbol{r})] S(\boldsymbol{k},\boldsymbol{k}_0)\, d\boldsymbol{k} = v(\boldsymbol{r})/2 + \int v_{\boldsymbol{R}'} \frac{\partial}{\partial \boldsymbol{R}_{\boldsymbol{R}'}} G_0(\boldsymbol{R}'-\boldsymbol{R})\, d\Sigma' \ . \tag{4.2.12}$$

This equality has the following meaning: when determining the value of $S(\boldsymbol{k}, \boldsymbol{k}_0)$ by (4.2.2) with the arbitrarily smooth function $v(\boldsymbol{r})$, then the left and right parts of equality (4.2.12) coincide with an accuracy to any order terms in h. In particular, if the solution of (4.2.1) is taken for v, then identity (4.2.12) takes the form coinciding with the Rayleigh equation (4.1.2):

$$\int S(\boldsymbol{k}, \boldsymbol{k}_0) q_k^{-1/2} \exp[\mathrm{i}\boldsymbol{k}\cdot\boldsymbol{r} - \mathrm{i}q_k h(\boldsymbol{r})]\, d\boldsymbol{k} = -q_0^{-1/2} \exp[\mathrm{i}\boldsymbol{k}_0\cdot\boldsymbol{r} + \mathrm{i}q_0 h(\boldsymbol{r})] \ . \tag{4.2.13}$$

Thus for the Dirichlet problem, formal expansions of SA in the integral-power series obtained (A) from the perturbation theory for integral equation (4.2.1) for density of the surface sources, and (B) from the Rayleigh equation (4.2.13), coincide.

Due to the result obtained here we can more reliably apply the Rayleigh hypothesis to calculate SA by the small perturbation method.

It is necessary to make a reservation here. Validity of the Rayleigh hypothesis for smooth roughness of the form $\varepsilon h(\boldsymbol{r})$ at rather small ε is established in Sect. 3.2, strictly speaking, only for the Dirichlet problem. However, general discussion of the Rayleigh hypothesis in Sect. 3.4 and calculations in Sect. 3.2 show that this result should be rather general and valid for a wide class of boundary conditions. Further into the present chapter we will constantly make use of this statement.

4.3 Neumann Problem

In the present section we apply the previously viewed procedure of calculating the integral-power expansions of SA for the Neumann problem when the boundary condition for the Helmholtz equation has the form

$$\left.\frac{\partial\psi}{\partial \boldsymbol{n}}\right|_{z=h(\boldsymbol{r})} = 0 \ . \tag{4.3.1}$$

For the sound waves the boundary satisfying condition (4.3.1) is called acoustically rigid or hard. Taking into account (2.1.10) it can also be written as

$$\left.\left(\frac{\partial}{\partial z} - \nabla h\cdot\nabla\right)\psi\right|_{z=h(\boldsymbol{r})} = 0 \ . \tag{4.3.2}$$

The equation for SA is obtained after substituting the representation for the field (2.2.10) into (4.3.2):

$$\left(\frac{\partial}{\partial z} - \nabla h\cdot\nabla\right) \times \left[q_0^{-1/2}\exp(\mathrm{i}\boldsymbol{k}_0\cdot\boldsymbol{r} + \mathrm{i}q_0 z) + \int q_k^{-1/2}\exp(\mathrm{i}\boldsymbol{k}\cdot\boldsymbol{r} - \mathrm{i}q_k z) S(\boldsymbol{k}, \boldsymbol{k}_0)\, d\boldsymbol{k}\right]_{z=h(\boldsymbol{r})} = 0 \ . \tag{4.3.3}$$

Calculations give

$$q_0^{1/2}\left(1-\frac{\boldsymbol{k}_0\cdot\nabla h}{q_0}\right)\exp[\mathrm{i}\boldsymbol{k}_0\cdot\boldsymbol{r}+\mathrm{i}q_0h(\boldsymbol{r})]-\int\left(\frac{1+\boldsymbol{k}\cdot\nabla h}{q_k}\right)$$
$$\times\, q_k^{1/2}\exp[\mathrm{i}\boldsymbol{k}\cdot\boldsymbol{r}-\mathrm{i}q_kh(\boldsymbol{r})]\cdot S(k,k_0)\,dk=0\ . \tag{4.3.4}$$

Here S should be sought in the form of expansion (4.1.4)

$$S=S^{(0)}+S^{(1)}+S^{(2)}+\cdots\ ,$$

where $S^{(n)}=\mathrm{O}(h^n)$ and all the exponentials of h entering into (4.3.3) should be expanded in the power series. As a result, we find

$$S^{(0)}(\boldsymbol{k},\boldsymbol{k}_0)=\delta(\boldsymbol{k}-\boldsymbol{k}_0)\ .$$

Equation

$$\int q_k^{1/2}S^{(1)}(\boldsymbol{k},\boldsymbol{k}_0)\exp(\mathrm{i}\boldsymbol{k}\cdot\boldsymbol{r})\,d\boldsymbol{k}=2q_0^{1/2}\left(\mathrm{i}q_0h-\frac{\boldsymbol{k}_0\cdot\nabla h}{q_0}\right)\exp(\mathrm{i}\boldsymbol{k}_0\cdot\boldsymbol{r})$$

appears for $S^{(1)}$. Whence

$$S^{(1)}(\boldsymbol{k},\boldsymbol{k}_0)=2q_0^{1/2}q_k^{-1/2}\int\left(\frac{\mathrm{i}q_0h-\boldsymbol{k}_0\cdot\nabla h}{q_0}\right)\exp[-\mathrm{i}(\boldsymbol{k}-\boldsymbol{k}_0)\cdot\boldsymbol{r}]\frac{d\boldsymbol{r}}{(2\pi)^2}\ . \tag{4.3.5}$$

Integrating by parts the term ∇h in (4.3.5) reduces to the replacement of operator ∇ by $\mathrm{i}(\boldsymbol{k}-\boldsymbol{k}_0)$, we finally come to the expression conveniently represented as

$$S^{(1)}(\boldsymbol{k},\boldsymbol{k}_0)=2\mathrm{i}(q_kq_{k_0})^{1/2}\frac{K^2-\boldsymbol{k}\cdot\boldsymbol{k}_0}{q_kq_{k_0}}h(\boldsymbol{k}-\boldsymbol{k}_0)\ , \tag{4.3.6}$$

where $h(\boldsymbol{k}-\boldsymbol{k}_0)$ is the roughness spectrum introduced in (4.1.7). For the second order correction the following equation:

$$\int q_k^{1/2}S^{(2)}(\boldsymbol{k},\boldsymbol{k}_0)\exp(\mathrm{i}\boldsymbol{k}\cdot\boldsymbol{r})\,d\boldsymbol{k}=\int\left(\mathrm{i}q_kh-\frac{\boldsymbol{k}\cdot\nabla h}{q_k}\right)\exp(\mathrm{i}\boldsymbol{k}\cdot\boldsymbol{r})q_k^{1/2}S^{(1)}_{(\boldsymbol{k},\boldsymbol{k}_0)}\,d\boldsymbol{k}\ .$$

originates. Its solution is

$$S^{(2)}(\boldsymbol{k},\boldsymbol{k}_0)=-2(q_kq_{k_0})^{-1/2}\int q_{k_1}^{-1}(K^2-\boldsymbol{k}\cdot\boldsymbol{k}_1)(K^2-\boldsymbol{k}_1\cdot\boldsymbol{k}_0)h(\boldsymbol{k}-\boldsymbol{k}_1)$$
$$\times\, h(\boldsymbol{k}_1-\boldsymbol{k}_0)\,d\boldsymbol{k}_1\ . \tag{4.3.7}$$

Hence the formula for SA in the case of the Neumann problem with an accuracy to the second order can be written as

$$S(\boldsymbol{k},\boldsymbol{k}_0) = \delta(\boldsymbol{k}-\boldsymbol{k}_0) + 2\mathrm{i}(q_k q_{k_0})^{1/2} B(\boldsymbol{k},\boldsymbol{k}_0) h(\boldsymbol{k}-\boldsymbol{k}_0)$$
$$- 2(q_k q_{k_0})^{1/2} \int B(\boldsymbol{k},\boldsymbol{k}_1) q_{k_1} B(\boldsymbol{k}_1,\boldsymbol{k}_0) h(\boldsymbol{k}-\boldsymbol{k}_1) h(\boldsymbol{k}_1-\boldsymbol{k}_0)\, d\boldsymbol{k}_1 \;, \tag{4.3.8}$$

where the dimensionless function B has the following form:

$$B(\boldsymbol{k},\boldsymbol{k}_0) = (q_k q_{k_0})^{-1} \cdot (K^2 - \boldsymbol{k}\cdot\boldsymbol{k}_0) \;. \tag{4.3.9}$$

It is obvious that this function is symmetric:

$$B(\boldsymbol{k},\boldsymbol{k}_0) = B(\boldsymbol{k}_0,\boldsymbol{k}) \;. \tag{4.3.10}$$

The first two terms in (4.3.8) clearly satisfy the reciprocity theorem (2.3.6):

$$S(\boldsymbol{k},\boldsymbol{k}_0) = S(-\boldsymbol{k}_0, -\boldsymbol{k}) \;.$$

This can be demonstrated for the third form in (4.3.8) also after substitution $\boldsymbol{k}' \to -\boldsymbol{k}'$.

Direct calculations show that the expression (4.3.8) satisfies the unitarity condition (2.3.10) to an accuracy of $\mathrm{O}(h^2)$.

The moments of SA are calculated by the direct averaging of (4.3.8) similar to that in Sect. 4.1. As a result

$$\bar{V}(\boldsymbol{k}) = 1 - 2q_k \int B^2(\boldsymbol{k}',\boldsymbol{k}) q_{k'} W(\boldsymbol{k}'-\boldsymbol{k})\, d\boldsymbol{k}' \tag{4.3.11}$$

and

$$\sigma(\boldsymbol{k},\boldsymbol{k}_0) = 4q_k q_{k_0} B^2(\boldsymbol{k},\boldsymbol{k}_0) W(\boldsymbol{k}-\boldsymbol{k}_0) \;, \tag{4.3.12}$$

where W is the roughness spectrum determined in (2.3.24)

Note that while integrating over $\boldsymbol{k}'$ in (4.3.11) there exists the singularity of the kind $q_{k'}^{-1}$ which is, generally speaking, integrable.

Consider the applicability condition of SPM for the problem viewed in the present section. It follows from (4.3.4) that in addition to the small Rayleigh parameter the following condition should be imposed also:

$$\boldsymbol{k}\cdot \nabla h/q_k \ll 1 \;.$$

In other words, the grazing angle for all spectra must greatly exceed the slope of roughness. Consequently, the grazing spectra for which $q \to 0$ in the given SPM scheme cannot be considered. The latter directly follows also from (4.3.8) applied to the sinusoidal surface of the kind (4.1.11). The amplitude of the first order spectrum is

$$S_1 = \mathrm{i}a(q_{k_0} q_{k_0+p})^{-1/2} [K^2 - k_0(k_0+p)] \;.$$

If $q_k = q_{k_0+p} \to 0$, then the spectrum amplitude tends to infinity, which is impossible in virtue of the energy conservation (4.1.17).

4.4 Scattering of Electromagnetic Waves at a Perfectly Conducting Boundary

In this section we examine the scattering problem for waves possessing polarization. This situation for monochromatic waves of frequency ω is completely described by Maxwell's equations

$$\mathrm{i}\omega \boldsymbol{H} = c\,\mathrm{curl}\,\boldsymbol{E}\ , \qquad \mathrm{i}\omega\varepsilon \boldsymbol{E} = -c\,\mathrm{curl}\,\boldsymbol{H}\ , \tag{4.4.1}$$

where ε is the dielectric permittivity of the medium. It can be easily shown using (4.4.1) that all components of the fields satisfy the Helmholtz equation. In this case the boundary condition is

$$\boldsymbol{E} \times (\boldsymbol{N} - \nabla h)|_{z=h(\boldsymbol{r})} = \boldsymbol{0}\ . \tag{4.4.2}$$

Here $\boldsymbol{N} = \boldsymbol{e}_z$ denotes the unit vector directed upward, vector $(\boldsymbol{N} - \nabla h)$ is proportional to the normal to the boundary according to (2.1.9), and condition (4.4.2) implies an absence of tangential components of the electric field on a perfectly conducting boundary.

The plane wave incident at the boundary can be represented in the following form:

$$\begin{cases} \boldsymbol{E}_1 = \boldsymbol{e}_1^+(\boldsymbol{k})\exp(\mathrm{i}\boldsymbol{k}\cdot\boldsymbol{r} + \mathrm{i}q_k z)\ , \\ \varepsilon^{-1/2}\boldsymbol{H}_1 = \boldsymbol{h}_1^+(\boldsymbol{k})\exp(\mathrm{i}\boldsymbol{k}\cdot\boldsymbol{r} + \mathrm{i}q_k z)\ , \end{cases} \tag{4.4.3}$$

where

$$\begin{aligned} \boldsymbol{e}_1^+(\boldsymbol{k}) &= (k^2\boldsymbol{N} - q_k\cdot\boldsymbol{k})/(K\cdot k)\ , \\ \boldsymbol{h}_1^+(\boldsymbol{k}) &= -\boldsymbol{N}\times\boldsymbol{k}/k\ . \end{aligned} \tag{4.4.4a}$$

In this case the wavenumber K is $K = \varepsilon^{1/2}\omega/c$. Direct substitution in (4.4.1) demonstrates that (4.4.3) is really the solution of Maxwell's equations, since

$$\mathrm{curl}\,\boldsymbol{E} = \mathrm{i}(\boldsymbol{k} + q_k\cdot\boldsymbol{N})\times\boldsymbol{e}_1^+(\boldsymbol{k})\exp(\mathrm{i}\boldsymbol{k}\cdot\boldsymbol{r} + \mathrm{i}q_k z) = \mathrm{i}K\varepsilon^{-1/2}\boldsymbol{H} \quad \text{etc.}$$

Solution (4.4.3) represents the vertically polarized wave with vector $\boldsymbol{E}$ lying in the plane of incidence and vector $\boldsymbol{H}$ in a horizontal one. The horizontally polarized wave (vector $\boldsymbol{E}$ is horizontal and vector $\boldsymbol{H}$ lies in the plane of incidence) is written as

$$\begin{cases} \boldsymbol{E}_2 = \boldsymbol{e}_2^+(\boldsymbol{k})\exp(\mathrm{i}\boldsymbol{k}\cdot\boldsymbol{r} + \mathrm{i}qz)\ , \\ \varepsilon^{-1/2}\boldsymbol{H}_2 = \boldsymbol{h}_2^+(\boldsymbol{k})\exp(\mathrm{i}\boldsymbol{k}\cdot\boldsymbol{r} + \mathrm{i}qz)\ , \end{cases}$$

where

$$\boldsymbol{e}_2^+(\boldsymbol{k}) = -\boldsymbol{h}_1^+(\boldsymbol{k})\ , \qquad \boldsymbol{h}_2^+(\boldsymbol{k}) = \boldsymbol{e}_1^+(\boldsymbol{k})\ . \tag{4.4.4b}$$

Correspondingly, the plane vertically and horizontally polarized waves which propagate downward and form the scattered field take on the form

$$\boldsymbol{E}_\sigma = \boldsymbol{e}_\sigma^-(\boldsymbol{k})\exp(\mathrm{i}\boldsymbol{k}\cdot\boldsymbol{r} - \mathrm{i}qz) , \qquad \varepsilon^{-1/2}\boldsymbol{H}_\sigma = \boldsymbol{h}_\sigma^-(\boldsymbol{k})\exp(\mathrm{i}\boldsymbol{k}\cdot\boldsymbol{r} - \mathrm{i}qz) ,$$

where indices $\sigma = 1$ and $\sigma = 2$ describe the vertical and the horizontal polarizations, as previously. Vectors $\boldsymbol{e}_\sigma^-$ and $\boldsymbol{h}_\sigma^-$ are obtained from (4.4.4) by replacement $q \to -q$:

$$\begin{aligned} &\boldsymbol{e}_1^-(\boldsymbol{k}) = (k^2\boldsymbol{N} + q\boldsymbol{k})/(Kk) , \qquad \boldsymbol{h}_1^-(\boldsymbol{k}) = \boldsymbol{h}_1^+(\boldsymbol{k}) = -\boldsymbol{N}\times\boldsymbol{k}/k ; \\ &\boldsymbol{e}_2^-(\boldsymbol{k}) = \boldsymbol{N}\times\boldsymbol{k}/k , \qquad \boldsymbol{h}_2^-(\boldsymbol{k}) = \boldsymbol{e}_1^-(\boldsymbol{k}) . \end{aligned} \tag{4.4.5}$$

According to Sect. 2.2 the scattering matrix is introduced by relations

$$\begin{aligned} \boldsymbol{E}_{\sigma_0} = {} & \boldsymbol{e}_{\sigma_0}^+(\boldsymbol{k}_0)q_0^{-1/2}\exp(\mathrm{i}\boldsymbol{k}_0\cdot\boldsymbol{r} + \mathrm{i}q_0 z) \\ & + \sum_\sigma \int d\boldsymbol{k} S_{\sigma\sigma_0}(\boldsymbol{k},\boldsymbol{k}_0)\boldsymbol{e}_\sigma^-(\boldsymbol{k})q^{-1/2}\exp(\mathrm{i}\boldsymbol{k}\cdot\boldsymbol{r} - \mathrm{i}qz) , \end{aligned} \tag{4.4.6a}$$

$$\begin{aligned} \varepsilon^{-1/2}\boldsymbol{H}_{\sigma_0} = {} & \boldsymbol{h}_{\sigma_0}^+(\boldsymbol{k}_0)q_0^{-1/2}\exp(\mathrm{i}\boldsymbol{k}_0\cdot\boldsymbol{r} + \mathrm{i}q_0 z) \\ & + \sum_\sigma \int d\boldsymbol{k} S_{\sigma\sigma_0}(\boldsymbol{k},\boldsymbol{k}_0)\boldsymbol{h}_\sigma^-(\boldsymbol{k})q^{-1/2}\exp(\mathrm{i}\boldsymbol{k}\cdot\boldsymbol{r} - \mathrm{i}qz) . \end{aligned} \tag{4.4.6b}$$

We start by formulating for SA the symmetry relations (the reciprocity theorem and the law of conservation of energy). Now these relations follow from equality

$$\int_{z=z_0} \boldsymbol{N}\cdot(\boldsymbol{E}^{(1)}\times\boldsymbol{H}^{(2)} - \boldsymbol{E}^{(2)}\times\boldsymbol{H}^{(1)})\,d\boldsymbol{r} = \boldsymbol{0} , \tag{4.4.7}$$

where z_0 is an arbitrary constant satisfying the condition $z_0 < \min h$, and $(\boldsymbol{E}^{(1)}, \boldsymbol{H}^{(1)})$ and $(\boldsymbol{E}^{(2)}, \boldsymbol{H}^{(2)})$ are arbitrary solutions of the boundary-value problem (4.4.1, 2). To prove (4.4.7) the Gauss formula (2.3.2) is used

$$\begin{aligned} &\int_\Omega \mathrm{div}(\boldsymbol{E}^{(1)}\times\boldsymbol{H}^{(2)} - \boldsymbol{E}^{(2)}\times\boldsymbol{H}^{(1)})\,d\boldsymbol{R} \\ &= \int_{z=h(\boldsymbol{r})} (\boldsymbol{E}^{(1)}\times\boldsymbol{H}^{(2)} - \boldsymbol{E}^{(2)}\times\boldsymbol{H}^{(1)})\,d\boldsymbol{\Sigma} - \int_{z=z_0} \boldsymbol{N}\cdot(\boldsymbol{E}^{(1)}\times\boldsymbol{H}^{(2)} - \boldsymbol{E}^{(2)}\times\boldsymbol{H}^{(1)})\,d\boldsymbol{r} . \end{aligned} \tag{4.4.8}$$

Here we integrate over the volume contained between the inhomogeneous boundary and the plane $z = z_0$. First of all, in virtue of the known identity

$$\mathrm{div}(\boldsymbol{E}\times\boldsymbol{H}) = \mathrm{curl}\,\boldsymbol{E}\cdot\boldsymbol{H} - \boldsymbol{E}\cdot\mathrm{curl}\,\boldsymbol{H}$$

and Maxwell's equations (4.4.1), we have

$$\mathrm{div}(\boldsymbol{E}^{(1)}\times\boldsymbol{H}^{(2)} - \boldsymbol{E}^{(2)}\times\boldsymbol{H}^{(1)}) = 0 .$$

Vectors $\boldsymbol{E}^{(1)}$ and $\boldsymbol{E}^{(2)}$ are normal to the boundary, so that

$$\boldsymbol{E}^{(1)} \times \boldsymbol{H}^{(2)} \cdot d\boldsymbol{\Sigma} = \boldsymbol{E}^{(2)} \times \boldsymbol{H}^{(1)} d\boldsymbol{\Sigma} = \boldsymbol{0} .$$

Thus (4.4.7) immediately follows from (4.4.8).

We substitute into (4.4.7) two solutions of the form (4.4.6) in the manner presented in Sect. 2.3 which correspond to values $(\sigma_0, \boldsymbol{k}_0) = (\sigma_1, \boldsymbol{k}_1)$ and $(\sigma_0, \boldsymbol{k}_0) = (\sigma_2, \boldsymbol{k}_2)$, respectively. The following equalities which can be immediately checked are convenient for calculations:

$$\begin{aligned} \boldsymbol{N} \cdot \boldsymbol{e}^+_{\sigma_1}(\boldsymbol{k}) \times \boldsymbol{h}^+_{\sigma_2}(-\boldsymbol{k}) &= -q_k/K \cdot \delta_{\sigma_1 \sigma_2} , \\ \boldsymbol{N} \cdot \boldsymbol{e}^-_{\sigma_1}(\boldsymbol{k}) \times \boldsymbol{h}^-_{\sigma_2}(-\boldsymbol{k}) &= q_k/K \delta_{\sigma_1 \sigma_2} \end{aligned} \quad (4.4.9a)$$

$$\begin{aligned} \boldsymbol{N} \cdot \boldsymbol{e}^+_{\sigma_1}(\boldsymbol{k}) \times \boldsymbol{h}^-_{\sigma_2}(-\boldsymbol{k}) &= -q_k/K \cdot (V_0)_{\sigma_1 \sigma_2} , \\ \boldsymbol{N} \cdot \boldsymbol{e}^-_{\sigma_1}(\boldsymbol{k}) \times \boldsymbol{h}^+_{\sigma_2}(-\boldsymbol{k}) &= q_k/K (V_0)_{\sigma_1 \sigma_2} . \end{aligned} \quad (4.4.9b)$$

Matrix $\hat{V}_0$ in (4.4.9) has the following form:

$$\hat{V}_0 = \begin{pmatrix} 1 & 0 \\ 0 & -1 \end{pmatrix} . \quad (4.4.10)$$

With account of (4.4.9a), (4.4.7) is reduced to the following:

$$\begin{aligned} \sum_{\sigma'} [&\boldsymbol{N} \cdot \boldsymbol{e}^+_{\sigma_1}(\boldsymbol{k}_1) \times \boldsymbol{h}^-_{\sigma'}(-\boldsymbol{k}_1) S_{\sigma' \sigma_2}(-\boldsymbol{k}_1, \boldsymbol{k}_2) \\ &+ \boldsymbol{N} \cdot \boldsymbol{e}^-_{\sigma'}(-\boldsymbol{k}_2) \times \boldsymbol{h}^+_{\sigma_2}(\boldsymbol{k}_2) S_{\sigma' \sigma_1}(-\boldsymbol{k}_2, \boldsymbol{k}_1)] = 0 . \end{aligned} \quad (4.4.11)$$

Building matrix $\hat{S}$ out of quantities $S_{\sigma_1 \sigma_2}$ according to (2.2.18)

$$\hat{S}(\boldsymbol{k}_1, \boldsymbol{k}_2) = \begin{pmatrix} S_{11} & S_{12} \\ S_{21} & S_{22} \end{pmatrix} (\boldsymbol{k}_1, \boldsymbol{k}_2) .$$

One can write (4.4.11) in matrix form

$$-\hat{V}_0 \cdot \hat{S}(-\boldsymbol{k}_1, \boldsymbol{k}_2) + [\hat{V}_0 \hat{S}(-\boldsymbol{k}_2, \boldsymbol{k}_1)]^{\mathrm{T}} = 0 ,$$

where the superscript "T" denotes transposition. Replacing $\boldsymbol{k}_1 \to -\boldsymbol{k}_1$, we finally write down the reciprocity theorem as

$$\hat{V}_0 \hat{S}(\boldsymbol{k}_1, \boldsymbol{k}_2) = \hat{S}^{\mathrm{T}}(-\boldsymbol{k}_2, -\boldsymbol{k}_1) \hat{V}_0 . \quad (4.4.12)$$

Now we obtain the unitarity relation. For this we note that when $(\boldsymbol{E}^{(1)}, \boldsymbol{H}^{(1)})$ is a solution of Maxwell's equations (4.4.1) then $(\boldsymbol{E}^{(1)*}, -\boldsymbol{H}^{(1)*})$ is a solution as well. Moreover, vector $\boldsymbol{E}^{(1)*}$ apparently satisfies the boundary condition (4.4.2). Replacement of $(\boldsymbol{E}^{(1)}, \boldsymbol{H}^{(1)}) \to (\boldsymbol{E}^{(1)*}, \boldsymbol{H}^{(1)*})$ in (4.4.7) gives

$$\int_{z=z_0} \boldsymbol{N} \cdot (\boldsymbol{E}^{(1)*} \times \boldsymbol{H}^{(2)} + \boldsymbol{E}^{(2)} \times \boldsymbol{H}^{(1)*}) \, d\boldsymbol{r} = 0 . \quad (4.4.13)$$

Then we substitute into (4.4.13) the same pair of solutions used in obtaining the reciprocity theorem and assume that $|\boldsymbol{k}_1| < K$, $|\boldsymbol{k}_2| < K$. After some calculations the following unitarity relation arises

$$\int\limits_{|\boldsymbol{k}'|<K} \hat{S}^{+}(\boldsymbol{k}',\boldsymbol{k}_1)\cdot\hat{S}(\boldsymbol{k}',\boldsymbol{k}_2)\,d\boldsymbol{k}' = \begin{pmatrix}1 & 0\\ 0 & 1\end{pmatrix}\delta(\boldsymbol{k}_1-\boldsymbol{k}_2)\ \ (|\boldsymbol{k}_1|,|\boldsymbol{k}_2|<K)\ , \tag{4.4.14}$$

where "+" denotes the Hermitian conjugation of 2×2 for matrix $\hat{S}$:

$$\hat{S}^{+}(\boldsymbol{k}_1,\boldsymbol{k}_2) = [\hat{S}^{*}(\boldsymbol{k}_1,\boldsymbol{k}_2)]^{\mathrm{T}} = \begin{pmatrix}S^{*}_{11} & S^{*}_{21}\\ S^{*}_{12} & S^{*}_{22}\end{pmatrix}(\boldsymbol{k}_1,\boldsymbol{k}_2)\ .$$

It is clear that (4.4.14) coincides with the unitarity relation (2.3.16) deduced by the law of conservation of energy.

The most natural basis for describing electromagnetic wave scattering corresponds to circularly polarized waves. The latter are invariant with respect to rotations of coordinates and insensitive to the "fictitious" depolarization effects of geometrical origin (for instance, the horizontally and vertically polarized waves partly transform into each other when reflecting from the slope plane). The electric field in vertically and horizontally polarized waves $\boldsymbol{E}_1$ and $\boldsymbol{E}_2$ used up to now, and the electric field in the right- and left-polarized waves $\boldsymbol{E}_1^{(\mathrm{c})}$ and $\boldsymbol{E}_2^{(\mathrm{c})}$ are related as

$$\boldsymbol{E}_1^{(\mathrm{c})} = (\boldsymbol{E}_1 - \mathrm{i}\boldsymbol{E}_2)/\sqrt{2}\ ,\qquad \boldsymbol{E}_2^{(\mathrm{c})} = (\boldsymbol{E}_1 + \mathrm{i}\boldsymbol{E}_2)/\sqrt{2}\ .$$

The same equalities exist for magnetic fields. The given transformation can be written in the form

$$\begin{pmatrix}\boldsymbol{E}_1^{(\mathrm{c})}\\ \boldsymbol{E}_2^{(\mathrm{c})}\end{pmatrix} = 2^{-1/2}\begin{pmatrix}1 & -\mathrm{i}\\ 1 & \mathrm{i}\end{pmatrix}\begin{pmatrix}\boldsymbol{E}_1\\ \boldsymbol{E}_2\end{pmatrix}\ . \tag{4.4.15}$$

Note that the matrix in (4.4.15) is unitary. Thus the transition to the circularly polarized waves implies summing up (4.4.6a) for $\sigma_0 = 1, 2$ with coefficients $2^{-1/2}$, $\mp\mathrm{i}2^{-1/2}$. Moreover, vector $\boldsymbol{e}_\sigma^-$ should be expressed by a relation inverse to (4.4.15)

$$\begin{pmatrix}\boldsymbol{e}_1^-\\ \boldsymbol{e}_2^-\end{pmatrix} = 2^{-1/2}\begin{pmatrix}1 & 1\\ \mathrm{i} & -\mathrm{i}\end{pmatrix}\begin{pmatrix}\boldsymbol{e}_1^{(\mathrm{c})}\\ \boldsymbol{e}_2^{(\mathrm{c})}\end{pmatrix}\ .$$

Hence, the scattering matrix $\hat{S}_{\mathrm{c}}$ in the basis of circularly polarized waves is obtained from the scattering matrix $\hat{S}$ for linearly polarized waves by the following unitary transformation:

$$\hat{S}_{\mathrm{c}}(\boldsymbol{k},\boldsymbol{k}_0) = \frac{1}{2}\begin{pmatrix}1 & \mathrm{i}\\ 1 & -\mathrm{i}\end{pmatrix}\hat{S}(\boldsymbol{k},\boldsymbol{k}_0)\begin{pmatrix}1 & 1\\ -\mathrm{i} & \mathrm{i}\end{pmatrix}\ . \tag{4.4.16}$$

The unitarity relation (4.4.14) for matrix $\hat{S}_c$ coincides with that for matrix $\hat{S}$. The reciprocity theorem (4.4.12) is rewritten in the simpler form

$$\hat{S}_c(\boldsymbol{k}_1, \boldsymbol{k}_2) = \hat{S}_c^{T}(-\boldsymbol{k}_2, -\boldsymbol{k}_1) \ . \tag{4.4.17}$$

Now we proceed to the calculation of SA by the small perturbation method. For this, according to the procedure used previously, solution (4.4.6a) should be substituted into the boundary condition (4.4.2) which yields equation

$$(\boldsymbol{N} - \nabla h) \times \boldsymbol{e}_{\sigma_0}^{+}(\boldsymbol{k}_0) q_0^{-1/2} \exp[\mathrm{i}\boldsymbol{k}_0 \cdot \boldsymbol{r} + \mathrm{i} q_0 h(\boldsymbol{r})]$$
$$+ \sum_{\sigma} \int S_{\sigma\sigma_0}(\boldsymbol{k}, \boldsymbol{k}_0)(\boldsymbol{N} - \nabla h) \times \boldsymbol{e}_{\sigma_0}^{-}(\boldsymbol{k}) q^{-1/2} \exp[\mathrm{i}\boldsymbol{k} \cdot \boldsymbol{r} - \mathrm{i} q h(\boldsymbol{r})]\, d\boldsymbol{k} = 0 \ . \tag{4.4.18}$$

The series expansion S in powers of h

$$S = S^{(0)} + S^{(1)} + S^{(2)} + \cdots$$

should be substituted into the previous equation. For the lower order (i.e., setting $h = 0$) we can easily find

$$\hat{S}^{(0)}(\boldsymbol{k}, \boldsymbol{k}_0) = \hat{V}_0 \delta(\boldsymbol{k} - \boldsymbol{k}_0) \ , \tag{4.4.19}$$

where $\hat{V}_0$ is given by (4.4.10). The following equation is the origin for $\hat{S}^{(1)}$:

$$\sum_{\sigma'} \int \boldsymbol{N} \times \boldsymbol{e}_{\sigma'}^{-}(\boldsymbol{k}') S_{\sigma'\sigma_0}^{(1)}(\boldsymbol{k}', \boldsymbol{k}_0) q_{k'}^{-1/2} \exp(\mathrm{i}\boldsymbol{k}' \cdot \boldsymbol{r})\, d\boldsymbol{k}'$$
$$+ [(\mathrm{i} q_0 h \boldsymbol{N} - \nabla h) \times \boldsymbol{e}_{\sigma_0}^{+}(\boldsymbol{k}_0) + (-1)^{\sigma_0}(\mathrm{i} q_0 h \boldsymbol{N} + \nabla h) \times \boldsymbol{e}_{\sigma_0}^{-}(\boldsymbol{k}_0)] q_0^{-1/2}$$
$$\times \exp(\mathrm{i}\boldsymbol{k}_0 \cdot \boldsymbol{r}) = 0 \ .$$

Multiply scalarly the latter by vector $\boldsymbol{e}_{\sigma}^{+}(\boldsymbol{k}) \exp(-\mathrm{i}\boldsymbol{k} \cdot \boldsymbol{r})/(2\pi)^2$ and integrate over $\boldsymbol{r}$, then factor the $h(\boldsymbol{k} - \boldsymbol{k}_0)$ results. Here the terms proportional to ∇h should be integrated by parts, which is equivalent to the replacement of operator ∇ by $\mathrm{i}(\boldsymbol{k} - \boldsymbol{k}_0)$. Direct calculation allows us to easily verify the following formulas:

$$\boldsymbol{e}_{\sigma}^{+}(\boldsymbol{k}) \cdot \boldsymbol{N} \times \boldsymbol{e}_{\sigma_0}^{\pm}(\boldsymbol{k}_0) = \begin{bmatrix} \mp q q_0 K^{-2} \boldsymbol{N} \cdot \boldsymbol{k} \times \boldsymbol{k}_0/(k k_0) & q K^{-1} (\boldsymbol{k}\boldsymbol{k}_0)/(k k_0) \\ \mp q_0 K^{-1} (\boldsymbol{k} \cdot \boldsymbol{k}_0)/(k k_0) & -\boldsymbol{N} \cdot \boldsymbol{k} \times \boldsymbol{k}_0/(k k_0) \end{bmatrix} ,$$

$$\boldsymbol{e}_{\sigma}^{+}(\boldsymbol{k}) \cdot (\boldsymbol{k} - \boldsymbol{k}_0) \times \boldsymbol{e}_{\sigma_0}^{\pm}(\boldsymbol{k}_0) = \begin{bmatrix} (\mp k^2 q_0 + k_0^2 q) K^{-2} \boldsymbol{N} \cdot \boldsymbol{k} \times \boldsymbol{k}_0/(k k_0) & k[(\boldsymbol{k} \cdot \boldsymbol{k}_0) - k_0^2]/(K k_0) \\ k_0[(\boldsymbol{k} \cdot \boldsymbol{k}_0) - k^2]/(K k) & 0 \end{bmatrix} .$$

Appropriate simple calculations using these formulas yield

$$\hat{S}^{(1)}(\boldsymbol{k}, \boldsymbol{k}_0) = 2\mathrm{i}(q q_0)^{1/2} B(\boldsymbol{k}, \boldsymbol{k}_0) h(\boldsymbol{k} - \boldsymbol{k}_0) \ , \tag{4.4.20}$$

where the dimensionless matrix $\hat{B}$ is

$$\hat{B}(\boldsymbol{k},\boldsymbol{k}_0)=\begin{bmatrix}\dfrac{[K^2(\boldsymbol{k}\cdot\boldsymbol{k}_0)-k^2k_0^2]}{qq_0kk_0} & \dfrac{K\boldsymbol{N}\cdot\boldsymbol{k}\times\boldsymbol{k}_0}{qkk_0}\\ \dfrac{K\boldsymbol{N}\cdot\boldsymbol{k}\times\boldsymbol{k}_0}{q_0kk_0} & \dfrac{-(\boldsymbol{k}\cdot\boldsymbol{k}_0)}{kk_0}\end{bmatrix} \tag{4.4.21}$$

Matrix $\hat{S}^{(2)}$ is calculated in a similar manner, and the result is

$$\hat{S}^{(2)}(\boldsymbol{k},\boldsymbol{k}_0)=-2(qq_0)^{1/2}\int q_{k'}\hat{B}(\boldsymbol{k},\boldsymbol{k}')\hat{V}_0\hat{B}(\boldsymbol{k}',\boldsymbol{k}_0)h(\boldsymbol{k}-\boldsymbol{k}')h(\boldsymbol{k}'-\boldsymbol{k}_0)\,d\boldsymbol{k}'\ . \tag{4.4.22}$$

Finally, matrix-valued SA with an accuracy to the h^2-order can be represented as

$$\hat{S}(\boldsymbol{k},\boldsymbol{k}_0)=\hat{V}_0\delta(\boldsymbol{k}-\boldsymbol{k}_0)+2\mathrm{i}(qq_0)^{1/2}\hat{B}(\boldsymbol{k},\boldsymbol{k}_0)h(\boldsymbol{k}-\boldsymbol{k}_0)$$
$$+\,2(qq_0)^{1/2}\int\hat{B}_2(\boldsymbol{k},\boldsymbol{k}_0;\xi)h(\boldsymbol{k}-\xi)h(\xi-\boldsymbol{k}_0)\,d\xi\ , \tag{4.4.23}$$

where

$$\hat{B}_2(\boldsymbol{k},\boldsymbol{k}_0;\xi)=-q_\xi\hat{B}(\boldsymbol{k},\xi)\hat{V}_0\hat{B}(\xi,\boldsymbol{k}_0) \tag{4.4.24}$$

and matrices $\hat{V}_0$ and $\hat{B}$ are given by (4.4.10) and (4.4.21), respectively. The structure of (4.4.23, 24) remains unchangeable in any basis. Consider the case of circularly polarized waves using transformation (4.4.16). Then

$$\hat{V}_0^{(c)}=\frac{1}{2}\begin{pmatrix}1 & \mathrm{i}\\ 1 & -\mathrm{i}\end{pmatrix}\begin{pmatrix}1 & 0\\ 0 & -1\end{pmatrix}\begin{pmatrix}1 & 1\\ -\mathrm{i} & \mathrm{i}\end{pmatrix}=\begin{pmatrix}0 & 1\\ 1 & 0\end{pmatrix}. \tag{4.4.25}$$

In other words, the clockwise and counterclockwise polarized waves turn into each other with the unit reflection coefficient when reflecting from the horizontal plane. Matrix B in this basis has the form

$$\hat{B}^{(c)}(\boldsymbol{k},\boldsymbol{k}_0)=(2qq_0kk_0)^{-1}$$
$$\times\begin{bmatrix}(K^2-qq_0)(\boldsymbol{k}\cdot\boldsymbol{k}_0)-k^2k_0^2+\mathrm{i}K(q-q_0)\boldsymbol{N}\cdot\boldsymbol{k}\times\boldsymbol{k}_0 & (K^2+qq_0)(\boldsymbol{k}\cdot\boldsymbol{k}_0)-k^2k_0^2+\mathrm{i}K(q+q_0)\boldsymbol{N}\cdot\boldsymbol{k}\times\boldsymbol{k}_0\\ (K^2+qq_0)(\boldsymbol{k}\cdot\boldsymbol{k}_0)-k^2k_0^2-\mathrm{i}K(q+q_0)\boldsymbol{N}\cdot\boldsymbol{k}\times\boldsymbol{k}_0 & (K^2-qq_0)(\boldsymbol{k}\cdot\boldsymbol{k}_0)-k^2k_0^2-\mathrm{i}K(q-q_0)\boldsymbol{N}\cdot\boldsymbol{k}\times\boldsymbol{k}_0\end{bmatrix}. \tag{4.4.26}$$

It obvious that the reciprocity theorem (4.4.17) is satisfied. To make (4.6.26) less cumbersome, matrix $\boldsymbol{B}$ can be represented as a linear combination of Pauli matrices:

$$\hat{\sigma}_0=\begin{pmatrix}1 & 0\\ 0 & 1\end{pmatrix},\quad \hat{\sigma}_1=\begin{pmatrix}0 & 1\\ 1 & 0\end{pmatrix},\quad \hat{\sigma}_2=\begin{pmatrix}0 & -\mathrm{i}\\ \mathrm{i} & 0\end{pmatrix},\quad \hat{\sigma}_3=\begin{pmatrix}1 & 0\\ 0 & -1\end{pmatrix}. \tag{4.4.27}$$

We set

$$\hat{B}^{(c)}(\boldsymbol{k}, \boldsymbol{k}_0) = \sum_{n=0}^{3} b_n(\boldsymbol{k}, \boldsymbol{k}_0) \cdot \hat{\sigma}_n \ .$$

Hence, matrix $\hat{B}^{(c)}$ is described by four dimensionless functions:

$$b_0(\boldsymbol{k}, \boldsymbol{k}_0) = (2qq_0kk_0)^{-1}[(K^2 - qq_0)(\boldsymbol{k}\cdot\boldsymbol{k}_0) - k^2k_0^2] \ , \tag{4.4.28a}$$

$$b_1(\boldsymbol{k}, \boldsymbol{k}_0) = (2qq_0kk_0)^{-1}[(K^2 + qq_0)(\boldsymbol{k}\cdot\boldsymbol{k}_0) - k^2k_0^2] \ , \tag{4.4.28b}$$

$$b_2(\boldsymbol{k}, \boldsymbol{k}_0) = -(2qq_0kk_0)^{-1}K(q + q_0)\boldsymbol{N}\cdot\boldsymbol{k} \times \boldsymbol{k}_0 \ , \tag{4.4.28c}$$

$$b_3(\boldsymbol{k}, \boldsymbol{k}_0) = \mathrm{i}(2qq_0kk_0)^{-1}K(q - q_0)\boldsymbol{N}\cdot\boldsymbol{k} \times \boldsymbol{k}_0 \ . \tag{4.4.28d}$$

The fact that correction of the second order (4.4.22) is also expressed through matrix $\hat{B}$ is related to the law of energy conservation (4.4.14). To demonstrate this representation (4.4.23) is substituted into (4.4.14) and the following relation is taken into account, which can be directly verified

$$\hat{V}_0 B(\boldsymbol{k}_1, \boldsymbol{k}_2) = B^{\mathrm{T}}(\boldsymbol{k}_2, \boldsymbol{k}_1)\hat{V}_0 \quad (|\boldsymbol{k}_1|, |\boldsymbol{k}_2| < K) \tag{4.4.29}$$

(due to the fact that matrix $\hat{B}$ is real, Hermitian conjugation is equivalent to transposition here). It can be easily verified also that in virtue of (4.4.29) the terms linear with respect to h reduce, which can be shown for the h^2-order terms with account of the following:

$$\begin{aligned} \hat{V}_0\hat{B}_2(\boldsymbol{k}_1, \boldsymbol{k}_2; \xi) &= -q_\xi \hat{V}_0 \hat{B}(\boldsymbol{k}_1, \xi)\hat{V}_0\hat{B}(\xi, \boldsymbol{k}_2) = -q_\xi \hat{B}^+(\xi, \boldsymbol{k}_1)B(\xi, \boldsymbol{k}_2) \ , \\ B_2^+(\boldsymbol{k}_2, \boldsymbol{k}_1; \xi)\hat{V}_0 &= -q_\xi^* \hat{B}^+(\xi, \boldsymbol{k}_1)\hat{V}_0\hat{B}^+(\boldsymbol{k}_2, \xi)\hat{V}_0 = -q_\xi^* \hat{B}^+(\xi, \boldsymbol{k}_1)B(\xi, \boldsymbol{k}_2) \ . \end{aligned} \tag{4.4.30}$$

Here it is necessary to make some comment of technical character. Relation (4.4.29) used in (4.4.30) is incorrect for $|\xi| > K$. However, using (4.4.21), one can directly verify that quantity q_ξ is in the even power in the product $\hat{B}(\boldsymbol{k}, \xi)V_0\hat{B}(\xi, \boldsymbol{k}_0)$, and, therefore, relations (4.4.30) are true for $|\xi| > K$ as well.

The reciprocity theorem (4.4.12) for SA (4.4.23) is also fulfilled, which can be easily shown taking into account the apparent equality: $B(-\boldsymbol{k}, -\boldsymbol{k}_0) = B(\boldsymbol{k}, \boldsymbol{k}_0)$.

4.5 Scattering of Sound Waves at the Interface Between Liquid Half-Spaces

This section regards the situation when both transmission and reflection of the scalar field at a rough boundary occur.

Let two half-spaces Ω_1 and Ω_2 (Fig. 2.2) separated by the rough interface $z = h(\boldsymbol{r})$ contain homogeneous liquids with densities ρ_1, ρ_2 and sound velocities

c_1, c_2. We take $\Psi^{(1)}$ and $\Psi^{(2)}$ for the acoustic field potentials satisfying the homogeneous Helmholtz equations

$$(\Delta + K_1^2)\Psi^{(1)} = 0 \ , \tag{4.5.1a}$$

$$(\Delta + K_2^2)\Psi^{(2)} = 0 \ , \tag{4.5.1b}$$

where $K_1 = \omega/c_1$ and $K_2 = \omega/c_2$ are the corresponding wave numbers. The boundary conditions have a sense of continuity at the boundary of normal velocities:

$$\frac{\partial \Psi^{(1)}}{\partial \boldsymbol{n}} = \frac{\partial \Psi^{(2)}}{\partial \boldsymbol{n}} \ , \qquad z = h(\boldsymbol{r}) \tag{4.5.2a}$$

and pressure

$$\varrho_1 \Psi^{(1)} = \varrho_2 \Psi^{(2)} \ , \qquad z = h(\boldsymbol{r}) \ . \tag{4.5.2b}$$

Normalizing of amplitudes for incident and scattered waves follows from the condition specified in Sect. 2.2: the vertical energy fluxes must be equal to the same constant for all waves propagating in the first and the second media.

According to the procedure presented in Sect. 2.2 we seek the solution of the scattering problem for the plane wave incident from the first (lower) half-space in the following form:

$$\begin{aligned}\Psi^{(1)} &= \exp(\mathrm{i}\boldsymbol{k}_0\cdot\boldsymbol{r} + \mathrm{i}q_0^{(1)}z)(\varrho_1 q_0^{(1)})^{-1/2} \\ &\quad + \int \exp(\mathrm{i}\boldsymbol{k}\cdot\boldsymbol{r} - \mathrm{i}q^{(1)}z)(\varrho_1 q^{(1)})^{-1/2} S^{11}(\boldsymbol{k},\boldsymbol{k}_0)\,d\boldsymbol{k} \ , \qquad z < \min h(\boldsymbol{r}) \ , \\ \Psi^{(2)} &= \int \exp(\mathrm{i}\boldsymbol{k}\cdot\boldsymbol{r} + \mathrm{i}q^{(2)}z)(\varrho_2 q^{(2)})^{-1/2} S^{21}(\boldsymbol{k},\boldsymbol{k}_0)\,d\boldsymbol{k} \ , \qquad z > \max h(\boldsymbol{r}) \ .\end{aligned} \tag{4.5.3a}$$

For the wave incident from the second half-space, the solution is

$$\begin{aligned}\Psi^{(1)} &= \int \exp(\mathrm{i}\boldsymbol{k}\cdot\boldsymbol{r} - \mathrm{i}q^{(1)}z)(\varrho_1 q^{(1)})^{-1/2} S^{12}(\boldsymbol{k},\boldsymbol{k}_0)\,d\boldsymbol{k} \ , \qquad z < \min h(\boldsymbol{r}) \ , \\ \Psi^{(2)} &= \exp(\mathrm{i}\boldsymbol{k}_0\cdot\boldsymbol{r} - \mathrm{i}q_0^{(2)}z)(\varrho_2 q_0^{(2)})^{-1/2} \\ &\quad + \int \exp(\mathrm{i}\boldsymbol{k}\cdot\boldsymbol{r} + \mathrm{i}q^{(2)}z)(\varrho_2 q^{(2)})^{-1/2} S^{22}(\boldsymbol{k},\boldsymbol{k}_0)\,d\boldsymbol{k} \ , \qquad z > \max h(\boldsymbol{r}) \ .\end{aligned} \tag{4.5.3b}$$

Here

$$q^{(1)} = q^{(1)}(k) = (K_1^2 - k^2)^{1/2} \ , \qquad q^{(2)} = q^{(2)}(k) = (K_2^2 - k^2)^{1/2}$$

and

$$q_0^{(1)} = q^{(1)}(k_0) \ , \qquad q_0^{(2)} = q^{(2)}(k_0)$$

with the usual rule of the square root branch choice: $\mathrm{Im}\{q^{(1,2)}\} \geqslant 0$, $\mathrm{Re}\{q^{(1,2)}\} \geqslant 0$. In this case the symmetry properties of SA follow from relation:

$$\varrho_1 \int\limits_{z=z_1<\min h(\boldsymbol{r})} \left(\Psi_2^{(1)} \cdot \frac{\partial \Psi_1^{(1)}}{\partial z} - \Psi_1^{(1)} \frac{\partial \Psi_2^{(1)}}{\partial z}\right) d\boldsymbol{r}$$

$$= \varrho_2 \int\limits_{z=z_2>\max h(\boldsymbol{r})} \left(\Psi_2^{(2)} \cdot \frac{\partial \Psi_1^{(2)}}{\partial z} - \Psi_1^{(2)} \frac{\partial \Psi_2^{(2)}}{\partial z}\right) d\boldsymbol{r}\ , \tag{4.5.4}$$

where $(\Psi_1^{(1)}, \Psi_2^{(1)})$ and $(\Psi_1^{(2)}, \Psi_2^{(2)})$ are arbitrary regular solutions of the problem (4.5.1, 2). Relation (4.5.4) results from the obvious chain of equalities:

$$0 = \varrho_1 \int\limits_{z_1<z<h(\boldsymbol{r})} (\Psi_2^{(1)} \Delta \Psi_1^{(1)} - \Psi_1^{(1)} \Delta \Psi_2^{(1)})\, d\boldsymbol{R}$$

$$+ \varrho_2 \int\limits_{h(\boldsymbol{r})<z<z_2} (\Psi_2^{(2)} \Delta \Psi_1^{(2)} - \Psi_1^{(2)} \Delta \Psi_2^{(2)})\, d\boldsymbol{R}$$

$$= \varrho_1 \int\limits_{z=z_1} \left(\Psi_2^{(1)} \frac{\partial \Psi_1^{(1)}}{\partial z} - \Psi_1^{(1)} \frac{\partial \Psi_2^{(1)}}{\partial z}\right) d\boldsymbol{r}$$

$$- \varrho_1 \int\limits_{z=h(\boldsymbol{r})} \left(\Psi_2^{(1)} \frac{\partial \Psi^{(1)}}{\partial \boldsymbol{n}} - \Psi_1^{(1)} \frac{\partial \Psi_2^{(1)}}{\partial \boldsymbol{n}}\right) d\Sigma$$

$$+ \varrho_2 \int\limits_{z=h(\boldsymbol{r})} \left(\Psi_2^{(2)} \frac{\partial \Psi_1^{(2)}}{\partial \boldsymbol{n}} - \Psi_1^{(2)} \frac{\partial \Psi_2^{(2)}}{\partial \boldsymbol{n}}\right) d\Sigma$$

$$- \varrho_2 \int\limits_{z=h_2} \left(\Psi_2^{(2)} \frac{\partial \Psi_1^{(2)}}{\partial z} - \Psi_1^{(2)} \frac{\partial \Psi_2^{(2)}}{\partial z}\right) d\boldsymbol{r}$$

$$= \varrho_1 \int\limits_{z=z_1} \left(\Psi_2^{(1)} \frac{\partial \Psi_1^{(1)}}{\partial z} - \Psi_1^{(1)} \frac{\partial \Psi_2^{(1)}}{\partial z}\right) d\boldsymbol{r}$$

$$- \rho_2 \int\limits_{z=z_2} \left(\Psi_2^{(2)} \frac{\partial \Psi_1^{(2)}}{\partial z} - \Psi_1^{(2)} \frac{\partial \Psi_2^{(2)}}{\partial z}\right) d\boldsymbol{r}$$

$$+ \int\limits_{z=h(\boldsymbol{r})} \left[\frac{\partial \Psi_1^{(1)}}{\partial \boldsymbol{n}} (\varrho_2 \Psi_2^{(2)} - \varrho_1 \Psi_2^{(1)}) - \frac{\partial \Psi_2^{(1)}}{\partial \boldsymbol{n}} (\varrho_2 \Psi_1^{(2)} - \varrho_1 \Psi_1^{(1)})\right] d\Sigma\ .$$

When passing to the last equality the boundary condition (4.5.2a) was used. Since the latter integral in this chain vanishes, then by the boundary condition (4.5.2b) we obtain (4.5.4).

Substituting all possible combinations of solutions of the form (4.5.4) with $\boldsymbol{k}_0 = \boldsymbol{k}_1$ and $\boldsymbol{k}_0 = \boldsymbol{k}_2$, we come to the following formulation of the reciprocity theorem:

$$S^{N_1N_2}(\boldsymbol{k}_1, \boldsymbol{k}_2) = S^{N_2N_1}(-\boldsymbol{k}_2, -\boldsymbol{k}_1) \ . \tag{4.5.5}$$

Further, if $(\Psi^{(1)}, \Psi^{(2)})$ is a solution for the boundary problem, then $(\Psi^{(1)*}, \Psi^{(2)*})$ also satisfies (4.5.1–3). Now substitute into (4.5.4) the result of complex conjugation of (4.5.3) with $\boldsymbol{k}_0 = \boldsymbol{k}_2$. As a result we obtain the unitarity relation of the form (4.3.17)

$$\sum_{N'=1,2} \int_{|\boldsymbol{k}'|<\omega/c_{N'}} S^{N'N_1}(\boldsymbol{k}', \boldsymbol{k}_1) \cdot [S^{N'N_2}(\boldsymbol{k}', \boldsymbol{k}_2)]^* \, d\boldsymbol{k}'$$
$$= \delta^{N_1N_2}\delta(\boldsymbol{k}_1 - \boldsymbol{k}_2) \ , \qquad (|\boldsymbol{k}_{1,2}| < \omega/c_{1,2}) \ . \tag{4.5.6}$$

The calculation pattern in obtaining (4.5.5, 6) is very close to that for relations (2.3.6, 10), and can be easily reproduced.

According to Sect. 2.2 one can comprise the following matrix, using a set of values $S^{N_1N_2}$:

$$\hat{S}(\boldsymbol{k}, \boldsymbol{k}_0) = \begin{bmatrix} S^{11}(\boldsymbol{k}, \boldsymbol{k}_0) & S^{12}(\boldsymbol{k}, \boldsymbol{k}_0) \\ S^{21}(\boldsymbol{k}, \boldsymbol{k}_0) & S^{22}(\boldsymbol{k}, \boldsymbol{k}_0) \end{bmatrix} .$$

The scattering amplitude $S^{N_1N_2}$ for small roughness can be easily determined by our usual scheme of applying the Rayleigh hypothesis. Namely, we shall calculate the field and its normal derivative at the boundary according to (4.5.3a, b) and seek the expression for $S^{N_1N_2}$ in the form of expansion in the small Rayleigh parameter

$$S^{N_1N_2} = S_0^{N_1N_2} + S_1^{N_1N_2} + S_2^{N_1N_2} + \cdots \ .$$

The exponential factors with elevations in (4.5.3) should be expanded in the power series as well. For the plane boundary $h = 0$ we easily obtain

$$\hat{S}_0(\boldsymbol{k}, \boldsymbol{k}_0) = \hat{V}_0(\boldsymbol{k})\delta(\boldsymbol{k} - \boldsymbol{k}_0) \ ,$$

where matrix $\hat{V}_0$ is

$$\hat{V}_0(k) = \begin{bmatrix} \mathscr{R}(k) & \mathscr{D}(k) \\ \mathscr{D}(k) & -\mathscr{R}(k) \end{bmatrix} . \tag{4.5.7}$$

$\mathscr{R}(k)$ and $\mathscr{D}(k)$ are the Fresnel reflection and transmission coefficients

$$\mathscr{R}(k) = (\varrho_2 q_k^{(1)} - \varrho_1 q_k^{(2)})(\varrho_2 q_k^{(1)} + \varrho_1 q_k^{(2)})^{-1} \ , \tag{4.5.8a}$$

$$\mathscr{D}(k) = 2(\varrho_1 \varrho_2 q_k^{(1)} q_k^{(2)})^{1/2} (\varrho_2 q_k^{(1)} + \varrho_1 q_k^{(2)})^{-1} \ . \tag{4.5.8b}$$

It is obvious that equality

$$\mathscr{R}^2 + \mathscr{D}^2 = 1$$

is fulfilled for any k.

Equations for S_1^{11} and S_1^{21}, obtained after substituting (4.5.3a) into (4.5.2), take on the following form:

$$\begin{aligned}
&\mathrm{i}\exp(\mathrm{i}\boldsymbol{k}_0\cdot\boldsymbol{r})h(\boldsymbol{r})(\varrho_1 q_0^{(1)})^{1/2}[1-\mathscr{R}(k_0)] + \int \exp(\mathrm{i}\boldsymbol{k}\cdot\boldsymbol{r})(\varrho_1/q^{(1)})^{1/2} S_1^{11}(\boldsymbol{k},\boldsymbol{k}_0)\,d\boldsymbol{k}\\
&\quad = \mathrm{i}\exp(\mathrm{i}\boldsymbol{k}_0\cdot\boldsymbol{r})h(\boldsymbol{r})(\varrho_2 q_0^{(2)})^{1/2}\mathscr{D}(k_0) + \int \exp(\mathrm{i}\boldsymbol{k}\cdot\boldsymbol{r})(\varrho_2/q^{(2)})^{1/2} S_1^{21}(\boldsymbol{k},\boldsymbol{k}_0)\,d\boldsymbol{k}\ ,\\
&\qquad \exp(\mathrm{i}\boldsymbol{k}_0\cdot\boldsymbol{r})(\varrho_1 q_0^{(1)})^{-1/2}[\mathrm{i}q_0^{(1)^2} h(\boldsymbol{r}) - \boldsymbol{k}_0\cdot\nabla h](1+\mathscr{R}(k_0))\\
&\qquad - \int \exp(\mathrm{i}\boldsymbol{k}\cdot\boldsymbol{r})(q^{(1)}/\rho_1)^{1/2} S_1^{11}(\boldsymbol{k},\boldsymbol{k}_0)\,d\boldsymbol{k}\\
&\quad = \exp(\mathrm{i}\boldsymbol{k}_0\cdot\boldsymbol{r})(\varrho_2 q_0^{(2)})^{-1/2}[\mathrm{i}q_0^{(2)^2} h(\boldsymbol{r}) - \boldsymbol{k}_0\cdot\nabla h]\mathscr{D}(k_0)\\
&\qquad + \int \exp(\mathrm{i}\boldsymbol{k}\cdot\boldsymbol{r})(q^{(2)}/\varrho_2)^{1/2} S_1^{21}(\boldsymbol{k},\boldsymbol{k}_0)\,d\boldsymbol{k}\ .
\end{aligned} \tag{4.5.9}$$

The analogous system originates also for S_1^{21} and S_1^{22}. Solving these equations is not complicated. To simplify calculations and formulas for subsequent approximations it is convenient to perform some linear transformations. That is, we introduce matrix $\hat{A}^{(\tau)}$ instead of some matrix $\hat{A}$ according to

$$\hat{A}^{(\tau)}(\boldsymbol{k},\boldsymbol{k}_0) = \hat{\tau}(k)\hat{A}(\boldsymbol{k},\boldsymbol{k}_0)\hat{\tau}^{-1}(k_0)\ , \tag{4.5.10}$$

where matrix $\hat{\tau}$ is

$$\hat{\tau}(k) = \hat{\tau}^{-1}(k) = 2^{-1/2}\begin{bmatrix} (1+\mathscr{R})^{1/2} & (1-\mathscr{R})^{1/2}\\ (1-\mathscr{R})^{1/2} & -(1+\mathscr{R})^{1/2}\end{bmatrix}\ . \tag{4.5.11}$$

The inverse transformation from matrix $\hat{A}^{(\tau)}$ to matrix $\hat{A}$ coincides with (4.5.10). The reciprocity theorem (4.5.5) is similar in both primary and τ-representation:

$$\hat{S}^{\mathrm{T}}(\boldsymbol{k},\boldsymbol{k}_0) = \hat{S}(-\boldsymbol{k}_0, -\boldsymbol{k})\ , \tag{4.5.12}$$

where the superscript "T" denotes transposition. Under the reference transformation matrix $\hat{V}_0$ becomes diagonal

$$\hat{V}_0^{(\tau)} = \begin{pmatrix} 1 & 0\\ 0 & -1\end{pmatrix}\ . \tag{4.5.13}$$

Simple, although tedious computations show that matrix $\hat{S}_1$ in τ-representation is diagonal as well:

$$\hat{S}_1^{(\tau)}(\boldsymbol{k},\boldsymbol{k}_0) = 2\mathrm{i}\begin{bmatrix} T_1(\boldsymbol{k},\boldsymbol{k}_0) & 0\\ 0 & T_2(\boldsymbol{k},\boldsymbol{k}_0)\end{bmatrix} h(\boldsymbol{k}-\boldsymbol{k}_0)\ , \tag{4.5.14}$$

where

$$T_1(\boldsymbol{k},\boldsymbol{k}_0) = (\varrho_1 q_k^{(2)} + \varrho_2 q_k^{(1)})^{-1/2}(\varrho_1 q_0^{(2)} + \varrho_2 q_0^{(1)})^{-1/2}$$
$$\times \{\varrho_2[K_1^2 - (\boldsymbol{k}\cdot\boldsymbol{k}_0)] - \varrho_1[K_2^2 - (\boldsymbol{k}\cdot\boldsymbol{k}_0]\} \qquad (4.5.15a)$$

and

$$T_2(\boldsymbol{k},\boldsymbol{k}_0) = (\varrho_1 q_k^{(2)} + \varrho_2 q_k^{(1)})^{-1/2}(\varrho_1 q_0^{(2)} + \varrho_2 q_0^{(1)})^{-1/2}$$
$$\times (\varrho_2 - \varrho_1)(q_k^{(1)} q_k^{(2)} q_0^{(1)} q_0^{(2)})^{1/2} \; . \qquad (4.5.15b)$$

Consider the case $\varrho_1/\varrho_2 \to 0$ when the boundary is hard for the waves incident from the first medium, and soft from the side of the second medium. In this case $\mathscr{R} = 1$, matrices $\hat{S}$ and $\hat{S}^{(\tau)}$ coincide, and

$$\hat{S}(\boldsymbol{k},\boldsymbol{k}_0) = 2\mathrm{i}\begin{bmatrix} (q_k^{(1)} q_0^{(1)})^{-1/2}[K_1^2 - (\boldsymbol{k}\cdot\boldsymbol{k}_0)] & 0 \\ 0 & (q^{(2)} q_0^{(2)})^{1/2} \end{bmatrix} h(\boldsymbol{k}-\boldsymbol{k}_0). \qquad (4.5.16)$$

Diagonal matrix elements in (4.5.16) coincide with (4.3.6), (4.1.6), which agrees in the physical sense (note, that in accordance with the direction of z axis the signum of elevations in (4.1.6) should be changed in this case).

Estimation of SA to the second order values in elevations is rather tedious and the result is the following (formulas for SA are given in τ-representation with the superscript "τ" in (4.5.17) supressed):

$$S_2^{11}(\boldsymbol{k},\boldsymbol{k}^0) = -2\int T_1(\boldsymbol{k},\boldsymbol{k}')T_1(\boldsymbol{k}',\boldsymbol{k}_0)h(\boldsymbol{k}-\boldsymbol{k}')h(\boldsymbol{k}'-\boldsymbol{k}_0)\,d\boldsymbol{k}' \; , \qquad (4.5.17a)$$

$$S_2^{22}(\boldsymbol{k},\boldsymbol{k}_0) = 2\int T_2(\boldsymbol{k},\boldsymbol{k}')T_2(\boldsymbol{k}',\boldsymbol{k}_0)h(\boldsymbol{k}-\boldsymbol{k}')h(\boldsymbol{k}'-\boldsymbol{k}_0)\,d\boldsymbol{k}' \; , \qquad (4.5.17b)$$

$$S_2^{12}(\boldsymbol{k},\boldsymbol{k}_0) = (\varrho_1 q_k^{(2)} + \varrho_2 q_k^{(1)})^{-1/2}(\varrho_1 q_0^{(2)} + \varrho_2 q_0^{(1)})^{-1/2}$$
$$\times \int (\varrho_1 q_{k'}^{(2)} + \varrho_2 q_{k'}^{(1)})^{-1}(\varrho_1\varrho_2 q_k^{(1)} q_k^{(2)})^{1/2}$$
$$\times [2(q_{k'}^{(1)} + q_{k'}^{(2)})\{\varrho_1[K_2^2 - (\boldsymbol{k}'\cdot\boldsymbol{k}_0)] - \varrho_2[K_1^2 - (\boldsymbol{k}'\cdot\boldsymbol{k}_0)]\}$$
$$+ (K_1^2 - K_2^2)(\varrho_1 q_{k'}^{(2)} + \varrho_2 q_{k'}^{(1)})]h(\boldsymbol{k}-\boldsymbol{k}')h(\boldsymbol{k}'-\boldsymbol{k}_0)\,d\boldsymbol{k}' \; . \qquad (4.5.17c)$$

The formula for S_2^{21} is similar to (4.5.17c) and can be obtained by virtue of the reciprocity theorem (4.5.12):

$$S_2^{21}(\boldsymbol{k},\boldsymbol{k}_0) = S_2^{12}(-\boldsymbol{k}_0,-\boldsymbol{k}) \; .$$

The result of calculating the matrix element S^{12} in the third approximation is also presented

$$\hat{S}_3^{12}(\boldsymbol{k},\boldsymbol{k}_0) = \mathrm{i}\int (S_2^{12}(\boldsymbol{k},\boldsymbol{k}')T_2(\boldsymbol{k}',\boldsymbol{k}_0) + T_1(k,k')S_2^{12}(\boldsymbol{k}',\boldsymbol{k}_0))\,d\boldsymbol{k}' \; . \qquad (4.5.18)$$

4.6 Electromagnetic Wave Scattering at the Interface Between Two Dielectrics

The present section considers the case when the waves propagating in two homogeneous media have polarization.

Let the interface $z = h(\boldsymbol{r})$ divide two homogeneous half-spaces with permittivities ε_1 (lower half-space) and ε_2 (upper half-space). For simplicity we assume both permeability equal to unity. In both the half-spaces the vertically and horizontally polarized waves propagate which transform into each other when scattering at the rough boundary. The corresponding solutions of Maxwell's equations were described in Sect. 4.4, see (4.4.6). In the present section we keep notation of Sect. 4.4. First consider the case when a plane wave is incident from the first half-space. The appropriate solutions for electric and magnetic fields are

$$\begin{aligned}\boldsymbol{E}^{(1)} &= \boldsymbol{e}_{\sigma_0}^{+(1)}(\boldsymbol{k}_0)q_0^{(1)-1/2}\exp(\mathrm{i}\boldsymbol{k}_0\cdot\boldsymbol{r}+\mathrm{i}q_0^{(1)}z)\\ &\quad+\sum_\sigma\int d\boldsymbol{k}S_{\sigma\sigma_0}^{11}(\boldsymbol{k},\boldsymbol{k}_0)\boldsymbol{e}_\sigma^{-(1)}(\boldsymbol{k})q^{(1)-1/2}\exp(\mathrm{i}\boldsymbol{k}\cdot\boldsymbol{r}-\mathrm{i}q^{(1)}z)\ ,\\ \varepsilon^{-1/2}\boldsymbol{H}^{(1)} &= \boldsymbol{h}_{\sigma_0}^{+(1)}(\boldsymbol{k}_0)q_0^{(1)-1/2}\exp(\mathrm{i}\boldsymbol{k}_0\cdot\boldsymbol{r}+\mathrm{i}q_0^{(1)}z)\\ &\quad+\sum_\sigma\int d\boldsymbol{k}S_{\sigma\sigma_0}^{11}(\boldsymbol{k},\boldsymbol{k}_0)\boldsymbol{h}_\sigma^{-(1)}(\boldsymbol{k})q^{(1)-1/2}\exp(\mathrm{i}\boldsymbol{k}\cdot\boldsymbol{r}-\mathrm{i}q^{(1)}z)\\ &\text{at } z<\min h(\boldsymbol{r})\ ,\end{aligned}\tag{4.6.1}$$

$$\begin{aligned}\boldsymbol{E}^{(2)} &= \sum_\sigma\int d\boldsymbol{k}S_{\sigma\sigma_0}^{21}(\boldsymbol{k},\boldsymbol{k}_0)\boldsymbol{e}_\sigma^{+(2)}(\boldsymbol{k})q^{(2)-1/2}\exp(\mathrm{i}\boldsymbol{k}\cdot\boldsymbol{r}+\mathrm{i}q^{(2)}z)\ ,\\ \varepsilon^{-1/2}\boldsymbol{H}^{(2)} &= \sum_\sigma\int d\boldsymbol{k}S_{\sigma\sigma_0}^{21}(\boldsymbol{k},\boldsymbol{k}_0)\boldsymbol{h}_\sigma^{+(2)}(\boldsymbol{k})q^{(2)-1/2}\exp(\mathrm{i}\boldsymbol{k}\cdot\boldsymbol{r}+\mathrm{i}q^{(2)}z)\\ &\text{at } z>\max h(\boldsymbol{r})\ .\end{aligned}\tag{4.6.2}$$

Indices σ_0, σ are referred to vertical ($\sigma = 1$) and horizontal ($\sigma = 2$) polarizations of waves, and superscripts in parenthesis denote the medium number. Formulas for the case of the plane wave incident from the second medium are written down quite analogously and are obtained from (4.6.1, 2) at formal replacements of indices $1 \rightleftarrows 2$, sign indices $+ \leftrightarrows -$: $\boldsymbol{e}^+ \rightleftarrows \boldsymbol{e}^-$, $\boldsymbol{h}^+ \rightleftarrows \boldsymbol{h}^-$ and change of sign of the vertical coordinate $z \to -z$.

According to (4.4.4, 5) we have

$$\begin{aligned}&\boldsymbol{e}_1^{\pm(N)}(\boldsymbol{k}) = (k^2\boldsymbol{N}\mp q_k^{(N)}\boldsymbol{k})/(K^{(N)}k)\ ,\qquad \boldsymbol{h}_1^{\pm(N)}(\boldsymbol{k}) = -\boldsymbol{N}\times\boldsymbol{k}/k\ ,\\ &\boldsymbol{e}_2^{\pm(N)}(\boldsymbol{k}) = \boldsymbol{N}\times\boldsymbol{k}/k\ ,\qquad \boldsymbol{h}_2^{\pm(N)}(\boldsymbol{k}) = (k^2\boldsymbol{N}\mp q_k^{(N)}\boldsymbol{k})/(K^{(N)}k)\ ,\end{aligned}\tag{4.6.3}$$

where $K^{(N)} = \varepsilon_N^{1/2}\omega/c$, $N = 1, 2$. (Using bold $\boldsymbol{N}$ designating unit vector along z-direction and superscript $N = 1, 2$ designating the number of half-space shouldn't lead to confusion)

The value of SA $S_{\sigma\sigma_0}^{NN_0}$ is determined by the boundary conditions which for the given problem are in continuity of the tangential components of the electric and magnetic fields and can be written in the form:

$$[\boldsymbol{n}, \boldsymbol{E}^{(1)}] = [\boldsymbol{n}, \boldsymbol{E}^{(2)}] \;, \qquad [\boldsymbol{n}, \boldsymbol{H}^{(1)}] = [\boldsymbol{n}, \boldsymbol{H}^{(2)}] \;, \qquad z = h(\boldsymbol{r}) \;, \tag{4.6.4}$$

where $\boldsymbol{n}$ is a vector normal to the boundary $z = h(\boldsymbol{r})$:

$$\boldsymbol{n} = \boldsymbol{N} - \nabla h \;. \tag{4.6.5}$$

Let $(\overset{1}{\boldsymbol{E}}, \overset{1}{\boldsymbol{H}})$ and $(\overset{2}{\boldsymbol{E}}, \overset{2}{\boldsymbol{H}})$ be a pair of arbitrary solutions of Maxwell's equations (4.4.1) satisfying the boundary conditions (4.6.4) as well. Due to these boundary conditions relation (4.4.7) is easily generalized for the case of two dielectric half-spaces

$$\int\limits_{z=z_1<\min h(\boldsymbol{r})} \boldsymbol{N}\cdot(\overset{1}{\boldsymbol{E}} \times \overset{2}{\boldsymbol{H}} - \overset{2}{\boldsymbol{E}} \times \overset{1}{\boldsymbol{H}})\, d\boldsymbol{r} = \int\limits_{z=z_2>\max h(\boldsymbol{r})} \boldsymbol{N}\cdot(\overset{1}{\boldsymbol{E}} \times \overset{2}{\boldsymbol{H}} - \overset{2}{\boldsymbol{E}} \times \overset{1}{\boldsymbol{H}})\, d\boldsymbol{r} \;. \tag{4.6.6}$$

To obtain the reciprocity theorem we should substitute for $(\overset{1}{\boldsymbol{E}}, \overset{1}{\boldsymbol{H}})$ and $\overset{2}{\boldsymbol{E}}, \overset{2}{\boldsymbol{H}})$ solutions of the type (4.6.1, 2) with $\boldsymbol{k}_0 = \boldsymbol{k}_1$ and $\boldsymbol{k}_0 = \boldsymbol{k}_2$ and the solution not being written for the plane wave incident from the second medium. Simple calculations with account of (4.4.9) yield

$$\begin{aligned} &\hat{\sigma}_3 \hat{S}^{11}(\boldsymbol{k}_1, \boldsymbol{k}_2) = \hat{S}^{11\mathrm{T}}(-\boldsymbol{k}_2, -\boldsymbol{k}_1)\hat{\sigma}_3 \;, \\ &\hat{\sigma}_3 \hat{S}^{22}(\boldsymbol{k}_1, \boldsymbol{k}_2) = \hat{S}^{22\mathrm{T}}(-\boldsymbol{k}_2, -\boldsymbol{k}_1)\hat{\sigma}_3 \;, \\ &\hat{\sigma}_3 \hat{S}^{12}(\boldsymbol{k}_1, \boldsymbol{k}_2) = \hat{S}^{21\mathrm{T}}(-\boldsymbol{k}_2, -\boldsymbol{k}_1)\hat{\sigma}_3 \;, \end{aligned} \tag{4.6.7}$$

(the superscript "T" denotes transposition). In (4.6.7) $\hat{S}^{NN_0}$ are 2×2 matrices and the Pauli matrix $\hat{\sigma}_3$ is given by (4.4.27). The 4×4 matrix $\hat{S}$ can be composed out of four matrices $\hat{S}^{NN_0}$ by (2.2.27):

$$\hat{S}(\boldsymbol{k}_1, \boldsymbol{k}_2) = \begin{pmatrix} \hat{S}^{11} & \hat{S}^{12} \\ \hat{S}^{21} & \hat{S}^{22} \end{pmatrix}(\boldsymbol{k}_1, \boldsymbol{k}_2) \;. \tag{4.6.8}$$

Then the reciprocity theorem is written as

$$\begin{pmatrix} \hat{\sigma}_3 & 0 \\ 0 & \hat{\sigma}_3 \end{pmatrix} \hat{S}(\boldsymbol{k}_1, \boldsymbol{k}_2) = \hat{S}^{\mathrm{T}}(-\boldsymbol{k}_2, -\boldsymbol{k}_1) \begin{pmatrix} \hat{\sigma}_3 & 0 \\ 0 & \hat{\sigma}_3 \end{pmatrix} . \tag{4.6.9}$$

It follows from (4.4.1) and the boundary conditions (4.6.4) that if $(\overset{2}{\boldsymbol{E}}, \overset{2}{\boldsymbol{H}})$ is a solution, then $(\overset{2}{\boldsymbol{E}}{}^*, -\overset{2}{\boldsymbol{H}}{}^*)$ is a solution, too. Therefore, instead of (4.6.6) it can be written down

$$\begin{aligned} &\int\limits_{z=z_1<\min h(\boldsymbol{r})} \boldsymbol{N}\cdot(\overset{1}{\boldsymbol{E}} \times \overset{2}{\boldsymbol{H}}{}^* + \overset{2}{\boldsymbol{E}}{}^* \times \overset{1}{\boldsymbol{H}})\, d\boldsymbol{r} \\ &\quad = \int\limits_{z=z_2>\max h(\boldsymbol{r})} \boldsymbol{N}\cdot(\overset{1}{\boldsymbol{E}} \times \overset{2}{\boldsymbol{H}}{}^* + \overset{2}{\boldsymbol{E}}{}^* \times \overset{1}{\boldsymbol{H}})\, d\boldsymbol{r} \;. \end{aligned} \tag{4.6.10}$$

The same set of solutions as in obtaining the reciprocity theorem is substituted for $(\overset{1}{E}, \overset{1}{H})$ and $(\overset{2}{E}, \overset{2}{H})$. Calculations quite analogous to the previous give the unitarity relation (2.3.17)

$$\int^{(h)} dk' \hat{S}^{+}(k', k_1) \cdot \hat{S}(k', k_2) = \delta(k_1 - k_2) , \tag{4.6.11}$$

where the superscript "+" denotes Hermitian conjugation. There is an assumption in (4.6.11) that the wave vectors k_1, k_2 and k' correspond to homogeneous waves. In terms of the 2×2 matrix (4.6.11) is

$$\int_{|k'|<K^{(1)}} \hat{S}^{11+}(k', k_1) \cdot \hat{S}^{11}(k', k_2) \, dk' + \int_{|k'|<K^{(2)}} \hat{S}^{21+}(k', k_1) \hat{S}^{21}(k', k_2) \, dk'$$

$$= \delta(k_1 - k_2) ,$$

for $|k_1| < K^{(1)}$ and $|k_2| < K^{(1)}$;

$$\int_{|k'|<K^{(1)}} \hat{S}^{11+}(k', k_1) \cdot \hat{S}^{12}(k', k_2) \, dk' + \int_{|k'|<K(2)} \hat{S}^{21+}(k', k_1) \cdot \hat{S}^{22}(k', k_2) \, dk' = 0 ,$$

for $|k_1| < K^{(1)}$ and $|k_2| < K^{(2)}$, etc. Both the unitarity relation (4.6.11) and the reciprocity theorem (4.6.9) each represent the set of 16 scalar relations.

Now consider calculation of SA in the small perturbation method. Taking into account the content of Sects. 3.2, 4, and 4.2 which show that the Rayleigh hypothesis can be applied for calculating expansion SA into the integral-power series in elevations, we shall proceed just as in the previous Sections of this Chapter. Namely, values of electric and magnetic fields at the boundary will be determined according to (4.6.1, 2) where z should be set equal to $h(r)$. Then the resulting relations are substituted into the boundary conditions (4.6.4) and SA is sought in the form of a power series in h; all exponentials containing elevations are expanded into the power series as well.

Calculations are very simple in essence, and require only time and throughness. Here we give only the results of the main stages: the reader can easily reconstruct omitted simple calculations.

For the plane boundary $h(r) = 0$

$$\hat{S} = \hat{V}_0 \delta(k - k_0) , \tag{4.6.12}$$

where $\hat{V}_0$ is the 4×4 matrix. Let a wave be incident at the boundary from the first medium (from below). The boundary conditions (4.6.4) become (index 0 by V is omitted for simplicity in the following formulas)

$$N \times [e_{\sigma_0}^{+(1)}(k_0) q_0^{(1)-1/2} + \sum_{\sigma} V_{\sigma\sigma_0}^{11} e_{\sigma}^{-(1)}(k_0) q_0^{(1)-1/2}]$$

$$= N \times \sum_{\sigma} V_{\sigma\sigma_0}^{21} e_{\sigma}^{+(2)}(k_0) q_0^{(2)-1/2} , \tag{4.6.13a}$$

$$\varepsilon_1^{1/2} N \times [h_{\sigma_0}^{+(1)}(k_0) q_0^{(1)-1/2} + \sum_{\sigma} V_{\sigma\sigma_0}^{11} h_{\sigma}^{-}(k_0) q_0^{(1)-1/2}]$$

$$= \varepsilon_2^{1/2} N \times \sum_{\sigma} V_{\sigma\sigma_0}^{21} h_{\sigma}^{+(2)}(k_0) q_0^{(2)-1/2} \tag{4.6.13b}$$

Let $\sigma_0 = 1$. Projecting (4.6.13a) on the vectors $\boldsymbol{N} \times \boldsymbol{k}_0/k_0$ and $\boldsymbol{k}_0/k_0$ we find

$$q_0^{(1)1/2} K_1 (1 - V_1^{11}) = q_0^{(2)1/2} K_2 V_{11}^{21},$$

$$q_0^{(1)-1/2} V_{21}^{11} = q_0^{(2)-1/2} V_{21}^{21} \ . \tag{4.6.14a}$$

Similarly for (4.6.13b) we have

$$q_0^{(1)1/2} V_{21}^{11} = -q_0^{(2)1/2} V_{21}^{21} \ ,$$

$$\varepsilon_1^{1/2} q_0^{(1)-1/2} (1 + V_{11}^{11}) = \varepsilon_2^{1/2} q_0^{(2)-1/2} V_{11}^{21} \ . \tag{4.6.14b}$$

Equations (4.6.14) split into two pairs of independent equations and are easily solved. The result is

$$V_{11}^{11} = a = (\varepsilon_2 q^{(1)} - \varepsilon_1 q^{(2)})/(\varepsilon_2 q^{(1)} + \varepsilon_1 q^{(2)}) \ ,$$

$$V_{11}^{21} = c = 2 \cdot (\varepsilon_1 \varepsilon_2 q^{(1)} q^{(2)})^{1/2}/(\varepsilon_2 q^{(1)} + \varepsilon_1 q^{(2)}) \ ,$$

$$V_{21}^{11} = V_{21}^{21} = 0 \ . \tag{4.6.15}$$

If the dielectric permittivity of one of the media takes negative values, then the denominator in (4.6.15) can vanish. This corresponds to origin of a surface wave, i.e., the wave propagates horizontally and decreases exponentially as $|z| \to \infty$. In this case, the scheme of solving the problem should be modified and this surface wave should be considered from the very beginning. Here we shall not analyze this situation and assume that absence of a surface wave.

The case $\sigma_0 = 2$ is examined quite analogously. Then (4.6.14) is replaced by the following relations:

$$q_0^{(1)1/2} K_1 V_{12}^{11} = -q_0^{(2)1/2} K_2 V_{12}^{21} \ ,$$

$$q_0^{(1)-1/2} (1 + V_{22}^{11}) = q_0^{(2)-1/2} V_{22}^{21} \ ,$$

$$q_0^{(1)1/2} (1 - V_{22}^{11}) = q_0^{(2)1/2} V_{22}^{21} \ ,$$

$$\varepsilon_1^{1/2} q_0^{(1)-1/2} V_{12}^{11} = \varepsilon_2^{1/2} q_0^{(2)-1/2} V_{12}^{21} \ . \tag{4.6.16}$$

The solution of these equations is

$$V_{22}^{11} = b = (q^{(1)} - q^{(2)})/(q^{(1)} + q^{(2)}) \ ,$$

$$V_{22}^{21} = d = 2(q^{(1)} q^{(2)})^{1/2}/(q^{(1)} + q^{(2)}) \ ,$$

$$V_{12}^{11} = V_{12}^{21} = 0 \ . \tag{4.6.17}$$

Calculations for the wave incident from the second medium and from above are quite similar. Hence for matrix $\hat{V}_0$ in (4.6.12) we have

$$\hat{V}_0 = \begin{bmatrix} a & 0 & c & 0 \\ 0 & b & 0 & d \\ c & 0 & -a & 0 \\ 0 & d & 0 & -b \end{bmatrix} . \tag{4.6.18}$$

Since, as is easily seen, $a^2 + c^2 = b^2 + d^2 = 1$, then matrix $\hat{V}_0$ is unitary. Since $\hat{V}_0(k) = \hat{V}_0^{\mathrm{T}}(k) = \hat{V}_0(-k)$, then the reciprocity theorem (4.6.9) is satisfied.

The scattering matrix in the first order in h can be represented as

$$\hat{S}_1(\boldsymbol{k}, \boldsymbol{k}_0) = \hat{\dot{S}}(\boldsymbol{k}, \boldsymbol{k}_0) h(\boldsymbol{k} - \boldsymbol{k}_0) ,$$

where $h(\boldsymbol{k} - \boldsymbol{k}_0)$ is the Fourier transform of the roughness shape (4.1.7). To calculate $\hat{S}_1$ one iteration of (4.6.4) should be performed. Taking into account relations (4.6.14, 16) greatly simplifies the appropriate calculations. As a result (4.6.14) is replaced by

$$\mathrm{i}\frac{\boldsymbol{N}\cdot\boldsymbol{k}\times\boldsymbol{k}_0}{kk_0} q_0^{(2)1/2} \frac{K_1^2 - K_2^2}{K_2} V_{11}^{21}(k_0) + q^{(1)1/2}\dot{S}_{21}^{11} = -q^{(2)1/2}\dot{S}_{21}^{21} ,$$

$$\mathrm{i}\frac{\boldsymbol{k}\cdot\boldsymbol{k}_0}{kk_0} q_0^{(2)1/2} \frac{K_1^2 - K_2^2}{K_2} V_{11}^{21}(k_0) + K_1 q^{(1)-1/2}\dot{S}_{11}^{11} = K_2 q^{(2)-1/2}\dot{S}_{11}^{21} ,$$

$$q^{(1)-1/2}\dot{S}_{21}^{11} = q^{(2)-1/2}\dot{S}_{21}^{21} ,$$

$$\mathrm{i}kk_0 q_0^{(2)-1/2} \frac{K_2^2 - K_1^2}{K_1} V_{11}^{21} + K_2 q^{(1)1/2}\dot{S}_{11}^{11} = -K_1 q^{(2)1/2}\dot{S}_{11}^{21} . \tag{4.6.19}$$

Finally, (4.6.16) is replaced by

$$q^{(1)1/2} K_1^{-1} \dot{S}_{12}^{11} = -q^{(2)1/2} K_2^{-1} \dot{S}_{12}^{21},$$

$$q^{(1)-1/2}\dot{S}_{22}^{11} = q^{(2)-1/2}\dot{S}_{22}^{21} ,$$

$$\mathrm{i}\frac{\boldsymbol{N}\cdot\boldsymbol{k}\times\boldsymbol{k}_0}{kk_0}(K_1^2 - K_2^2) q_0^{(2)-1/2} V_{22}^{21}(k_0) + K_1 q^{(1)-1/2}\dot{S}_{12}^{11} = K_2 q^{(2)-1/2}\dot{S}_{12}^{21} .$$

$$-\mathrm{i}\frac{\boldsymbol{k}\cdot\boldsymbol{k}_0}{kk_0} q_0^{(2)-1/2}(K_1^2 - K_2^2) V_{22}^{21}(k_0) + q^{(1)1/2}\dot{S}_{22}^{11} = -q^{(2)1/2}\dot{S}_{22}^{21} . \tag{4.6.20}$$

Each system (4.6.19, 20) splits into pairs of independent equations: the 1st and the 3rd, the 2nd and the 4th. These equations are all easily solved and the result can be represented in the following form:

$$\dot{S}_{\sigma\sigma_0}^{NN_0}(\boldsymbol{k}, \boldsymbol{k}_0) = 2\mathrm{i} q_k^{(N)1/2} q_{k_0}^{(N_0)1/2} B_{\sigma\sigma_0}^{NN_0}(\boldsymbol{k}, \boldsymbol{k}_0) , \tag{4.6.21}$$

where

$$\hat{B}(\boldsymbol{k},\boldsymbol{k}_0)=\hat{d}(k)\left[\begin{pmatrix}\hat{\sigma}_3 & \hat{\sigma}_3\\ \hat{\sigma}_3 & \hat{\sigma}_3\end{pmatrix}\frac{\boldsymbol{k}\cdot\boldsymbol{k}_0}{kk_0}+\begin{pmatrix}\hat{\sigma}_1 & \hat{\sigma}_1\\ \hat{\sigma}_1 & \hat{\sigma}_1\end{pmatrix}\frac{\boldsymbol{N}\cdot\boldsymbol{k}\times\boldsymbol{k}_0}{kk_0}\right]\hat{d}(k_0)$$

$$-\hat{d}_1(k)\begin{bmatrix}1&0&1&0\\0&0&0&0\\1&0&1&0\\0&0&0&0\end{bmatrix}\hat{d}_1(k_0)\ . \qquad (4.6.22)$$

The Pauli matrices $\hat{\sigma}_1$, $\hat{\sigma}_3$ are given by (4.4.27). Here $\hat{d}(k)$ and $\hat{d}_1(k)$ are diagonal 4×4 matrices of the form

$$\hat{d}(k)=(\varepsilon_2-\varepsilon_1)^{1/2}\,\mathrm{diag}\left(\frac{\varepsilon_1^{1/2}q^{(2)}}{\varepsilon_2q^{(1)}+\varepsilon_1q^{(2)}},\frac{\omega/c}{q^{(1)}+q^{(2)}},-\frac{\varepsilon_2^{1/2}q^{(1)}}{\varepsilon_2q^{(1)}+\varepsilon_1q^{(2)}},\frac{\omega/c}{q^{(1)}+q^{(2)}}\right)\ , \qquad (4.6.23a)$$

$$\hat{d}_1(k)=(\varepsilon_2-\varepsilon_1)^{1/2}\,\mathrm{diag}\left(\frac{\varepsilon_2^{1/2}k}{\varepsilon_2q^{(1)}+\varepsilon_1q^{(2)}},0,\frac{\varepsilon_1^{1/2}k}{\varepsilon_2q^{(1)}+\varepsilon_1q^{(2)}},0\right)\ . \qquad (4.6.23b)$$

Since $\hat{d}(\boldsymbol{k})=\hat{d}(-\boldsymbol{k})$ and $d_1(\boldsymbol{k})=d_1(-\boldsymbol{k})$ it is clear that the reciprocity theorem (4.6.9) is fulfilled with respect to matrix $\hat{S}$.

Transition to the basis of the circularly polarized waves is performed by formulas similar to (4.4.16).

Consider the matrix $\hat{S}^{11}$ in the limit $\varepsilon_2\to\infty$. Proceeding from formulas (4.6.21–23) one can easily see that

$$\hat{B}^{11}(\boldsymbol{k},\boldsymbol{k}_0)=\begin{pmatrix}K_1q_k^{(1)-1} & 0\\ 0 & 1\end{pmatrix}\begin{bmatrix}\boldsymbol{k}\cdot\boldsymbol{k}_0/(kk_0) & \boldsymbol{N}\cdot\boldsymbol{k}\times\boldsymbol{k}_0/(kk_0)\\ \boldsymbol{N}\cdot\boldsymbol{k}\times\boldsymbol{k}_0/(kk_0) & -\boldsymbol{k}\cdot\boldsymbol{k}_0/(kk_0)\end{bmatrix}$$

$$\times\begin{pmatrix}K_1q_0^{(1)-1} & 0\\ 0 & 1\end{pmatrix}-\begin{pmatrix}k/q_k^{(1)} & 0\\ 0 & 0\end{pmatrix}\begin{pmatrix}1 & 0\\ 0 & 0\end{pmatrix}\begin{pmatrix}k_0/q_0^{(1)} & 0\\ 0 & 0\end{pmatrix}\ .$$

This relation coincides, in fact, with (4.4.21), as it must be.

Now consider the two-dimensional problem $h=h(x)$ and the case of the horizontally polarized wave when the electric field has only a y-component: $\boldsymbol{E}=(0,\psi,0)$. In virtue of (4.4.1) we have

$$\boldsymbol{H}=-\mathrm{i}\frac{c}{\omega}\left(-\frac{\partial\psi}{\partial z},0,\frac{\partial\psi}{\partial x}\right)\ ,\qquad (\boldsymbol{n}\times\boldsymbol{H})_y=-\mathrm{i}\frac{c}{\omega}\frac{\partial\psi}{\partial\boldsymbol{n}}\ .$$

Therefore in this case the boundary conditions (4.6.4) can be written as

$$\psi^{(1)}=\psi^{(2)}\ ,\qquad \frac{\partial\psi^{(1)}}{\partial\boldsymbol{n}}=\frac{\partial\psi^{(2)}}{\partial\boldsymbol{n}}\ . \qquad (4.6.24)$$

Formulas (4.6.24) coincide with the boundary conditions (4.5.2) at $\varrho_1=\varrho_2$ describing the case of the acoustic wave scattering at the interface between two liquid half-spaces with equal density. According to formulas (4.5.11, 14, 15) we have at $\varrho_1 = \varrho_2$:

$$\hat{\tau} = (q^{(1)} + q^{(2)})^{-1/2} \begin{pmatrix} q^{(1)1/2} & q^{(2)1/2} \\ q^{(2)1/2} & -q^{(1)1/2} \end{pmatrix} ,$$

$$\hat{\hat{S}}_1^{(\tau)} = -2\mathrm{i} \begin{bmatrix} \dfrac{K_2^2 - K_1^2}{(q^{(1)} + q^{(2)})^{1/2}(q_0^{(1)} + q_0^{(2)})^{1/2}} & 0 \\ 0 & 0 \end{bmatrix} .$$

Passing from τ-representation to the initial one by (4.5.10)

$$\hat{\hat{S}} = \hat{\tau}\hat{\hat{S}}^{(\tau)}\hat{\tau}$$

we have

$$\hat{\hat{S}}^{NN_0} = 2\mathrm{i} q_k^{(N)1/2} q_{k_0}^{(N_0)1/2} (K_1^2 - K_2^2)(q_k^{(1)} + q_k^{(2)})^{-1}(q_0^{(1)} + q_0^{(2)})^{-1} . \tag{4.6.25}$$

It is easily seen that (4.6.25) coincides with quantities $\hat{\hat{S}}_{22}^{NN_0}$ determined by (4.6.21–23) if taking into account that in the two-dimensional case we can set $\boldsymbol{k} \cdot \boldsymbol{k}_0 / k k_0 = 1$.

Let roughness have the form $h(\boldsymbol{r}) = H = \text{const}$. In this case scattering reduces to the specular reflection without changing the horizontal component of the wave vector $\boldsymbol{k} = \boldsymbol{k}_0$. Let $H \to 0$. Using transformation properties (2.4.4) we come to the equation, see also (5.2.15) below,

$$\mathrm{i}H \begin{bmatrix} 2q^{(1)}\hat{V}_0^{11}(k) & (q^{(1)} - q^{(2)})\hat{V}_0^{12}(k) \\ (q^{(1)} - q^{(2)})\hat{V}_0^{21}(k) & 2q^{(2)}\hat{V}_0^{22}(k) \end{bmatrix}$$

$$= 2\mathrm{i}H \begin{bmatrix} q^{(1)}\hat{B}^{11}(\boldsymbol{k},\boldsymbol{k}) & q^{(1)1/2}q^{(2)1/2}\hat{B}^{12}(\boldsymbol{k},\boldsymbol{k}) \\ q^{(2)1/2}q^{(1)1/2}\hat{B}^{21}(\boldsymbol{k},\boldsymbol{k}) & q^{(2)}\hat{B}^{22}(\boldsymbol{k},\boldsymbol{k}) \end{bmatrix} ,$$

whence

$$\hat{V}_0^{NN_0}(k) = \frac{2(q^{(N)}q^{(N_0)})^{1/2}}{(-1)^{N+1}q^{(N)} + (-1)^{N_0+1}q^{(N_0)}} \hat{B}^{NN_0}(\boldsymbol{k},\boldsymbol{k}) . \tag{4.6.26}$$

Validity of relation (4.6.26) can be proved by direct calculations as well.

5. Kirchhoff-Tangent Plane Approximation

Along with the perturbation theory the geometrical optics (quasiclassical) approximation represents another general approach in analyzing the scattering processes. The method may be employed when characteristic spatial scales of irregularities somewhat exceed the radiation wavelength. The approach is based on the assumption that the wave field locally has the structure of a plane wave whose amplitude and wave vector vary slowly from point to point. For scattering from a rough boundary this ansatz is formulated as follows: at each boundary point the incident wave is reflected in such a way as if the boundary is a plane. One can say that the boundary is approximated at each point by the tangent plane. Based on this representation, the scattered field and its normal derivative at the boundary can be easily expressed through the incident field. When these values are known it is possible to reconstruct the scattered field in the total space and, in particular, calculate SA, for instance, by formula of the kind (3.2.6). The above method was first proposed by *Brekhovskikh* [5.1, 2] and referred to as the tangent plane approximation or the Kirchhoff approximation (for the statistical case this approximation was developed by *Isakovich* [5.3]).

There is another more formal approach to realize this idea. It suggests writing the integral equation for the surface sources density in such a form when the first iteration of this equation for the plane boundary yields an exact solution for the problem. Then it can be expected that for a boundary that does not deviate much from the plane locally, the first iteration of the reference integral equation still gives a good result. The desired form of the integral equation for surface sources is obtained, particularly under the assumption that the scattered field is generated by the distribution of dipoles at the rough boundary, see formula (3.1.40). In this case the directivity diagrams are "eights" by figure, with axes running along local normals to the boundary (Fig. 5.1). It is obvious that for the plane the incident field singly reradiated by these surface dipoles is not then reradiated once more, since radiating and receiving occurs this time via zeros in the directivity diagrams of the boundary dipoles.

For definiteness in what follows we shall call the first approach to constructing the solution – the tangent plane approximation and the second one – the Kirchhoff approximation. However, this terminology is rather arbitrary. Both methods have a close base and usually lead to consistent results, what we shall see in the following sections of this chapter.

Section 5.2 also discusses the relation between the Kirchhoff approximation and the small perturbation method considered in Chap. 4.

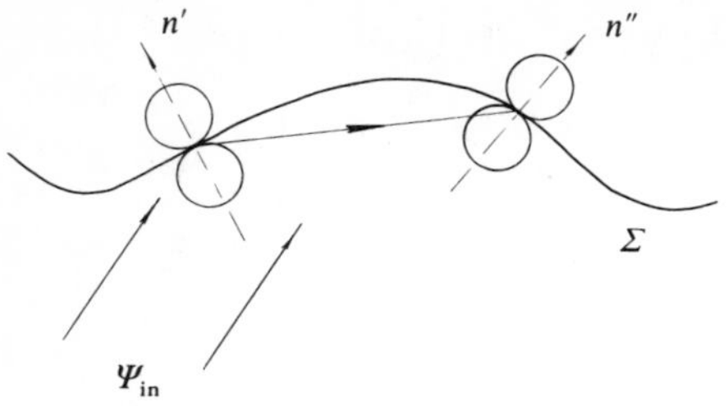

Fig. 5.1. Reradiation of the incident field by surface dipole sources

Corrections to the Kirchhoff-tangent plane approximation for the determinated case are analyzed in Sect. 5.3. They are due to the local diffraction effects and multiple reflections and shadowings as well.

The statistical problem is examined in Sect. 5.4. The theoretical approach to it consists in averaging the exact solution of the problem represented as an iterative series entailed by the integral equation for surface sources. Statistical ensemble of elevations is assumed as Gaussian and the result of averaging is given by diagrams. Such a diagrammatic approach was first applied in the works by *Zipfel* and *DeSanto* [5.4]; see also [5.5]. Section 5.4 presents different diagrammatic techniques by evaluating higher order diagrams with the help of several small parameters [5.6].

The last Sect. 5.5 proceeds to the two-scale model suggested first by *Kur'yanov* [5.7] which is widely used in practice. Its point is the following: The boundary roughness is assumed to consist of superposition of large- and small-scale components. Scattering at large-scale components (forming the so-called underlying surface) is treated in the Kirchhoff-tangent plane approximation, and small-scale components are taken into account by the perturbation theory. Hence such a pattern covers scattering in directions close to specular which is mainly due to the underlying surface and scattering at large angles caused by Bragg's scattering at small-scale components.

5.1 Tangent Plane Approximation

We start with the scalar problem. Let the plane wave

$$\Psi_{\text{in}} = q_0^{-1/2} \exp(\mathrm{i}\boldsymbol{k}_0 \cdot \boldsymbol{r} + \mathrm{i} q_0 z) \tag{5.1.1}$$

be incident onto the rough boundary.

The field at each point of space can be reconstructed with the help of general formula (3.1.27). Determining the field *outside* the boundary Σ as

$$\Psi|_{\boldsymbol{R} \in \Omega_2} = f_2 = 0 \ ,$$

see Fig. 2.2, we pass to formula (3.1.29). This time, however, we determine it as

$$\Psi|_{\boldsymbol{R}\in\Omega_2} = f_2 = q_0^{-1/2}\exp(\mathrm{i}\boldsymbol{k}_0\cdot\boldsymbol{r} + \mathrm{i}q_0 z) = \Psi_{\mathrm{in}}|_{\boldsymbol{R}\in\Omega_2} \ .$$

The total field is represented as a sum of the incident and the scattered fields:

$$\Psi = \Psi_{\mathrm{in}} + \Psi_{\mathrm{sc}} \ .$$

Then instead of (3.1.29) we obtain

$$\int \Psi_{\mathrm{sc}}|_{z=h(\boldsymbol{r})}\cdot\frac{\partial}{\partial\boldsymbol{n}}G_0(\boldsymbol{R}-\boldsymbol{R}_*)\,d\Sigma - \int\frac{\partial\Psi_{\mathrm{sc}}}{\partial\boldsymbol{n}}\bigg|_{z=h(\boldsymbol{r})}G_0(\boldsymbol{R}-\boldsymbol{R}_*)\,d\Sigma$$

$$= \begin{cases}\Psi_{\mathrm{sc}}(\boldsymbol{R}_*), & \boldsymbol{R}_*\in\Omega_1 \ ,\\ 0 \ , & \boldsymbol{R}_*\in\Omega_2 \ .\end{cases} \tag{5.1.2b}$$

Formula (5.1.2a) is much like (3.2.3) and using representation (3.2.4) for the Green's function gives, as in the beginning of Sect. 3.2, the following formula for SA:

$$S(\boldsymbol{k},\boldsymbol{k}_0) = \frac{1}{2}\int\left\{(q_k + \boldsymbol{k}\cdot\nabla h)\Psi_{\mathrm{sc}}|_{z=h(\boldsymbol{r})} + \mathrm{i}[1+(\nabla h)^2]^{1/2}\frac{\partial\Psi_{\mathrm{sc}}}{\partial\boldsymbol{n}}\bigg|_{z=h(\boldsymbol{r})}\right\}$$

$$\times\, q_k^{-1/2}\exp[-\mathrm{i}\boldsymbol{k}\cdot\boldsymbol{r} + \mathrm{i}q_k h(\boldsymbol{r})]\,d\boldsymbol{r}/(2\pi)^2 \ . \tag{5.1.3}$$

Taking into account (2.1.10) we can write

$$S(\boldsymbol{k},\boldsymbol{k}_0) = \frac{1}{2}\int\left[(q_k + \boldsymbol{k}\cdot\nabla h)\Psi_{\mathrm{sc}}|_{z=h(\boldsymbol{r})} + \mathrm{i}\left(\frac{\partial\Psi_{\mathrm{sc}}}{\partial z} - \nabla\Psi_{\mathrm{sc}}\cdot\nabla h\right)\bigg|_{z=h(\boldsymbol{r})}\right]$$

$$\times\, q_k^{-1/2}\exp[-\mathrm{i}\boldsymbol{k}\cdot\boldsymbol{r} + \mathrm{i}q_k h(\boldsymbol{r})]\,d\boldsymbol{r}/(2\pi)^2 \ . \tag{5.1.4}$$

Hence, SA in formula (5.1.3) is expressed through the *scattered* (not total) field at the rough boundary under the assumption that the incident field has the form (5.1.1).

In reflection from the slope plane

$$h(\boldsymbol{r}) = \boldsymbol{a}\cdot\boldsymbol{r} \ ,$$

where

$$\boldsymbol{a} = \nabla h \tag{5.1.5}$$

the incident and reflected fields are related as

$$\Psi_{\mathrm{sc}} = R(q_{on})\Psi_{\mathrm{in}} \ , \qquad z = h(\boldsymbol{r}) \ . \tag{5.1.6}$$

Here q_{on} is the projection of the three-dimensional wave vector of the incident wave $q_0\boldsymbol{N} + \boldsymbol{k}_0$ on the unit normal vector to the boundary (here and below $\boldsymbol{N}$ is the unit upward vector). Reflection coefficient $\mathscr{R}$ for isotropic media can be considered to be a function of q_{on}. It is clear that $R(q_{on})$ coincides with the coefficient $V_0(k)$ in expansion (4.1.1) for $k = (K^2 - q_{on}^2)^{1/2}$ (this reasoning becomes more complicate for multi-componental fields – see following). The scattered field for the plane boundary is just a reflected wave of the form

$$\Psi_{sc} = \mathscr{R}(q_{on}) \exp(i\tilde{\boldsymbol{k}} \cdot \boldsymbol{r} - iq_{\tilde{k}} z) .$$

Wave vector components for the reflected wave are calculated by the formulas

$$\tilde{\boldsymbol{k}} = \boldsymbol{k}_0 + 2(1 + a^2)^{-1}(q_0 - \boldsymbol{a} \cdot \boldsymbol{k}_0)\boldsymbol{a} ,$$

$$q_{\tilde{k}} = -q_0 + 2(1 + a^2)^{-1}(q_0 - \boldsymbol{a} \cdot \boldsymbol{k}_0) .$$

Thus

$$q_{\tilde{k}} + \boldsymbol{a} \cdot \tilde{\boldsymbol{k}} = q_0 - \boldsymbol{a} \cdot \boldsymbol{k}_0$$

whence relation

$$\boldsymbol{a} = (\tilde{\boldsymbol{k}} - \boldsymbol{k}_0)/(q + q_0) \tag{5.1.7}$$

follows.

Taking into account (5.1.7) we easily find

$$i\left(\frac{\partial \Psi_{sc}}{\partial z} - \nabla h \cdot \nabla \Psi_{sc}\right)\Bigg|_{z=h(\boldsymbol{r})} = \mathscr{R}(q_{on})(q_0 - \boldsymbol{k}_0 \cdot \nabla h)\Psi_{in} \tag{5.1.8}$$

Substituting (5.1.6, 8, 1) into (5.1.4) yields the following formula for SA:

$$S(\boldsymbol{k}, \boldsymbol{k}_0) = (qq_0)^{-1/2}/2 \int \mathscr{R}(q_{on})(q + q_0 + (\boldsymbol{k} - \boldsymbol{k}_0) \cdot \nabla h)$$
$$\times \exp[-i(\boldsymbol{k} - \boldsymbol{k}_0) \cdot \boldsymbol{r} + i(q + q_0)h(\boldsymbol{r})]\, d\boldsymbol{r}/(2\pi)^2 . \tag{5.1.9}$$

Now consider the Dirichlet problem when $\mathscr{R} = 1$. We use the following apparent transformation:

$$\nabla h \exp[i(q + q_0)h(\boldsymbol{r})] = [i(q + q_0)]^{-1} \nabla \{\exp[i(q + q_0)h(\boldsymbol{r})]\}$$

and integrate in (5.1.9) by parts, transferring the action of operator ∇ to $\exp[i(\boldsymbol{k} - \boldsymbol{k}_0) \cdot \boldsymbol{r}]$ and changing the sign. As a result this operation is equivalent to the following replacement in (5.1.9):

$$\nabla h \to (\boldsymbol{k} - \boldsymbol{k}_0)/(q + q_0) . \tag{5.1.10}$$

Thus for the Dirichlet problem we obtain

$$S(\boldsymbol{k}, \boldsymbol{k}_0) = -\frac{(qq_0)^{-1/2}}{(q + q_0)}(K^2 + qq_0 - \boldsymbol{k} \cdot \boldsymbol{k}_0)$$
$$\times \int \exp[-i(\boldsymbol{k} - \boldsymbol{k}_0) \cdot \boldsymbol{r} + i(q + q_0)h(\boldsymbol{r})]\, d\boldsymbol{r}/(2\pi)^2 . \tag{5.1.11}$$

This formula can be easily transformed into that which will be used further

$$S(\boldsymbol{k},\boldsymbol{k}_0) = -\frac{2(qq_0)^{1/2}}{(q+q_0)}\cdot\left[1+\frac{(\boldsymbol{k}-\boldsymbol{k}_0)^2+(q-q_0)^2}{4qq_0}\right]$$
$$\times\int \exp[-\mathrm{i}(\boldsymbol{k}-\boldsymbol{k}_0)\cdot\boldsymbol{r}+\mathrm{i}(q+q_0)h(\boldsymbol{r})]\frac{d\boldsymbol{r}}{(2\pi)^2}\ . \quad (5.1.12)$$

Since for the Neumann problem, $\mathscr{R} = +1$, then the resulting formula in the tangent plane approximation for the rigid boundary differs from (5.1.11) only in the sign.

The tangent plane approximation, similar to the geometric optics approximation, is the more accurate when the wavelength is less, see in addition Sects. 5.3, 4. For the wavenumber $K = \omega/c$, which is large compared to the characteristic horizontal scale of function $h(\boldsymbol{r})$, integral (5.1.9) can be calculated by the stationary phase method (the one-dimensional version of this method for analytical profiles was already used in Sect. 3.2). The stationary point $\boldsymbol{r} = \boldsymbol{r}_s$ is determined from the condition that the gradient of phase in the exponent in (5.1.9) must vanish which yields

$$\nabla h(r_s) = (\boldsymbol{k}-\boldsymbol{k}_0)/(q+q_0)\ . \quad (5.1.13)$$

This relation coincides with (5.1.10). From the physical point of view, (5.1.13) represents the equation for the specular points position $\boldsymbol{r}_s$. According to (5.1.7) it is just the plane with a slope determined by the right-hand side of (5.1.13) which ensures transforming the incident wave into a reflected one at specular reflection.

Returning to the general case, replace the function in (5.1.9) by its value at stationary point

$$\mathscr{R} = \mathscr{R}(q_{0n}^{(s)}) = \mathscr{R}(\boldsymbol{k},\boldsymbol{k}_0)\ ,$$

where by (5.1.13) the vector of the unit normal to the boundary is

$$\boldsymbol{n} = \left[\boldsymbol{N}-\frac{(\boldsymbol{k}-\boldsymbol{k}_0)}{(q+q_0)}\right]\Big/\left[1+\frac{(\boldsymbol{k}-\boldsymbol{k}_0)^2}{(q+q_0)^2}\right]^{-1/2} \quad (5.1.14)$$

whence

$$q_{0n}^{(s)} = (q_0\boldsymbol{N}+\boldsymbol{k}_0)\cdot\boldsymbol{n} = 2^{-1/2}(K^2+qq_0-\boldsymbol{k}\cdot\boldsymbol{k}_0)^{1/2}\ .$$

As a result, we obtain the following formula for calculating SA in the tangent plane approximation:

$$S(\boldsymbol{k},\boldsymbol{k}_0) = \mathscr{R}(\boldsymbol{k},\boldsymbol{k}_0)\frac{(qq_0)^{-1/2}}{(q+q_0)}(K^2+qq_0-\boldsymbol{k}\cdot\boldsymbol{k}_0)$$
$$\times\int \exp[-\mathrm{i}(\boldsymbol{k}-\boldsymbol{k}_0)\cdot\boldsymbol{r}+\mathrm{i}(q+q_0)h(\boldsymbol{r})]\frac{d\boldsymbol{r}}{(2\pi)^2}\ . \quad (5.1.15)$$

The present reasoning can be easily generalized for the boundary between two media. In fact diagonal matrix element S^{11} and S^{22} for scalar fields, see Sect. 2.2, can be calculated by formula (5.1.15). For example, let us examine the boundary between two liquid half-spaces already considered in Sect. 4.5. The reflection coefficient $\mathscr{R}$ in (5.1.15) is given by (4.5.8a) where $q^{(1)}$ and $q^{(2)}$ are the projections of three-dimensional wave vectors on the local unit normal, so that

$$q_n^{(1)} = (q_0^{(1)}\boldsymbol{N} + \boldsymbol{k}_0)\cdot\boldsymbol{n} = 2^{-1/2}\cdot(K_1^2 + q^{(1)}q_0^{(1)} - \boldsymbol{k}\cdot\boldsymbol{k}_0)^{1/2} \,,$$

$$q_n^{(2)} = [K_2^2 - K_1^2 + (q_n^{(1)})^2]^{1/2} = 2^{-1/2}(2K_2^2 - K_1^2 + q^{(1)}q^{(0)} - \boldsymbol{k}\cdot\boldsymbol{k}_0)^{1/2} \,, \tag{5.1.16}$$

and

$$\mathscr{R}(\boldsymbol{k}, \boldsymbol{k}_0) = \mathscr{R}^{11}(\boldsymbol{k}, \boldsymbol{k}_0) = \frac{(\varrho_2 q_n^{(1)} - \varrho_1 q_n^{(2)})}{(\varrho_2 q_n^{1)} + \varrho_1 q_n^{(2)})} \,.$$

Hence the formula for matrix element S^{11} can be represented in the form

$$S^{11}(\boldsymbol{k}, \boldsymbol{k}_0) = \mathscr{R}^{11}(\boldsymbol{k}, \boldsymbol{k}_0)\frac{(q^{(1)}q_0^{(1)})^{-1/2}}{(q^{(1)} + q_0^{(1)})}(K_1^2 + q^{(1)}q_0^{(1)} - \boldsymbol{k}\cdot\boldsymbol{k}_0)$$
$$\times \int \exp[-\mathrm{i}(\boldsymbol{k} - \boldsymbol{k}_0)\cdot\boldsymbol{r} + \mathrm{i}(q^{(1)} + q_0^{(1)})h(\boldsymbol{r})]\frac{d\boldsymbol{r}}{(2\pi)^2} \,. \tag{5.1.17}$$

Correspondingly, the expression for matrix element S^{22} is

$$S^{22}(\boldsymbol{k}, \boldsymbol{k}_0) = \mathscr{R}^{22}(\boldsymbol{k}, \boldsymbol{k}_0)\frac{(q^{(2)}q_0^{(2)})^{-1/2}}{(q^{(2)} + q_0^{(2)})}(K_2^2 + q^{(2)}q_0^{(2)} - \boldsymbol{k}\cdot\boldsymbol{k}_0)$$
$$\times \int \exp[-\mathrm{i}(\boldsymbol{k} - \boldsymbol{k}_0)\cdot\boldsymbol{r} - \mathrm{i}(q^{(2)} + q_0^{(2)})h(\boldsymbol{r})]\frac{d\boldsymbol{r}}{(2\pi)^2} \,, \tag{5.1.18}$$

where

$$q_n^{(2)} = 2^{-1/2}(K_2^2 + q^{(2)}q_0^{(2)} - \boldsymbol{k}\cdot\boldsymbol{k}_0)^{1/2} \,,$$

$$q_n^{(1)} = 2^{-1/2}(2K_1^2 - K_2^2 + q^{(2)}q_0^{(2)} - \boldsymbol{k}\cdot\boldsymbol{k}_0)^{1/2} \,,$$

and

$$\mathscr{R}^{22}(\boldsymbol{k}, \boldsymbol{k}_0) = \frac{(\varrho_1 q_n^{(2)} - \varrho_2 q_n^{(1)})}{(\varrho_1 q_n^{(2)} + \varrho_2 q_n^{(1)})} \,.$$

Now let us consider calculating the nondiagonal matrix element S^{21}, describing wave transmission from the first to the second medium. Based on the same arguments concerning the local field structure, we can write for the refracted field Ψ_{sc} and its normal derivative at the boundary on the second medium side

$$\Psi_{\mathrm{sc}}\Big|_{z=h(\boldsymbol{r})} = \mathscr{D}\cdot\left(\frac{\varrho_1 q_n^{(1)}}{\varrho_2 q_n^{(2)}}\right)^{1/2} \Psi_{\mathrm{in}}\Big|_{z=h(\boldsymbol{r})} ,$$

$$\frac{\partial \Psi_{\mathrm{sc}}}{\partial \boldsymbol{n}}\Big|_{z=h(\boldsymbol{r})} = \mathrm{i}(q_{\boldsymbol{k}}^{(2)}\boldsymbol{N} + \tilde{\boldsymbol{k}})\cdot\boldsymbol{n}\Psi_{\mathrm{sc}}|_{z=h(\boldsymbol{r})} , \tag{5.1.19}$$

where

$$\Psi_{\mathrm{in}} = (\varrho_1 q_0^{(1)})^{-1/2}\exp[\mathrm{i}\boldsymbol{k}_0\cdot\boldsymbol{r} + \mathrm{i}q_0^{(1)}h(\boldsymbol{r})]$$

is the incident field. In (5.1.19) $(q_{\boldsymbol{k}}^{(2)}\boldsymbol{N} + \tilde{\boldsymbol{k}})$ is the wave vector of the refracted wave and $\mathscr{D}$ is the transmission coefficient, see (4.5.8a):

$$\mathscr{D} = 2\frac{\varrho_1 q_n^{(1)}\varrho_2 q_n^{(2)})^{1/2}}{(\varrho_2 q_n^{(1)} + \varrho_1 q_n^{(2)})} , \tag{5.1.20}$$

where $q_n^{(1)}$ and $q_n^{(2)}$ are projections of the three-dimensional wave vectors of the incident and the refracted waves on the unit normal to the boundary. The field at each point of the second medium is estimated by (5.1.2a):

$$\Psi_{\mathrm{sc}}(\boldsymbol{R}_*) = -\int \Psi_{\mathrm{sc}}\Big|_{z=h(\boldsymbol{r})}\frac{\partial}{\partial \boldsymbol{n}}G_0(\boldsymbol{R}-\boldsymbol{R}_*)\,d\Sigma$$

$$+\int\frac{\partial\Psi_{\mathrm{sc}}}{\partial \boldsymbol{n}}\Big|_{z=h(\boldsymbol{r})} G_0(\boldsymbol{R}-\boldsymbol{R}_*)\,d\Sigma , \qquad \boldsymbol{R}_*\in\Omega_2 \tag{5.1.21}$$

(since normal $\boldsymbol{n}$ is inward with respect to the second medium, (5.1.21) differs from (5.1.2a) in sign). By definition of SA (4.5.3a) we obtain, similarly to the previous

$$S^{21}(\boldsymbol{k},\boldsymbol{k}_0) = (q_{\boldsymbol{k}}^{(2)}q_0^{(1)})^{-1/2}/2\int[q_{\boldsymbol{k}}^{(2)} + q_{\tilde{\boldsymbol{k}}}^{(2)} - (\boldsymbol{k}+\tilde{\boldsymbol{k}})\cdot\nabla h](q_n^{(1)}/q_n^{(2)})^{1/2}$$

$$\times\,\mathscr{D}\exp[-\mathrm{i}(\boldsymbol{k}-\boldsymbol{k}_0)\cdot\boldsymbol{r} + \mathrm{i}(-q_{\boldsymbol{k}}^{(2)} + q_0^{(1)})h(\boldsymbol{r})]\frac{d\boldsymbol{r}}{(2\pi)^2} . \tag{5.1.22}$$

In estimating the integral by the stationary phase method the stationary point is obtained from

$$\nabla h(\boldsymbol{r}_{\mathrm{s}}) = (\boldsymbol{k}-\boldsymbol{k}_0)/(-q_{\boldsymbol{k}}^{(2)} + q_0^{(1)}) . \tag{5.1.23}$$

Substituting this value of ∇h into the integrand in (5.1.22), we get after simple calculations

$$S^{21}(\boldsymbol{k},\boldsymbol{k}_0) = \mathscr{D}(\boldsymbol{k},\boldsymbol{k}_0)\left(\frac{q_n^{(1)}}{q_n^{(2)}}\right)^{1/2}(q_{\boldsymbol{k}}^{(2)}q_0^{(1)})^{-1/2}\frac{(K_2^2 - q_{\boldsymbol{k}}^{(2)}q_0^{(1)} - \boldsymbol{k}\cdot\boldsymbol{k}_0)}{(q_{\boldsymbol{k}}^{(2)} - q_0^{(1)})}$$

$$\times\int\exp[-\mathrm{i}(\boldsymbol{k}-\boldsymbol{k}_0)\cdot\boldsymbol{r} + \mathrm{i}(-q_{\boldsymbol{k}}^{(2)} + q_0^{(1)})h(\boldsymbol{r})]\frac{d\boldsymbol{r}}{(2\pi)^2} , \tag{5.1.24}$$

where $q_n^{(1)}$, $q_n^{(2)}$ in (5.1.20, 24) should be calculated with respect to the vector of normal at the stationary point

$$\boldsymbol{n} = \left(\boldsymbol{N} - \frac{(\boldsymbol{k}-\boldsymbol{k}_0)}{(-q_k^{(2)}+q_0^{(1)})}\right)\cdot\left(1 + \frac{(\boldsymbol{k}-\boldsymbol{k}_0)^2}{(-q_k^{(2)}+q_0^{(1)})^2}\right)^{-1/2} .$$

Expressions for $q_n^{(1)}$, $q_n^{(2)}$ are obtained similarly to (5.1.16) and have the following form:

$$\begin{aligned} q_n^{(1)} &= (K_1^2 - q_k^{(2)}q_0^{(1)} - \boldsymbol{k}\cdot\boldsymbol{k}_0)(K_1^2 + K_2^2 - 2q_k^{(2)}q_0^{(1)} - 2\boldsymbol{k}\cdot\boldsymbol{k}_0)^{-1/2} , \\ q_n^{(2)} &= (K_2^2 - q_k^{(2)}q_0^{(1)} - \boldsymbol{k}\cdot\boldsymbol{k}_0)(K_1^2 + K_2^2 - 2q_k^{(2)}q_0^{(1)} - 2\boldsymbol{k}\cdot\boldsymbol{k}_0)^{-1/2} . \end{aligned} \tag{5.1.25}$$

As a result we obtain

$$\begin{aligned} S^{21}(\boldsymbol{k},\boldsymbol{k}_0) &= \mathscr{D}(\boldsymbol{k},\boldsymbol{k}_0)\frac{(q^{(2)}q_0^{(1)})^{-1/2}}{(q^{(2)}-q_0^{(1)})}(K_2^2 - q^{(2)}q_0^{(1)} - \boldsymbol{k}\cdot\boldsymbol{k}_0)^{1/2} \\ &\quad\times (K_1^2 - q^{(2)}q_0^{(1)} - \boldsymbol{k}\cdot\boldsymbol{k}_0)^{1/2} \\ &\quad\times \int \exp[-\mathrm{i}(\boldsymbol{k}-\boldsymbol{k}_0)\cdot\boldsymbol{r} + \mathrm{i}(-q^{(2)}+q_0^{(1)})h(\boldsymbol{r})]\, d\boldsymbol{r}/(2\pi)^2 , \end{aligned} \tag{5.1.26}$$

where $\mathscr{D}$ is calculated according to (5.1.20, 25). Changing the sign for h and replacing the wave numbers $K_1 \rightleftarrows K_2$ and superscripts (1) $\rightleftarrows$ (2) yields the formula for S^{12}:

$$\begin{aligned} S^{12}(\boldsymbol{k},\boldsymbol{k}_0) &= \mathscr{D}(\boldsymbol{k},\boldsymbol{k}_0)\frac{(q^{(1)}q_0^{(2)})^{-1/2}}{(-q^{(1)}+q_0^{(2)})}(K_1^2 - q^{(1)}q_0^{(2)} - \boldsymbol{k}\cdot\boldsymbol{k}_0)^{1/2} \\ &\quad\times (K_2^2 - q^{(1)}q_0^{(2)} - \boldsymbol{k}\cdot\boldsymbol{k}_0)^{1/2} \\ &\quad\times \int \exp[-\mathrm{i}(\boldsymbol{k}-\boldsymbol{k}_0)\cdot\boldsymbol{r} + \mathrm{i}(q^{(1)}-q_0^{(2)})h(\boldsymbol{r})]\frac{d\boldsymbol{r}}{(2\pi)^2} , \end{aligned} \tag{5.1.27}$$

where this time

$$\begin{aligned} q_n^{(2)} &= (K_2^2 - q^{(1)}q_0^{(2)} - \boldsymbol{k}\cdot\boldsymbol{k}_0)(K_1^2 + K_2^2 - 2q^{(1)}q_0^{(2)} - 2\boldsymbol{k}\cdot\boldsymbol{k}_0)^{-1/2} , \\ q_n^{(1)} &= (K_1^2 - q^{(1)}q_0^{(2)} - \boldsymbol{k}\cdot\boldsymbol{k}_0)(K_1^2 + K_2^2 - 2q^{(1)}q_0^{(2)} - 2\boldsymbol{k}\cdot\boldsymbol{k}_0)^{-1/2} , \end{aligned} \tag{5.1.28}$$

and $\mathscr{D}(\boldsymbol{k},\boldsymbol{k}_0)$ is calculated, as previously, by (5.1.20).

One should thoroughly choose the root branches in formulas (5.1.25–28) which are related to conditions $\mathrm{Re}\{q_n^{(1,2)}\} \geqslant 0$ and $\mathrm{Im}\{q_n^{(1,2)}\} \geqslant 0$. The phenomenon arising here is apparent for the trivial case $h(\boldsymbol{r}) = 0$. Therefore, formulas (5.1.26, 27) are written down with an accuracy to the constant phase factor.

It can be easily shown that the reciprocity theorem (4.5.5) is satisfied in this approximation.

Now consider application of the tangent plane method for the multicomponental fields, taking as an example the reflection of electromagnetic waves from a perfectly conducting plane. This problem has already been considered in Sect. 4.4 in the framework of the small perturbation method. Here it is convenient to treat it in the basis of circularly polarized waves. For the multicomponental fields SA is given by a formula of the kind (5.1.15):

$$S_{\sigma\sigma_0}(\boldsymbol{k},\boldsymbol{k}_0) = \mathscr{R}_{\sigma\sigma_0}(\boldsymbol{k},\boldsymbol{k}_0)\frac{(qq_0)^{-1/2}}{(q+q_0)}(K^2 + qq_0 - \boldsymbol{k}\cdot\boldsymbol{k}_0)$$

$$\times \int \exp[-\mathrm{i}(\boldsymbol{k}-\boldsymbol{k}_0)\cdot\boldsymbol{r} + \mathrm{i}(q+q_0)h(\boldsymbol{r})]\, d\boldsymbol{r}/(2\pi)^2 \ . \tag{5.1.29}$$

The only difference is that the reflection coefficient acquired indices and $\mathscr{R}_{\sigma\sigma_0}$ equals the ratio of the reflected σth polarization wave amplitude to that of the incident σ_0 polarization wave. It was shown in Sect. 4.4 that in reflecting from the perfectly conducting plane the clockwise polarized wave turns into the counterclockwise polarized wave with the same amplitude and vice versa. However, it should be taken into account that the amplitude of the clock counterclockwise polarized wave acquires a certain additional phase factor when passing into the coordinates related to the plane with the given slope. Imagine the plane originated by the vectors $\boldsymbol{E}_1 = \mathrm{Re}\,\boldsymbol{E}^{(c)}$ and $\boldsymbol{E}_2 = \mp\mathrm{Im}\,\boldsymbol{E}^{(c)}$. For the circularly polarized wave with unit amplitude the projection of vector $\boldsymbol{N} = \boldsymbol{e}_z$ on this plane must be directed along the vector $\boldsymbol{E}_1$, see (4.4.15). However, the vector of unit normal to the slope boundary $\boldsymbol{n}$ have projections on vectors $\boldsymbol{E}_1$ and $\boldsymbol{E}_2$ equal to $(\boldsymbol{E}_1\cdot\boldsymbol{n})$ and $(\boldsymbol{E}_2\cdot\boldsymbol{n})$, respectively. Therefore, in the system of coordinates related to the slope plane, the initial clockwise polarized wave gets an additional phase vector a_1:

$$a_1 = \frac{(\boldsymbol{E}_1\cdot\boldsymbol{n}) - \mathrm{i}(\boldsymbol{E}_2\cdot\boldsymbol{n})}{[(\boldsymbol{E}_1\cdot\boldsymbol{n})^2 + (\boldsymbol{E}_2\cdot\boldsymbol{n})^2]^{1/2}} \ .$$

Since for the incident wave

$$\boldsymbol{E}_1 = \frac{k_0^2\boldsymbol{N} - q_0\boldsymbol{k}_0}{Kk_0} \ , \qquad \boldsymbol{E}_2 = \boldsymbol{N}\times\boldsymbol{k}_0/k_0 \ ,$$

see (4.4.4, 15), and $\boldsymbol{n}$ is given by (5.1.14), as a result of direct calculations we find

$$a_1 = \frac{k_0^2 q + q_0\boldsymbol{k}\cdot\boldsymbol{k}_0 - \mathrm{i}K\boldsymbol{N}\cdot\boldsymbol{k}\times\boldsymbol{k}_0}{k_0(K^2 + qq_0 - \boldsymbol{k}\cdot\boldsymbol{k}_0)^{1/2}(K^2 - qq_0 + \boldsymbol{k}\cdot\boldsymbol{k}_0)^{1/2}} \ .$$

In reflecting the clockwise polarized wave from the slope plane, its amplitude equal to a_1 does not change, but the wave becomes a counterclockwise polarized one. When returning to the initial coordinate system this counterclockwise wave takes an additional phase factor equal to

$$a_2 = \frac{(\boldsymbol{E}_1 \cdot \boldsymbol{n}) - \mathrm{i}(\boldsymbol{E}_2 \cdot \boldsymbol{n})}{[(\boldsymbol{E}_1 \cdot \boldsymbol{n})^2 + (\boldsymbol{E}_2 \cdot \boldsymbol{n})^2]^{1/2}} ,$$

where $\boldsymbol{E}_1$ and $\boldsymbol{E}_2$ correspond to the reflected wave and take the form

$$\boldsymbol{E}_1 = \frac{k^2 \boldsymbol{N} + q\boldsymbol{k}}{Kk} , \qquad \boldsymbol{E}_2 = \boldsymbol{N} \times \boldsymbol{k}/k .$$

Calculations yield

$$a_2 = \frac{k^2 q_0 + q\boldsymbol{k} \cdot \boldsymbol{k}_0 - \mathrm{i}K\boldsymbol{N} \cdot \boldsymbol{k} \times \boldsymbol{k}_0}{k(K^2 + qq_0 - kk_0)^{1/2}(K^2 - qq_0 + \boldsymbol{k} \cdot \boldsymbol{k}_0)^{1/2}} .$$

As a result we have

$$\mathscr{R}_{21}(\boldsymbol{k}, \boldsymbol{k}_0) = a_1 a_2 = \frac{K^2 + qq_0\boldsymbol{k} \cdot \boldsymbol{k}_0 - k^2 k_0^2 - \mathrm{i}K(q + q_0)\boldsymbol{N} \cdot \boldsymbol{k} \times \boldsymbol{k}_0}{kk_0(K^2 + qq_0 - \boldsymbol{k} \cdot \boldsymbol{k}_0)} . \tag{5.1.30}$$

Coefficient $\mathscr{R}_{12}$ can be obtained from (5.1.30) by the formal replacement $\mathrm{i} \to -\mathrm{i}$. Bearing in mind that for reflecting from the perfectly conducting plane $\mathscr{R}_{11} = \mathscr{R}_{22} = 0$ (depolarization does not exist), we obtain the following formula for SA in the tangent plane approximation (in the basis of circularly polarized waves):

$$S^{(c)}(\boldsymbol{k}, \boldsymbol{k}_0) = \begin{bmatrix} 0 & (K^2 + qq_0)(\boldsymbol{k} \cdot \boldsymbol{k}_0) - k^2 k_0^2 + \mathrm{i}K(q + q_0)\boldsymbol{N} \cdot \boldsymbol{k} \times \boldsymbol{k}_0 \\ (K^2 + qq_0)(\boldsymbol{k} \cdot \boldsymbol{k}_0) - k^2 k_0^2 - \mathrm{i}K(q + q_0)\boldsymbol{N} \cdot \boldsymbol{k} \times \boldsymbol{k}_0 & 0 \end{bmatrix}$$
$$\times (2kk_0 qq_0)^{-1} \frac{2(qq_0)^{1/2}}{(q + q_0)}$$
$$\times \int \exp[-\mathrm{i}(\boldsymbol{k} - \boldsymbol{k}_0) \cdot \boldsymbol{r} + \mathrm{i}(q + q_0)h(\boldsymbol{r})]\, d\boldsymbol{r}/(2\pi)^2 . \tag{5.1.31}$$

In notations of (4.4.27, 28) we can write (5.1.31) as

$$S^{(c)}(\boldsymbol{k}, \boldsymbol{k}_0) = [b_1(\boldsymbol{k}, \boldsymbol{k}_0)\hat{\sigma}_1 + b_2(\boldsymbol{k}, \boldsymbol{k}_0)\hat{\sigma}_2] \cdot \frac{2(qq_0)^{1/2}}{(q + q_0)}$$
$$\times \int \exp[-\mathrm{i}(\boldsymbol{k} - \boldsymbol{k}_0) \cdot \boldsymbol{r} + \mathrm{i}(q + q_0)h(\boldsymbol{r})]\, d\boldsymbol{r}/(2\pi)^2 . \tag{5.1.31}$$

There is a general comment with respect to formulas for SA obtained in this section: strictly speaking, the integrals included in these formulae should be calculated by the stationary phase method. If the stationary points for the corresponding integrands are absent, i.e., there are no specular points reflecting the wave in a given direction, then replacements of the kind (5.1.13, 22) are of no sense and the accuracy of the corresponding formula is expected to reduce.

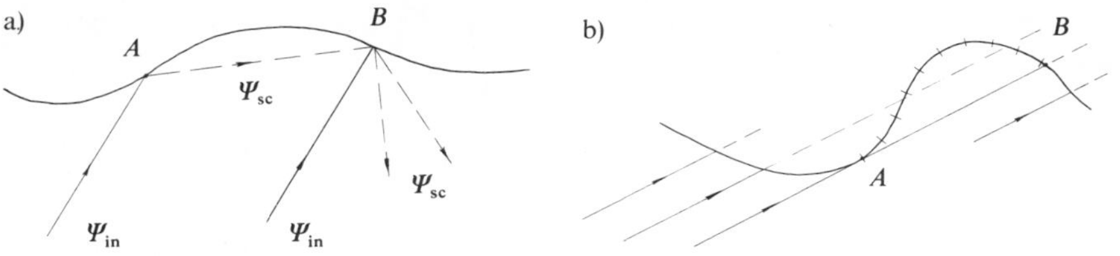

Fig. 5.2a,b. Situations with multiple reflections (**a**) and shadowings (**b**)

Numerical experiments confirm this conclusion: the tangent plane approximation does not give very good results for directions far from specular.

The tangent plane approximation needs a rather smooth rough boundary, so that it can be approximated by the tangent plane at any point.

This approximation is limited also to a requirement of absence of multiple reflections, see Fig. 5.2. It is apparent that for an incident wave arriving at the point B on the rough surface after reflection at the point A, the structure of the incident wave field at B changes and corresponds to the superposition of two waves; parameters of one of the latter are under the effect of the rough boundary shape far from point B, see Fig. 5.2a.

A requirement of absence of shadowing plays the same restricting role. Figure 5.2b represents the case in which the boundary part AB is shadowed by other irregularities and the structure of the incident field there also changes (upon neglecting the diffraction effects the field at the shadowed site can be assumed as zero).

Thus, the tangent plane method is applied under the assumption that multiple wave reflections/shadowings do not exist. This condition is fulfilled, when, on one hand, the slopes of irregularities are sufficiently small, and, on the other hand, when the grazing angle of the incident wave is rather large and exceeds the characteristic slope of irregularities. The second requirement is in small diffraction effects, the appropriate quantity criterion and will be considered in Sect. 5.3.

5.2 Kirchhoff Approximation

In the present section we consider an approximate solution of the scattering problem which is related to the iteration procedure of solving the appropriate integral equation for surface sources.

We start from the Dirichlet problem. According to the results of Sect. 3.1 the scattered field can be sought in the form

$$\Psi_{sc}(\boldsymbol{R}_*) = \int \nu(\boldsymbol{R}) \frac{\partial}{\partial \boldsymbol{R}_{\boldsymbol{R}}} G_0(\boldsymbol{R} - \boldsymbol{R}_*)\, d\Sigma \ , \qquad \boldsymbol{R} \in \Sigma \ , \tag{5.2.1}$$

where $\boldsymbol{R}_*$ is the point of observation and ν is the density of surface sources (dipoles). The latter should satisfy (3.1.42):

$$\nu(\boldsymbol{R})/2 + \int \nu(\boldsymbol{R}')\frac{\partial}{\partial \boldsymbol{n}_{\boldsymbol{R}'}} G_0(\boldsymbol{R}' - \boldsymbol{R})\, d\Sigma' = -\Psi_{\text{in}}(\boldsymbol{R}) \; ; \qquad \boldsymbol{R}, \boldsymbol{R}' \in \Sigma \; . \tag{5.2.2}$$

For the first approximation one can neglect the integral term in this equation and set

$$\nu(\boldsymbol{R}) \approx \nu^{(0)}(\boldsymbol{R}) = -2\Psi_{\text{in}}(\boldsymbol{R}) \; . \tag{5.2.3}$$

Substituting (5.2.3) into (5.2.1) we obtain the formula for the scattered field in the Kirchhoff approximation. Let the wave of the form

$$\Psi_{\text{in}} = q_0^{-1/2} \exp(\mathrm{i}\boldsymbol{k}_0 \cdot \boldsymbol{r} + \mathrm{i}q_0 z)$$

be incident on the rough surface then

$$\Psi_{\text{sc}}(\boldsymbol{R}_*) = -2q_0^{-1/2} \int \exp[\mathrm{i}\boldsymbol{k}_0 \cdot \boldsymbol{r} + \mathrm{i}q_0 h(\boldsymbol{r})] \frac{\partial}{\partial \boldsymbol{n}_{\boldsymbol{R}}} G_0(\boldsymbol{R} - \boldsymbol{R}_*)\, d\Sigma \; , \quad \boldsymbol{R} = (\boldsymbol{r}, h(\boldsymbol{r})) \; . \tag{5.2.4}$$

Taking into account (2.1.10, 11) and using the representation of the Green's function (2.1.24) for $z_* < \min h(\boldsymbol{r})$ yield

$$\begin{aligned} d\Sigma \frac{\partial}{\partial \boldsymbol{n}_{\boldsymbol{R}}} G_0(\boldsymbol{R} - \boldsymbol{R}_*) &= d\boldsymbol{r}\left(\frac{\partial}{\partial z} - \nabla h \cdot \nabla\right) G_0(\boldsymbol{R} - \boldsymbol{R}_*) \\ &= d\boldsymbol{r}(q_k + \boldsymbol{k} \cdot \nabla h) \int \exp[-\mathrm{i}\boldsymbol{k} \cdot (\boldsymbol{r} - \boldsymbol{r}_*) + \mathrm{i}q_k(z - z_*)]\, d\boldsymbol{k}/(8\pi^2) \; . \end{aligned} \tag{5.2.5}$$

By definition of SA (2.2.14) we obtain

$$\begin{aligned} S(\boldsymbol{k}, \boldsymbol{k}_0) = &-(qq_0)^{-1/2} \int (q + \boldsymbol{k} \cdot \nabla h) \\ &\times \exp[-\mathrm{i}(\boldsymbol{k} - \boldsymbol{k}_0) \cdot \boldsymbol{r} + \mathrm{i}(q + q_0)h(\boldsymbol{r})]\, d\boldsymbol{r}/(2\pi)^2 \; . \end{aligned}$$

We integrate by parts the term proportional to ∇h in the way similar to the one used in Sect. 5.1. This operation reduces to replacement (5.1.10) and, as a result

$$\begin{aligned} S(\boldsymbol{k}, \boldsymbol{k}_0) = &-\frac{(qq_0)^{-1/2}}{(q + q_0)}(K^2 + qq_0 - \boldsymbol{k} \cdot \boldsymbol{k}_0) \\ &\times \int \exp[-\mathrm{i}(\boldsymbol{k} - \boldsymbol{k}_0) \cdot \boldsymbol{r} + \mathrm{i}(q + q_0)h(\boldsymbol{r})] \frac{d\boldsymbol{r}}{(2\pi)^2} \; . \end{aligned}$$

This expression coincides with (5.1.11) obtained by the tangent plane method in the previous section.

In the case of the Neumann problem the scattered field can also be sought in the form (5.2.1) with $v(\boldsymbol{R})$ satisfying (3.1.37). The latter differs from (5.2.2) in the sign of the integral term and the right-hand side. Therefore, for the Neumann problem in the Kirchhoff approximation instead of (5.2.3) we have

$$v(\boldsymbol{R}) \approx 2\Psi_{\text{in}}(\boldsymbol{R})$$

and the resulting formula for SA differs from the Dirichlet case only in sign. We came to the same result in Sect. 5.1 using the tangent plane method. For further consideration we represent the appropriate formula for SA in the following form:

$$S(\boldsymbol{k},\boldsymbol{k}_0) = \left[1 - \frac{(\boldsymbol{k}-\boldsymbol{k}_0)^2 + (q-q_0)^2}{4(K^2 - \boldsymbol{k}\cdot\boldsymbol{k}_0)}\right]\frac{(K^2 - \boldsymbol{k}\cdot\boldsymbol{k}_0)}{qq_0}$$
$$\times \frac{2(qq_0)^{1/2}}{(q+q_0)}\int \exp[-\mathrm{i}(\boldsymbol{k}-\boldsymbol{k}_0)\cdot\boldsymbol{r} + \mathrm{i}(q+q_0)h(\boldsymbol{r})]\frac{d\boldsymbol{r}}{(2\pi)^2} \ . \tag{5.2.6}$$

Now we consider scattering the electromagnetic waves from the perfectly conducting surface. In this case the scattered field is generated physically by currents flowing over this surface. Vector potential $\boldsymbol{A}$ of the electromagnetic field generated by currents with density $\boldsymbol{j}$ satisfies the well-known equation which immediately follows from Maxwell's equations:

$$(\Delta + K^2)\boldsymbol{A} = -4\pi\boldsymbol{j}/c \ .$$

Whence

$$\boldsymbol{A} = -\frac{4\pi}{c}\int \boldsymbol{j}_{\boldsymbol{R}'} G_0(\boldsymbol{R}-\boldsymbol{R}')\,d\boldsymbol{R}'$$

and the magnetic field for our case is given by

$$\boldsymbol{H}_{\text{sc}}(\boldsymbol{R}_*) = \operatorname{curl}\boldsymbol{A}(\boldsymbol{R}_*) = -\frac{4\pi}{c}\int \boldsymbol{j}_{\boldsymbol{R}'} \times \nabla_{\boldsymbol{R}'} G_0(\boldsymbol{R}'-\boldsymbol{R}_*)\,d\Sigma_{\boldsymbol{R}'} \ , \tag{5.2.7}$$

where $\boldsymbol{j}$ is the current surface density and $d\Sigma$ is the element of surface area. Tending the observation point from inside the region to the point $\boldsymbol{R}$ at the boundary and in virtue of simple consequence of (3.1.35) we find

$$\boldsymbol{H}_{\text{sc}}(\boldsymbol{R}) = -\frac{2\pi}{c}\boldsymbol{j}_{\boldsymbol{R}} \times \boldsymbol{n}_{\boldsymbol{R}} - \frac{4\pi}{c}\int \boldsymbol{j}_{\boldsymbol{R}'} \times \nabla_{\boldsymbol{R}'} G_0(\boldsymbol{R}'-\boldsymbol{R})\,d\Sigma' \ , \tag{5.2.8}$$

where $\boldsymbol{n}_{\boldsymbol{R}}$ is the vector of external normal at the point $\boldsymbol{R}$. On the other hand, for magnetic field circulation over contour $ABCD$ shown in Fig. 5.3 and taking into

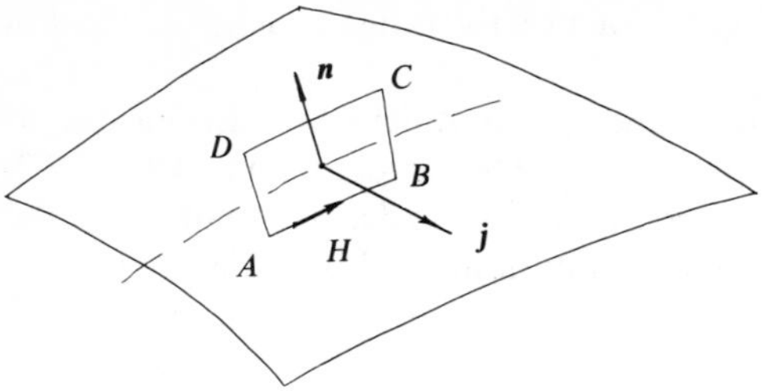

Fig. 5.3. On derivation of relationship between surface current and magnetic field at the surface

account that the total field at the site CD (i.e., inside the ideal conductor) is zero, and in virtue of equation

$$\operatorname{curl} \boldsymbol{H} = \frac{4\pi}{c} \cdot \boldsymbol{j} + \frac{\varepsilon}{c} \partial \boldsymbol{E} / \partial t$$

we find the following relation between surface current and magnetic field at the surface:

$$\boldsymbol{j} = -\frac{c}{(4\pi)} \boldsymbol{n} \times \boldsymbol{H} = -\frac{c}{(4\pi)} \boldsymbol{n} \times (\boldsymbol{H}_{\text{in}} + \boldsymbol{H}_{\text{sc}}) \ .$$

Substituting (5.2.8) into this relation yields the following equation for $\boldsymbol{j}$:

$$\boldsymbol{j}_{\boldsymbol{R}} = -\frac{c}{2\pi} \boldsymbol{n}_{\boldsymbol{R}} \times \boldsymbol{H}_{\text{in}}(\boldsymbol{R}) + 2\boldsymbol{n}_{\boldsymbol{R}} \times \int \boldsymbol{j}_{\boldsymbol{R}'} \times \nabla_{\boldsymbol{R}'} G_0(\boldsymbol{R}' - \boldsymbol{R}) \, d\Sigma' \ . \qquad (5.2.9)$$

When the ideal conductor surface is plane the integral in (5.2.9) becomes zero exactly. In the Kirchhoff approximation the integral term in (5.2.9) is neglected. Then

$$\boldsymbol{j} = -\frac{c}{2\pi} \boldsymbol{n} \times \boldsymbol{H}_{\text{in}} \ .$$

Substituting this expression for $\boldsymbol{j}$ into (5.2.7) we find the following relation for the magnetic field of the scattered wave:

$$\boldsymbol{H}_{\text{sc}}(\boldsymbol{R}_*) = 2 \int [\boldsymbol{n}_{\boldsymbol{R}'} \times \boldsymbol{H}_{\text{in}}(\boldsymbol{R}')] \times \nabla_{\boldsymbol{R}'} G_0(\boldsymbol{R}' - \boldsymbol{R}_*) \, d\Sigma' \ .$$

Taking the plane wave of σ_0th polarization, see (4.4.6),

$$\varepsilon^{-1/2} \boldsymbol{H}_{\text{in}} = \boldsymbol{h}^{+}_{\sigma_0}(\boldsymbol{k}_0) q_0^{-1/2} \exp(\mathrm{i}\boldsymbol{k}_0 \cdot \boldsymbol{r} + \mathrm{i} q_0 z)$$

using expansion in plane waves (2.1.24) for G_0, we obtain the expression for the magnetic field of the scattered wave

$$\varepsilon^{1/2} \boldsymbol{H}_{\rm sc}(\boldsymbol{R}_*) = \int (qq_0)^{-1/2} [\boldsymbol{N} - \nabla h(\boldsymbol{r}') \times \boldsymbol{h}^+_{\sigma_0}(\boldsymbol{k}_0)] \times (q\boldsymbol{N} - \boldsymbol{k}) \; . \\ \times \exp[-\mathrm{i}(\boldsymbol{k} - \boldsymbol{k}_0)\cdot \boldsymbol{r}' + \mathrm{i}(q + q_0) h(\boldsymbol{r}')] \\ \times \frac{d\boldsymbol{r}'}{(2\pi)^2} \exp(\mathrm{i}\boldsymbol{k}\cdot\boldsymbol{r}_* - \mathrm{i}qz_*) q^{-1/2}\, d\boldsymbol{k}$$

(here, as usual, $\boldsymbol{N} = \boldsymbol{e}_z$ is the unit vertical vector). Comparing this expansion to (4.4.6) gives

$$\sum_{\sigma'} S_{\sigma'\sigma_0}(\boldsymbol{k}, \boldsymbol{k}_0) \boldsymbol{h}^-_{\sigma'}(\boldsymbol{k}) = (qq_0)^{-1/2} \int [\boldsymbol{N} - \nabla h(\boldsymbol{r}') \times \boldsymbol{h}^+_{\sigma_0}(\boldsymbol{k}_0)] (q\boldsymbol{N} - \boldsymbol{k}) \\ \times \exp[-\mathrm{i}(\boldsymbol{k} - \boldsymbol{k}_0)\cdot \boldsymbol{r}' + \mathrm{i}(q + q_0) h(\boldsymbol{r}')] \frac{d\boldsymbol{r}'}{(2\pi)^2} \; . \tag{5.2.10}$$

According to the procedure presented in Sect. 5.1, the term proportional to $\nabla h(\boldsymbol{r}')$ is integrated by parts which is equivalent to replacement (5.1.10). In addition we multiply (5.2.10) by $\boldsymbol{h}^-_\sigma(\boldsymbol{k})$ and apply orthogonality of vectors $\boldsymbol{h}^-$, $\boldsymbol{h}^-_{\sigma_1}(\boldsymbol{k}) \cdot \boldsymbol{h}^-_{\sigma_2}(\boldsymbol{k}) = \delta_{\sigma_1 \sigma_2}$. As a result

$$S_{\sigma\sigma_0}(\boldsymbol{k}, \boldsymbol{k}_0) = \boldsymbol{h}^-_\sigma(\boldsymbol{k}) \cdot \left[(\boldsymbol{k} - q\boldsymbol{N}) \times \left\{ \left[\boldsymbol{N} - \frac{(\boldsymbol{k} - \boldsymbol{k}_0)}{(q + q_0)} \right] \times \boldsymbol{h}^+_{\sigma_0}(\boldsymbol{k}_0) \right\} \right] \\ \times (qq_0)^{-1/2} \int \exp[-\mathrm{i}(\boldsymbol{k} - \boldsymbol{k}_0)\cdot \boldsymbol{r} + \mathrm{i}(q + q_0) h(\boldsymbol{r})] \frac{d\boldsymbol{r}}{(2\pi)^2} \; . \tag{5.2.11}$$

Simple calculations lead to the following result:

$$\hat{S}(\boldsymbol{k}, \boldsymbol{k}_0) = (2qq_0kk_0)^{-1} \cdot \begin{bmatrix} (K^2 + qq_0)\boldsymbol{k}\cdot\boldsymbol{k}_0 - k^2 k_0^2 & K(q+q_0)\boldsymbol{N}\cdot\boldsymbol{k}\times\boldsymbol{k}_0 \\ K(q+q_0)\boldsymbol{N}\cdot\boldsymbol{k}\times\boldsymbol{k}_0 & -(K^2 + qq_0)\boldsymbol{k}\cdot\boldsymbol{k}_0 + k^2 k_0^2 \end{bmatrix} \\ \times \frac{2(qq_0)^{1/2}}{(q + q_0)} \int \exp[-\mathrm{i}(\boldsymbol{k} - \boldsymbol{k}_0)\cdot \boldsymbol{r} + \mathrm{i}(q + q_0) h(\boldsymbol{r})] \frac{d\boldsymbol{r}}{(2\pi)^2} \; . \tag{5.2.12}$$

In this formula SA is written down in initial basis of vertically and horizontally polarized waves. The transition to the basis of circularly polarized waves with the help of (4.4.16) yields exactly (5.1.31).

Now compare the Kirchhoff approximation and the small perturbation method. The problems considered in Chap. 4 confirm that the expression for SA accurate within the order of h can be represented in the form

$$S(\boldsymbol{k}, \boldsymbol{k}_0) = V_0(k)\delta(\boldsymbol{k} - \boldsymbol{k}_0) + 2\mathrm{i}(qq_0)^{1/2} B(\boldsymbol{k}, \boldsymbol{k}_0) h(\boldsymbol{k} - \boldsymbol{k}_0) \; . \tag{5.2.13}$$

For instance, in virtue of (4.1.10) we have for the Dirichlet problem

$$V_0(k) = -1 \; ; \qquad B(\boldsymbol{k}, \boldsymbol{k}_0) = -1 \tag{5.2.14a}$$

and for the Neumann problem by (4.3.8)

$$V_0(k) = 1 \; ; \qquad B(\boldsymbol{k}, \boldsymbol{k}_0) = \frac{K^2 - \boldsymbol{k} \cdot \boldsymbol{k}_0}{q q_0} \; . \tag{5.2.14b}$$

The coefficients V_0 and B in the general case are related as

$$V_0(k) = B(\boldsymbol{k}, \boldsymbol{k}) \; . \tag{5.2.15}$$

To prove (5.2.15) we consider the case of the plane boundary elevated above the level $z = 0$ by H: $h(\boldsymbol{r}) = H = \text{const}$. According to (2.4.2) the following exact expression originates for SA:

$$S(\boldsymbol{k}, \boldsymbol{k}_0) = V_0(k) \exp(2\mathrm{i}qH)\delta(\boldsymbol{k} - \boldsymbol{k}_0) = V_0(k)[1 + 2\mathrm{i}qH + \mathrm{O}(H^2)]\delta(\boldsymbol{k} - \boldsymbol{k}_0) \; . \tag{5.2.16}$$

On the other hand, we have by (5.2.13)

$$S(\boldsymbol{k}, \boldsymbol{k}_0) = V_0(k)\delta(\boldsymbol{k} - \boldsymbol{k}_0) + 2\mathrm{i}qHB(\boldsymbol{k}, \boldsymbol{k}_0)\delta(\boldsymbol{k} - \boldsymbol{k}_0) + \mathrm{O}(H^2) \; .$$

Comparing the H order terms in these two expressions gives (5.2.15).

It follows from the problems presented in Sects. 5.1, 2 that the formula for SA in the general case in the tangent plane approximation can be represented as

$$S(\boldsymbol{k}, \boldsymbol{k}_0) = g(\boldsymbol{k}, \boldsymbol{k}_0) B(\boldsymbol{k}, \boldsymbol{k}_0) \frac{2(q q_0)^{1/2}}{(q + q_0)} \times \int \exp[-\mathrm{i}(\boldsymbol{k} - \boldsymbol{k}_0) \cdot \boldsymbol{r} + \mathrm{i}(q + q_0) h(\boldsymbol{r})] \, d\boldsymbol{r}/(2\pi)^2 \; . \tag{5.2.17}$$

According to (5.1.12), for the Dirichlet problem

$$g(\boldsymbol{k}, \boldsymbol{k}_0) = 1 + [(\boldsymbol{k} - \boldsymbol{k}_0)^2 + (q - q_0)^2]/4 q q_0 \tag{5.2.18a}$$

and in virtue of (5.2.6) for the Neumann problem

$$g(\boldsymbol{k}, \boldsymbol{k}_0) = 1 - \frac{(\boldsymbol{k} - \boldsymbol{k}_0)^2 + (q - q_0)^2}{4(K^2 - \boldsymbol{k}\boldsymbol{k}_0)} \; . \tag{5.2.18b}$$

For $h(\boldsymbol{r}) = \text{const}$ formula (5.2.17) transforms into the exact expression (5.2.16) and by (5.2.15)

$$g(\boldsymbol{k}, \boldsymbol{k}) = 1 \; . \tag{5.2.19}$$

Now we consider in the framework of the Kirchhoff approximation the case of small roughness (small Rayleigh parameter), when the exponential in (5.2.17) can be expanded in series in h and limited to the first order term. This obviously results in

$$S(\boldsymbol{k},\boldsymbol{k}_0) = V_0(k)\delta(\boldsymbol{k}-\boldsymbol{k}_0) + 2\mathrm{i}(qq_0)^{1/2}g(\boldsymbol{k},\boldsymbol{k}_0)B(\boldsymbol{k},\boldsymbol{k}_0)h(\boldsymbol{k}-\boldsymbol{k}_0) + \mathrm{O}(h^2) \ . \quad (5.2.20)$$

Relations (5.2.13, 20) *are not equal* due to the factor $g(\boldsymbol{k},\boldsymbol{k}_0)$ in (5.2.20). This difference is not significant for scattering in directions close to specular: $\boldsymbol{k} \approx \boldsymbol{k}_0$ for by (5.2.19) $g \approx 1$, and is significant for large scattering angles. For the last case the Kirchhoff approximation does not hold as was mentioned in Sect. 5.1. However, if function $h(\boldsymbol{r})$ is smooth, with small gradients and changes rather slowly in the scale of a wavelength, then scattering at large angles does not practically occur, since the integral in (5.2.17) turns out to be exponentially small. In such a case one can use in (5.2.17) approximation $g \approx 1$.

The curious situation arises concerning the problem of scattering electromagnetic waves at a perfectly conducting surface. In the basis of horizontally and vertically polarized waves, matrix $\hat{B}(\boldsymbol{k},\boldsymbol{k}_0)$ determined by formula (4.4.21), and the matrix in (5.2.12) differ, and are equal only at $\boldsymbol{k} = \boldsymbol{k}_0$. However, in the basis of circularly polarized waves, nondiagonal matrix elements coincide for all arguments and diagonal ones corresponding to depolarization effects do not, see (4.4.26) and (5.1.31).

Hence, the results obtained by the Kirchhoff approximation and the small perturbation method agree in the general case only for small and smooth roughness. Application of the small perturbation method is limited to the small Rayleigh parameter. However, in this case characteristic horizontal scale of roughness can be compared to the wavelength, and scattering at any angles can be considered (including inhomogeneous waves). The Kirchhoff approximation is applicable for an arbitrarily large Rayleigh parameter, but irregularities must be sufficiently smooth so that scattering at large angles does not occur.

Now we proceed to the statistical case. Averaging formula (5.2.17) and taking into account the definition (2.4.6) and (5.2.15, 19) yields

$$\bar{V}(k) = V_0(k)\overline{\exp[2\mathrm{i}qh(\boldsymbol{r})]} \ . \quad (5.2.21)$$

The result of averaging the exponential for spatially homogeneous statistics is $\boldsymbol{r}$-independent. It is clear that (5.2.21) can be obtained under the assumption that the statistical ensemble consists of horizontal planes only. Here the phase factor which is due to the double wave run from $z = 0$ to $z = h$ is only averaged.

To obtain simple and tractable formulas we suppose in what follows that the ensemble of elevations is Gaussian. For any Gaussian quantity $h(\boldsymbol{r})$, $\bar{h} = 0$ the relation

$$\overline{\exp(h)} = \exp(\overline{h^2}/2) \quad (5.2.22)$$

holds.

Moreover, any linear transformation of Gaussian field

$$\tilde{h}(\boldsymbol{r}) = \int T(\boldsymbol{r},\boldsymbol{r}')h(\boldsymbol{r}')\,d\boldsymbol{r}' \quad (5.2.23)$$

gives Gaussian field $\tilde{h}$ again. In particular, (5.2.22, 23) yield

$$\overline{\exp[\int T(\boldsymbol{r},\boldsymbol{r}')h(\boldsymbol{r}')\,d\boldsymbol{r}]} = \exp[\tfrac{1}{2}\int T(\boldsymbol{r},\boldsymbol{r}_1)T(\boldsymbol{r},\boldsymbol{r}_2)\overline{h(\boldsymbol{r}_1)h(\boldsymbol{r}_2)}\,d\boldsymbol{r}_1\,d\boldsymbol{r}_2] \ . \tag{5.2.24}$$

Formula (5.2.21) for the Gaussian ensemble of elevations turns into

$$\bar{V}(k) = V_0(k)\exp(-2q^2\sigma^2) \ , \tag{5.2.25}$$

where

$$\sigma^2 = \overline{h^2} \tag{5.2.26}$$

is the mean-square elevation. The mean reflection coefficient tends rapidly to zero with an increasing Rayleigh parameter, which is due to the nearly even distribution of the phase in (5.2.21) at $(0, 2\pi)$.

Now we consider the second moment for SA. According to (2.4.9, 12) after simple calculations we have

$$\sigma(\boldsymbol{k},\boldsymbol{k}_0) = |g(\boldsymbol{k},\boldsymbol{k}_0)B(\boldsymbol{k},\boldsymbol{k}_0)|^2 4qq_0(q+q_0)^{-2}\int \exp[-\mathrm{i}(\boldsymbol{k}-\boldsymbol{k}_0)\cdot\boldsymbol{r}]$$
$$\times \{\exp[-(q+q_0)^2 D(\boldsymbol{r})] - \exp[-(q+q_0)^2 D(\infty)]\}\frac{d\boldsymbol{r}}{(2\pi)^2} \ , \tag{5.2.27}$$

where

$$D(\boldsymbol{r}) = \tfrac{1}{2}\overline{[h(\boldsymbol{r}+\varrho) - h(\varrho)]^2} = \sigma^2 - \tilde{W}(\boldsymbol{r})$$

is the structure function of elevations and $\tilde{W}$ is the correlation function, see (4.1.24). The integral in (5.2.27) can be calculated for the large Rayleigh parameter by the stationary phase method. The stationary point is $\boldsymbol{r}_s = 0$ and function $D(\boldsymbol{r})$ is approximated as

$$D(\boldsymbol{r}) \approx \sigma^2 x^2/2l_x^2 + \sigma^2 y^2/2l_y^2 \ . \tag{5.2.28}$$

Relation (5.2.28) determines in fact horizontal scales l_x and l_y. The quantities

$$\delta_x^2 = \sigma^2/l_x^2 = \overline{(\partial h/\partial x)^2} \ , \qquad \delta_y^2 = \sigma^2/l_y^2 = \overline{(\partial h/\partial y)^2} \tag{5.2.29}$$

represent the mean-square slopes of irregularities in x- and y-directions (these coordinates are chosen so that the cross term in (5.2.28) proportional to the product xy is absent). Substitution of (5.2.28) into (5.2.27) gives the standard Gaussian integral. As a result we have

$$\sigma(\boldsymbol{k},\boldsymbol{k}_0) = |g(\boldsymbol{k},\boldsymbol{k}_0)B(\boldsymbol{k},\boldsymbol{k}_0)|^2 4qq_0(q+q_0)^{-2}\left[(q+q_0)^{-2}P\left(\frac{\boldsymbol{k}-\boldsymbol{k}_0}{q+q_0}\right)\right] , \tag{5.2.30}$$

where

$$P(\theta_x, \theta_y) = (2\pi\delta_x\delta_y)^{-1} \exp\left(\frac{-\theta_x^2}{2\delta_x^2} + \frac{-\theta_y^2}{2\delta_y^2}\right)$$

is probability density of the slopes. It was shown in Sect. 5.1, see (5.1.13), that the argument of function P in (5.2.30) is equal to the boundary slope at the specular point. Thus, all approximations in deriving formula (5.2.30) reduce to the assumption that the statistical ensemble consists of planes with given slope distribution. Thus for roughness with small slopes the scattered field is concentrated around specular direction $\boldsymbol{k} \approx \boldsymbol{k}_0$ within a narrow cone with width of the order of δ_x, δ_y.

5.3 Corrections to the Kirchhoff Approximation: Deterministic Case

In the present and the next section we consider corrections to the Kirchhoff approximation taking the Dirichlet problem as the example. It was seen in Sect. 5.2 that the Kirchhoff approximation originates as the first approximation in solving the integral equation (5.2.2) for dipole surface sources by iteration. Calculating corrections to the Kirchhoff approximation is related to the estimation of the following iteration. In the present section we apply the integral equation for the monopole surface source density which is also widely used for the Dirichlet problem.

According to (3.1.32) the following representation exists for the field:

$$\Psi(\boldsymbol{R}_*) = \Psi_{\text{in}}(\boldsymbol{R}_*) - \int \frac{\partial \Psi}{\partial \boldsymbol{n}_{\boldsymbol{R}'}} G_0(\boldsymbol{R}' - \boldsymbol{R}_*)\, d\Sigma' \ . \tag{5.3.1}$$

Here $\boldsymbol{R}_* \in \Omega$ is the point of observation and $\boldsymbol{R}' \in \Sigma$ is the point belonging to the rough boundary. We fix the point $\boldsymbol{R} \in \Sigma$; let $\boldsymbol{n}_{\boldsymbol{R}}$ be the constant vector of unit normal at this point. Applying operator $\boldsymbol{n}_{\boldsymbol{R}} \cdot \nabla_{\boldsymbol{R}_*}$ to (5.3.1) we find

$$\frac{\partial \Psi(\boldsymbol{R}_*)}{\partial \boldsymbol{n}_{\boldsymbol{R}}} = \frac{\partial \Psi_{\text{in}}(\boldsymbol{R}_*)}{\partial \boldsymbol{n}_{\boldsymbol{R}}} + \int \frac{\partial \Psi}{\partial \boldsymbol{n}_{\boldsymbol{R}'}} \cdot \frac{\partial G_0(\boldsymbol{R}' - \boldsymbol{R}_*)}{\partial \boldsymbol{n}_{\boldsymbol{R}}}\, d\Sigma' \ . \tag{5.3.2}$$

Let the point $\boldsymbol{R}_* \in \Omega$ tend to the point $\boldsymbol{R} \in \Sigma$ along the normal $\boldsymbol{n}_{\boldsymbol{R}}$. The limiting transition is performed using the same considerations as used in obtaining (3.1.35). As a result

$$\frac{1}{2} \frac{\partial \Psi}{\partial \boldsymbol{n}_{\boldsymbol{R}}} = \frac{\partial \Psi_{\text{in}}}{\partial \boldsymbol{n}_{\boldsymbol{R}}} + \int \frac{\partial \Psi}{\partial \boldsymbol{n}_{\boldsymbol{R}'}} \frac{\partial G_0(\boldsymbol{R}' - \boldsymbol{R})}{\partial \boldsymbol{n}_{\boldsymbol{R}}}\, d\Sigma' \ . \tag{5.3.3}$$

Since $\boldsymbol{R} \in \Sigma$ is an arbitrary point belonging to the rough boundary then (5.3.3) is the integral equation for the normal derivative of the total field at the boundary. The iterative solution of (5.3.3) is

$$\frac{\partial \Psi}{\partial \boldsymbol{n}_{\boldsymbol{R}}} = 2\frac{\partial \Psi_{\text{in}}}{\partial \boldsymbol{n}_{\boldsymbol{R}}} + 4\int \frac{\partial \Psi_{\text{in}}(\boldsymbol{R}_1)}{\partial \boldsymbol{n}_{\boldsymbol{R}_1}} \cdot \frac{\partial}{\partial \boldsymbol{n}_{\boldsymbol{R}}} G_0(\boldsymbol{R}_1 - \boldsymbol{R})\, d\Sigma_{\boldsymbol{R}_1} + \cdots . \tag{5.3.4}$$

Let the incident field correspond to the point source located at the point $\boldsymbol{R}_0$: $\Psi_{\text{in}} = G_0(\boldsymbol{R} - \boldsymbol{R}_0)$. Substituting solution (5.3.4) into (5.3.1) yields

$$\begin{aligned} G(\boldsymbol{R}_*, \boldsymbol{R}_0) = {} & G_0(\boldsymbol{R}_* - \boldsymbol{R}_0) - 2\int \frac{\partial G_0(\boldsymbol{R}_1 - \boldsymbol{R}_0)}{\partial \boldsymbol{n}_{\boldsymbol{R}_1}} G_0(\boldsymbol{R}_1 - \boldsymbol{R}_*)\, d\Sigma_{\boldsymbol{R}_1} \\ & - 4\int \frac{\partial G_0(\boldsymbol{R}_1 - \boldsymbol{R}_0)}{\partial \boldsymbol{n}_{\boldsymbol{R}_1}} \cdot \frac{\partial G_0(\boldsymbol{R}_1 - \boldsymbol{R}_2)}{\partial \boldsymbol{n}_{\boldsymbol{R}_2}} \\ & \times G_0(\boldsymbol{R}_2 - \boldsymbol{R}_*)\, d\Sigma_{\boldsymbol{R}_1}\, d\Sigma_{\boldsymbol{R}_2} - \cdots . \end{aligned} \tag{5.3.5}$$

It is of interest to compare (5.3.5) with the analogous solution obtained from (5.2.2). Substituting iterative solution (3.1.43) of this equation into (5.2.1) we find

$$\begin{aligned} G(\boldsymbol{R}_*, \boldsymbol{R}_0) = {} & G_0(\boldsymbol{R}_* - \boldsymbol{R}_0) - 2\int G_0(\boldsymbol{R}_1 - \boldsymbol{R}_0) \frac{\partial}{\partial \boldsymbol{n}_{\boldsymbol{R}_1}} G_0(\boldsymbol{R}_1 - \boldsymbol{R}_*)\, d\Sigma_{\boldsymbol{R}_1} \\ & + 4\int G_0(\boldsymbol{R}_1 - \boldsymbol{R}_0) \frac{\partial}{\partial \boldsymbol{n}_{\boldsymbol{R}_1}} G_0(\boldsymbol{R}_1 - \boldsymbol{R}_2) \frac{\partial}{\partial \boldsymbol{n}_{\boldsymbol{R}_2}} \\ & \times G_0(\boldsymbol{R}_2 - \boldsymbol{R}_*)\, d\Sigma_{\boldsymbol{R}_1}\, d\Sigma_{\boldsymbol{R}_2} - \cdots . \end{aligned} \tag{5.3.6}$$

In (5.3.6) we apply the reciprocity theorem (2.3.7) and replace arguments $\boldsymbol{R}_0 \rightleftarrows \boldsymbol{R}_*$. In addition, we replace in the second term in (5.3.6) integration variables $\boldsymbol{R}_1 \rightleftarrows \boldsymbol{R}_2$. Then (5.3.6) takes the form

$$\begin{aligned} G_0(\boldsymbol{R}_0, \boldsymbol{R}_*) = {} & G_0(\boldsymbol{R}_* - \boldsymbol{R}_0) - 2\int G_0(\boldsymbol{R}_1 - \boldsymbol{R}_*) \frac{\partial}{\partial \boldsymbol{n}_{\boldsymbol{R}_1}} G_0(\boldsymbol{R}_1 - \boldsymbol{R}_0)\, d\Sigma_{\boldsymbol{R}_1} \\ & - 4\int G_0(\boldsymbol{R}_2 - \boldsymbol{R}_*) \frac{\partial}{\partial \boldsymbol{n}_{\boldsymbol{R}_2}} G_0(\boldsymbol{R}_1 - \boldsymbol{R}_2) \frac{\partial}{\partial \boldsymbol{n}_{\boldsymbol{R}_1}} \\ & \times G_0(\boldsymbol{R}_1 - \boldsymbol{R}_0)\, d\Sigma_{\boldsymbol{R}_1}\, d\Sigma_{\boldsymbol{R}_2} - \cdots . \end{aligned}$$

The oddness of function ∇G_0 is taken into account in the last term. In virtue of this relation the iterative solutions of (5.3.3 and 2.2) coincide in all orders.

There exist two types of corrections to the density of monopoles $\mu(\boldsymbol{R}) = 2\partial \Psi_{\text{in}}/\partial \boldsymbol{n}_{\boldsymbol{R}}$ at the point $\boldsymbol{R} \in \Sigma$ which arise relating to expansion (5.3.4). The first ones are the local corrections related to the contribution of the points lying in the immediate vicinity of the point $\boldsymbol{R}$. The second type are nonlocal corrections related to the effects of multiple reflections and shadowings, see the end of Sect. 5.1.

First consider local corrections. We fix some point O at the boundary and choose the coordinates so that this point is at the origin of the coordinates and the boundary in the vicinity of the point can be approximated as

$$h(x, y) \approx x^2/2\varrho_1 + y^2/2\varrho_2 \ , \tag{5.3.7}$$

where ϱ_1 and ϱ_2 are the radii of curvature. Let the incident field be the plane wave propagating in the plane (x, z):

$$\Psi_{\text{in}} = q_0^{-1/2} \exp(\mathrm{i}k_0 x + \mathrm{i}q_0 z) \ . \tag{5.3.8}$$

According to (5.3.4) in the Kirchhoff approximation the value $\partial\Psi/\partial\boldsymbol{n}$ at the point O is

$$\left[\frac{\partial\Psi}{\partial\boldsymbol{n}}(\mathrm{O})\right]_1 = 2\mathrm{i}q_0^{1/2} \ . \tag{5.3.9}$$

Respectively, the second order correction at this point has the form

$$\begin{aligned}\left[\frac{\partial\Psi}{\partial\boldsymbol{n}}(0)\right]_2 &= 4\int \frac{\partial\Psi_{\text{in}}}{\partial\boldsymbol{n}_{\boldsymbol{R}_1}}\frac{\partial}{\partial\boldsymbol{n}_{\boldsymbol{R}}} G_0(\boldsymbol{R}_1 - \boldsymbol{R})\,d\Sigma_{\boldsymbol{R}_1} \\ &= -\frac{\mathrm{i}}{\pi} q_0^{-1/2} \int \left(q_0 - k_0\frac{\partial h}{\partial x_1}\right)\exp(\mathrm{i}k_0 x_1 + \mathrm{i}q_0 h)\frac{h}{R_1}\frac{\mathrm{i}KR_1 - 1}{R_1^2} \\ &\quad \times \exp(\mathrm{i}KR_1)\,dx_1\,dy_1 \ ,\end{aligned} \tag{5.3.10}$$

where $h = h(x_1, y_1)$ and $R_1 = (x_1^2 + y_1^2 + h^2)^{1/2}$. In calculating the integral (5.3.10) we approximately set

$$R_1 \approx (x_1^2 + y_1^2)^{1/2} \quad \text{and} \quad \exp(\mathrm{i}q_0 h) \approx 1$$

and use approximation (5.3.7) for $h(x_1, y_1)$. Using in (5.3.10) the polar coordinates $x_1 = r\cos\varphi$ and $y_1 = r\sin\varphi$, we have

$$\begin{aligned}\left[\frac{\partial\Psi}{\partial\boldsymbol{n}}(\mathrm{O})\right]_2 &= -\frac{\mathrm{i}}{2\pi q_0^{1/2}}\int_0^{2\pi} d\varphi\left(\frac{\cos^2\varphi}{\varrho_1} + \frac{\sin^2\varphi}{\varrho_2}\right) \\ &\quad \times \int_0^\infty \exp[\mathrm{i}(K + k_0\cos\varphi)r](\mathrm{i}Kr - 1)\,dr \ .\end{aligned} \tag{5.3.11}$$

When integrating over r in (5.3.11) keep in mind that K has an infinitely small positive imaginary part, so that substitutions at $r = \infty$ turn to zero. We have

$$\begin{aligned}\left[\frac{\partial\Psi}{\partial\boldsymbol{n}}(\mathrm{O})\right]_2 &= -\frac{q_0^{1/2}}{2\pi}\int_0^{2\pi} d\varphi[K(K + k_0\cos\varphi)^{-2} - (K + k_0\cos\varphi)^{-1}] \\ &\quad \times [\varrho_2^{-1} + (\varrho_1^{-1} - \varrho_2^{-1})\cos^2\varphi]\,d\varphi \ .\end{aligned} \tag{5.3.12}$$

Integral (5.3.12) is easily calculated using differentiation of the following formula:

$$(2\pi)^{-1} \int_0^{2\pi} (p + q\cos\varphi)^{-1}\, d\varphi = (p^2 - q^2)^{-1/2} \ .$$

Taking into account (5.3.9) the result can be represented as

$$\left[\frac{\partial \Psi}{\partial \boldsymbol{n}}(\mathrm{O})\right]_2 = \left[\frac{\partial \Psi}{\partial \boldsymbol{n}}(\mathrm{O})\right]_1 \left(1 + \frac{\mathrm{i}}{2K\varrho_1 \sin^3 \chi_0} + \frac{\mathrm{i}}{2K\varrho_2 \sin \chi_0}\right) , \qquad (5.3.13)$$

where χ_0 is the local grazing angle for the incident wave at the point O: $k_0 = K\cos\chi_0$. Thus, if ϱ_1 and ϱ_2 are of the similar order, then the Kirchhoff approximation becomes applicable under the condition

$$K\varrho \sin^3 \chi_0 \gg 1 \ . \qquad (5.3.14)$$

Condition (5.3.14) can be rather restrictive for small grazing angles.

With estimation (5.3.13) the approximated solution of (5.3.3) can be written in the form

$$\frac{\partial \Psi}{\partial \boldsymbol{n}_{\boldsymbol{R}}} = 2\frac{\partial \Psi_{\mathrm{in}}}{\partial \boldsymbol{n}_{\boldsymbol{R}}}\left(1 - \frac{\mathrm{i}}{2k\varrho_1 \sin^3 \chi_0} - \frac{\mathrm{i}}{2k\varrho_2 \sin \chi_0}\right)^{-1} .$$

Since the correction terms are small, we can write with the same accuracy

$$\frac{\partial \Psi}{\partial \boldsymbol{n}_{\boldsymbol{R}}} = 2\frac{\partial \Psi_{\mathrm{in}}}{\partial \boldsymbol{n}_{\boldsymbol{R}}} \exp\left(\frac{\mathrm{i}}{2K\varrho_1 \sin^3 \chi_0} + \frac{\mathrm{i}}{2K\varrho_2 \sin \chi_0}\right) . \qquad (5.3.15)$$

This result is used in Sect. 6.3.

Now consider the role of nonlocal corrections, following in principle the work by *Liszka* and *McCoy* [5.8]. Each integral in (5.3.5) in the limit $K \to \infty$ can be estimated by the stationary phase method. The phase factor in the nth term in series (5.3.5) generated by the product of the Green's function G_0 is

$$\Phi = K(|\boldsymbol{R}_0 - \boldsymbol{R}_1| + |\boldsymbol{R}_1 - \boldsymbol{R}_2| + \cdots + |\boldsymbol{R}_n - \boldsymbol{R}_*|) \ . \qquad (5.3.16)$$

Value Φ/K is the length of the broken line connecting the points $\boldsymbol{R}_0$ and $\boldsymbol{R}_*$ under the condition that n vertices of the broken line belong to the boundary. The stationary phase condition

$$\frac{\partial \phi}{\partial \boldsymbol{R}_k} = 0 \ , \qquad k = 1, 2, \ldots, n \qquad (5.3.17)$$

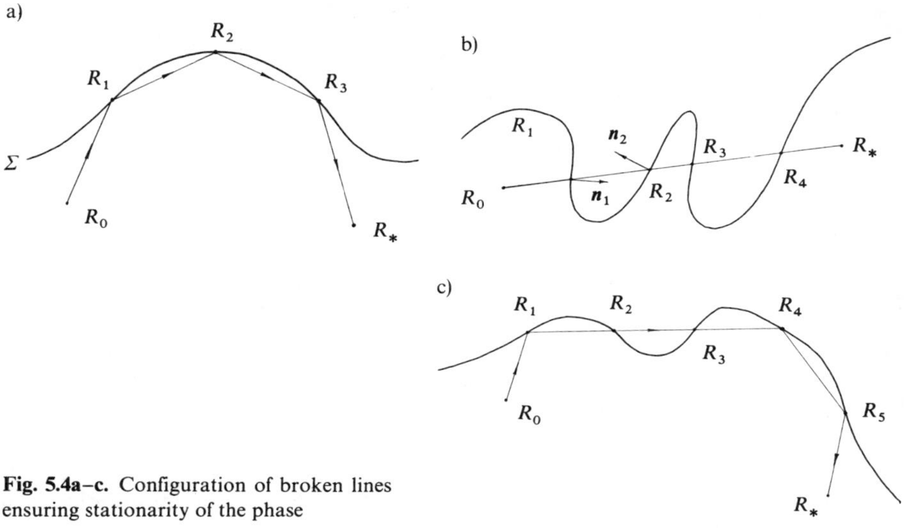

Fig. 5.4a–c. Configuration of broken lines ensuring stationarity of the phase

implies a minimum of this broken line length. Figure 5.4a,b shows main configurations ensuring this minimality. The broken line in Fig. 5.4a corresponds to the trajectory of the real ray undergoing multiple reflections from the boundary. Equations (5.3.17) imply equality of the incident and reflection angles at the points $\boldsymbol{R}_k$. The straight line $\boldsymbol{R}_0\boldsymbol{R}_*$ in Fig. 5.4b also ensures minimality of broken line length, although this straight line does not correspond to any real ray, since some parts of this line are outside the region Ω. Besides that, various combinations of configurations of the form 5.4a, b are possible, one of which is shown in Fig. 5.4c. These configurations do not correspond to real rays either.

First of all consider the case of shadowings, see Fig. 5.4b. The iterative solution of (5.3.3) can be represented in the form

$$\mu_{m+1}(\boldsymbol{R}) = 2\frac{\partial \Psi_{\text{in}}}{\partial \boldsymbol{n}_{\boldsymbol{R}}} + 2\int \mu_m(\boldsymbol{R}_1)\frac{\partial}{\partial \boldsymbol{n}_{\boldsymbol{R}}} G_0(\boldsymbol{R}_1 - \boldsymbol{R})\, d\Sigma_{\boldsymbol{R}_1} \ , \qquad m = 0, 1, \dots \ , \tag{5.3.18}$$

where

$$\mu_0 = 0 \quad \text{and} \quad \frac{\partial \Psi(\boldsymbol{R})}{\partial \boldsymbol{n}} = \lim_{m\to\infty} \mu_m(\boldsymbol{R}) \ .$$

Thus in the $(m+1)$th approximation, density of monopoles is generated by the incident wave and monopole sources originating as a result of the mth iteration. For the plane boundary the integral term in (5.3.3) obviously vanishes. Let the incident field be a plane wave of the form (5.3.8). In this case density of monopole sources at the boundary is calculated by the formula

$$\frac{\partial \Psi}{\partial \boldsymbol{n}} = \mu_1 = 2\frac{\partial \Psi_{\text{in}}}{\partial \boldsymbol{n}} = 2\mathrm{i}q_0^{-1/2}\boldsymbol{\ell}\cdot\boldsymbol{n}\exp(\mathrm{i}\boldsymbol{\ell}\cdot\boldsymbol{R}) \ , \tag{5.3.19}$$

where

$$\boldsymbol{\ell} = \boldsymbol{k}_0 + q_0\boldsymbol{N}$$

is the three-dimensional wave vector of the incident wave. These sources generate the reflected plane wave with inverse sign of amplitude. It is clear from (5.3.1) that *outside* region Ω these monopoles generate as well the plane wave with inverse sign of amplitude propagating in the same direction as the incident one (5.3.8). Hence the total field outside the region Ω vanishes, which completely corresponds to the extinction theorem (3.1.29b).

Now turn to Fig. 5.4b and consider the case of rather short waves. According to the phase minimality (5.3.16) besides the incident wave, monopoles with density $\mu(\boldsymbol{R}_2)$ located near the point $\boldsymbol{R}_2$ contribute the density of monopoles at the point $\boldsymbol{R}_1$. Similarly, density of monopoles at point $\boldsymbol{R}_3$ is affected by monopoles in the vicinity of point $\boldsymbol{R}_1$ and $\boldsymbol{R}_2$, and the point $\boldsymbol{R}_4$ is influenced by points $\boldsymbol{R}_1$, $\boldsymbol{R}_2$ and $\boldsymbol{R}_3$. The size of the vicinity of points $\boldsymbol{R}_m$, $m = 1, 2, \ldots$, contributing to the monopoles density at points $\boldsymbol{R}_{m+1}, \boldsymbol{R}_{m+2}, \ldots$ decreases with increasing K. Therefore at rather large K the boundary Σ in vicinity of points $\boldsymbol{R}_1, \boldsymbol{R}_2, \ldots$ can be considered plane. We saw above that in this case monopoles in the vicinity of point $\boldsymbol{R}_1$ and having density $\mu(\boldsymbol{R}_1)$, see (5.3.19), "extinct" the plane incident wave at points $\boldsymbol{R}_2$, $\boldsymbol{R}_3$, etc.,

$$\begin{aligned} 2\int \mu(R_1)\cdot\frac{\partial}{\partial \boldsymbol{n}_{\boldsymbol{R}_2}} G_0(\boldsymbol{R}_1-\boldsymbol{R}_2)\,d\Sigma_{\boldsymbol{R}_1} &= 4\int \frac{\partial \Psi_{\text{in}}}{\partial \boldsymbol{n}_{\boldsymbol{R}_1}}\frac{\partial}{\partial \boldsymbol{n}_{\boldsymbol{R}_2}} G_0(\boldsymbol{R}_1-\boldsymbol{R}_2)\,d\Sigma_{\boldsymbol{R}_1} \\ &= -2\frac{\partial \Psi_{\text{in}}}{\partial \boldsymbol{n}_{\boldsymbol{R}_2}} \end{aligned} \tag{5.3.20}$$

and correspondingly,

$$\mu_2(\boldsymbol{R}_2) = 0 \ .$$

Validity of equality (5.3.20) can be proved directly by calculating the integral by the stationary phase method. Monopoles at the point $\boldsymbol{R}_2$ affect the monopoles at points $\boldsymbol{R}_3$ and $\boldsymbol{R}_4$ in a similar way. However, Fig. 5.4b shows that projection of external normal $\boldsymbol{n}_{\boldsymbol{R}_2}$ on the wave vector $\boldsymbol{k}$ changes the sign: $\boldsymbol{k}\cdot\boldsymbol{n}_{\boldsymbol{R}_2} < 0$. Therefore monopoles with density

$$\mu_1(\boldsymbol{R}_2) = 2\frac{\partial \Psi_{\text{in}}}{\partial \boldsymbol{n}_{\boldsymbol{R}_2}} = 2\mathrm{i}q_0^{-1/2}(\boldsymbol{k}\cdot\boldsymbol{n}_{\boldsymbol{R}_2})\exp(\mathrm{i}\boldsymbol{k}\cdot\boldsymbol{R}_2)$$

generate at points $\boldsymbol{R}_3$ and $\boldsymbol{R}_4$, sources which differ from (5.3.20) in the sign:

$$4\int \frac{\partial \Psi_{\text{in}}}{\partial \boldsymbol{n}_{\boldsymbol{R}_2}} \frac{\partial}{\partial \boldsymbol{n}_{\boldsymbol{R}_m}} G_0(\boldsymbol{R}_2 - \boldsymbol{R}_m)\, d\Sigma_{\boldsymbol{R}_2} = 2\frac{\partial \Psi_{\text{in}}}{\partial \boldsymbol{n}_{\boldsymbol{R}_m}} \; ; \qquad m = 3, 4 \; . \tag{5.3.21}$$

Therefore,

$$\mu_2(\boldsymbol{R}_3) = \mu_1(\boldsymbol{R}_3) = 2\frac{\partial \Psi_{\text{in}}}{\partial \boldsymbol{n}_{\boldsymbol{R}_3}} \; .$$

Density $\mu_2(\boldsymbol{R}_4)$ is calculated analogously to $\mu_2(\boldsymbol{R}_3)$. However, the sources $\mu_1(\boldsymbol{R}_3)$ contributed into $\mu_2(\boldsymbol{R}_4)$ by (5.3.20) should be additionally taken into account. As a result

$$\mu_2(\boldsymbol{R}_4) = 0 \; .$$

Developing these considerations we find

$$\mu_3(\boldsymbol{R}_2) = \mu_3(\boldsymbol{R}_3) = 0 \; ;$$

but

$$\mu_3(\boldsymbol{R}_4) = -2\partial \Psi_{\text{in}}/\partial \boldsymbol{n}_{\boldsymbol{R}_4} \; .$$

only in the following approximation do we find

$$\mu_4(\boldsymbol{R}_2) = \mu_4(\boldsymbol{R}_3) = \mu_4(\boldsymbol{R}_4) = 0 \; .$$

This result describes the expected effect of shadowing of points $\boldsymbol{R}_2$, $\boldsymbol{R}_3$ and $\boldsymbol{R}_4$ by point $\boldsymbol{R}_1$. However, to obtain this effect four iterations of the primary equation are needed. Obviously, the same situation exists in the general case for an arbitrary number M of the crossings the boundary by the incident wave vector. The incident wave and monopoles originating at the first point "extinct" monopoles subsequently at point $\boldsymbol{R}_2, \ldots, \boldsymbol{R}_M$ with each new iteration. However to "extinct" all these monopoles M iterations (5.3.18) should be fulfilled in the general case.

Regarding the configuration in Fig. 5.4a it is clear that the rays many times reflected from the boundary contribute into the field, just as the once reflected rays do.

Thus to correctly take into account the shadowing effects and multiple reflections, a rather large number of terms in expansion (5.3.4) should be considered. This number is larger, when the grazing angle for the incident wave is less.

5.4 Corrections to the Kirchhoff Approximation: Statistical Case

As we have seen, the criteria for applicability of the Kirchhoff approximation besides the condition (5.3.14) contain the requirement of absence of shadowings and multiple reflections, see Sect. 5.1. However, when considering the statistical rough surfaces it is difficult to attribute the exact sense to these requirements and they are of euristic character. To obtain corrections to the Kirchhoff approximation directly in the framework of the statistical problem is the more consistent approach. The estimation of corrections gives the corresponding applicability criteria in terms of statistical characteristics of the ensemble of elevations.

In the present section we construct the expression for SA in the form of an iterative series and then average this expression for the Gauss ensemble of elevations. It is essential that the obtained result is at least formally precise and the approximation is to be done at the terminal stage – when calculating the appropriate series. The content of this section is based on the work [5.6].

By (4.2.2) SA is expressed through surface density of dipoles $v(\boldsymbol{r})$ as follows:

$$S(\boldsymbol{k},\boldsymbol{k}_0) = q^{-1/2}/(8\pi^2) \int v(\boldsymbol{r})(q + \boldsymbol{k}\cdot\nabla h)\exp[-\mathrm{i}\boldsymbol{k}\cdot\boldsymbol{r} + \mathrm{i}qh(\boldsymbol{r})]\,d\boldsymbol{r}\ , \tag{5.4.1}$$

where, in virtue of (4.2.3),

$$\begin{aligned} q_0^{1/2} v(\boldsymbol{r}) = & -2\exp[\mathrm{i}\boldsymbol{k}_0\cdot\boldsymbol{r} + \mathrm{i}q_0 h(\boldsymbol{r})] + 4\int \exp[\mathrm{i}\boldsymbol{k}_0\cdot\boldsymbol{r}_1 + \mathrm{i}q_0 h(\boldsymbol{r}_1)] \\ & \times d\Sigma_1 \frac{\partial}{\partial \boldsymbol{n}_{\boldsymbol{R}_1}} G_0(\boldsymbol{R}_1 - \boldsymbol{R}) - 8\int \exp[\mathrm{i}\boldsymbol{k}_0\cdot\boldsymbol{r}_1 + \mathrm{i}q_0 h(\boldsymbol{r}_1)] \\ & \times d\Sigma_1 \frac{\partial}{\partial \boldsymbol{n}_{\boldsymbol{R}_1}} G_0(\boldsymbol{R}_1 - \boldsymbol{R}_2)\, d\Sigma_2 \frac{\partial}{\partial \boldsymbol{n}_{\boldsymbol{R}_2}} G_0(\boldsymbol{R}_2 - \boldsymbol{R}) + \cdots \end{aligned} \tag{5.4.2}$$

and

$$\boldsymbol{R}_m = [\boldsymbol{r}_m, h(\boldsymbol{r}_m)]\ , \qquad m = 1, 2, \ldots\ ; \qquad \boldsymbol{R} = [\boldsymbol{r}, h(\boldsymbol{r})]\ .$$

Substitution of (5.4.2) into (5.4.1) gives the required formally precise representation of SA in the form of an iterative series:

$$\begin{aligned} (2\pi)^2(qq_0)^{1/2} S(\boldsymbol{k},\boldsymbol{k}_0) = & \int \exp[\mathrm{i}\boldsymbol{k}_0\cdot\boldsymbol{r}_1 + \mathrm{i}q_0 h(\boldsymbol{r}_1)][q + \boldsymbol{k}\cdot\nabla h(\boldsymbol{r}_1)] \\ & \times \exp[-\mathrm{i}\boldsymbol{k}\cdot\boldsymbol{r}_1 + \mathrm{i}qh(\boldsymbol{r}_1)]\, d\boldsymbol{r}_1 \\ & - 2\int \exp[\mathrm{i}\boldsymbol{k}_0\cdot\boldsymbol{r}_1 + \mathrm{i}q_0 h(\boldsymbol{r}_1)]\, d\boldsymbol{r}_1 \frac{\partial}{\partial \boldsymbol{n}_1} G_0(\boldsymbol{R}_1 - \boldsymbol{R}_2) \end{aligned}$$

$$\times [q + \boldsymbol{k} \cdot \nabla h(\boldsymbol{r}_2)] \exp[-\mathrm{i}\boldsymbol{k} \cdot \boldsymbol{r}_2 + \mathrm{i}qh(\boldsymbol{r}_2)]\, d\boldsymbol{r}_2$$

$$+ 4 \int \exp[\mathrm{i}\boldsymbol{k}_0 \cdot \boldsymbol{r}_1 + \mathrm{i}q_0 h(\boldsymbol{r}_1)]\, d\boldsymbol{r}_1 \frac{\partial}{\partial \boldsymbol{n}_1} G_0(\boldsymbol{R}_1 - \boldsymbol{R}_2)$$

$$\times d\boldsymbol{r}_2 \frac{\partial}{\partial \boldsymbol{n}_2} G_0(\boldsymbol{R}_2 - \boldsymbol{R}_3)(q + \boldsymbol{k} \cdot \nabla h(\boldsymbol{r}_3))$$

$$\times \exp[-\mathrm{i}\boldsymbol{k} \cdot \boldsymbol{r}_3 + \mathrm{i}qh(\boldsymbol{r}_3)]\, d\boldsymbol{r}_3 - \cdots . \tag{5.4.3}$$

For convenient statistical averaging in this series, the Green's function G_0 should be given as a three-dimensional Fourier transformation (2.1.19). Taking into account also that

$$d\Sigma \frac{\partial}{\partial \boldsymbol{n}} = d\boldsymbol{r} \cdot \left(\frac{\partial}{\partial z} - \nabla h \cdot \nabla \right) \tag{5.4.4}$$

see Sect. 4.2, we find

$$d\Sigma_m \frac{\partial G_0(\boldsymbol{R}_m - \boldsymbol{R}_{m+1})}{\partial \boldsymbol{n}_m} = -\mathrm{i} d\boldsymbol{r}_m (q_m - \boldsymbol{k}_m \cdot \nabla h)$$

$$\times \int \frac{\exp[-\mathrm{i}\boldsymbol{k}_m \cdot (\boldsymbol{r}_m - \boldsymbol{r}_{m+1}) - \mathrm{i}q_m (h_m - h_{m+1})]}{(K^2 - k_m^2 - q_m^2 + \mathrm{i}\varepsilon)}$$

$$\times \frac{d\boldsymbol{k}_m \, dq_m}{(2\pi)^3} . \tag{5.4.5}$$

Here $h_m = h(\boldsymbol{r}_m)$; we emphasize that quantities $\boldsymbol{k}_m$ and q_m in (5.4.5) are independent integration variables (in contrast to $\boldsymbol{k}_0$, $\boldsymbol{k}$, and q_0, q). (The term $\mathrm{i}\varepsilon$, $\varepsilon > 0$ reminds us that K has the vanishing positive real part). After substitution of (5.4.5) into (5.4.3) the terms of the obtained series will contain factors ∇h_m and we can get rid of them through integrating by parts using the following transformation:

$$\nabla h_m \cdot \exp[-\mathrm{i}(q_m - q_{m-1}) h_m]$$

$$= \cdot [-\mathrm{i}(q_m - q_{m-1})]^{-1} \cdot \nabla [\exp(-\mathrm{i}(q_m - q_{m-1}) h_m)] .$$

The results of this operation reduces to the formal replacement

$$\nabla h_m \rightarrow (\boldsymbol{k}_m - \boldsymbol{k}_{m-1})/(q_m - q_{m-1}) ,$$

cf. (5.1.10). It is convenient to write the expressions obtained after these simple transformations using the diagrammatic notation

$$\frac{i}{\pi}(qq_0)^{1/2}S(\boldsymbol{k},\boldsymbol{k}_0) =$$ 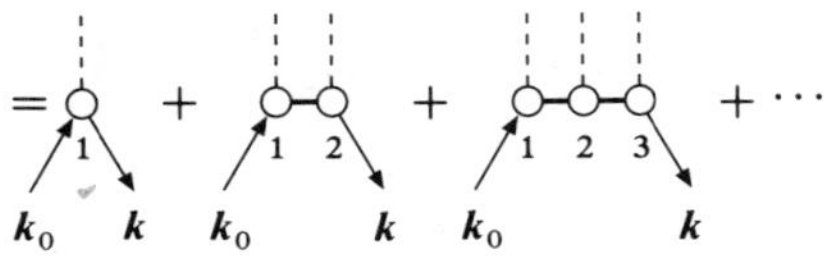 $$+ \cdots . \tag{5.4.6}$$

These diagrams are decoded according to the following rules:

a) the nth circle (vertex) is appropriated for the integration variables $\boldsymbol{r}_n$, $\boldsymbol{k}_n$, q_n and the factor

$$\frac{2i}{(2\pi)^3}[q_n + \boldsymbol{k}_n\cdot(\boldsymbol{k}_n - \boldsymbol{k}_{n-1})/(q_n - q_{n-1})]\exp[-i(\boldsymbol{k}_n - \boldsymbol{k}_{n-1})\cdot\boldsymbol{r}_n] \ . \tag{5.4.7}$$

However, in the last vertex at $n = M$ in the diagram of the Mth order (containing M vertexes), the integration over $\boldsymbol{k}_M$ and q_M is not to be done, and one should set $\boldsymbol{k}_m = \boldsymbol{k}$ and $q_m = -q$.

b) the dashed line in the nth vertex is appropriate for the factor

$$\exp[i(q_{n-1} - q_n)h(\boldsymbol{r}_n)]$$

(here also $q_M = -q$).

c) the solid bold line connecting the nth and $(n + 1)$th vertexes is appropriate to the factor (propagator)

$$(K^2 - k_n^2 - q_n^2 + i\varepsilon)^{-1} \ . \tag{5.4.8}$$

The entering and leaving arrows in the diagram and values of arguments $\boldsymbol{k}_0$, $\boldsymbol{k}$ attached to them just indicate horizontal components of wave vectors of incident and scattered waves and can be omitted. The vertexes are numerated from the left to the right.

The Kirchhoff – tangent plane approximation corresponds to considering the first diagram of series (5.4.6). Really decoding the diagram according to the above formulated rules we can easily find

$$S(\boldsymbol{k},\boldsymbol{k}_0) = -i\pi(qq_0)^{-1/2}$$ 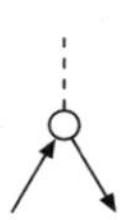 $$= -(qq_0)^{-1/2}\left[q + \frac{\boldsymbol{k}\cdot(\boldsymbol{k} - \boldsymbol{k}_0)}{(q + q_0)}\right]$$

$$\times \int \exp[-i(\boldsymbol{k} - \boldsymbol{k}_0)\cdot\boldsymbol{r} + i(q + q_0)h(\boldsymbol{r})]\frac{d\boldsymbol{r}}{(2\pi)^2}$$

This formula coincides with (5.1.11) what can be easily verified.

Further, we limit ourselves to the consideration of the first statistical moment SA, that is, of calculating the mean reflection coefficient (Sect. 2.4).

Elevation of the free surface corresponds only to the dashed lines, i.e., reduced to the factor

$$Q = \exp\left[i\sum_{n=1}^{M} t_n h(\boldsymbol{r}_n)\right] ,$$

where $t_n = q_{n-1} - q_n$. Now we consider the case of Gaussian and space homogeneous ensemble of elevations. Then, in virtue of (5.2.24) we have

$$\bar{Q} = \exp\left[-\frac{1}{2} \sum_{n,l=1}^{M} t_n t_l \tilde{W}(\boldsymbol{r}_n - \boldsymbol{r}_l) \right] ,$$

where $\tilde{W}(\boldsymbol{r}) = \overline{h(\boldsymbol{r} + \varrho)h(\varrho)}$ is the correlation function of elevations determined by (4.1.24). This value can be also demonstrated graphically with connecting each two vertexes (the nth and lth) with line

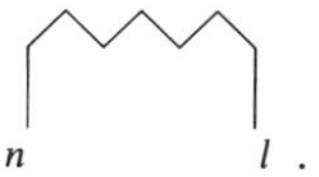

The straight sections of this line correspond to the factors $\exp(-t_n^2\sigma^2/2)$ and $\exp(-t_l^2\sigma^2/2)$, where $\sigma^2 = \tilde{W}(0)$, and the section of broken line corresponds to the factor $\exp[-t_n t_l \tilde{W}(\boldsymbol{r}_n - \boldsymbol{r}_l)]$. We have in assumed notation

$$\frac{\mathrm{i}}{\pi}(qq_0)^{1/2}\bar{S}(\boldsymbol{k}, \boldsymbol{k}_0) =$$

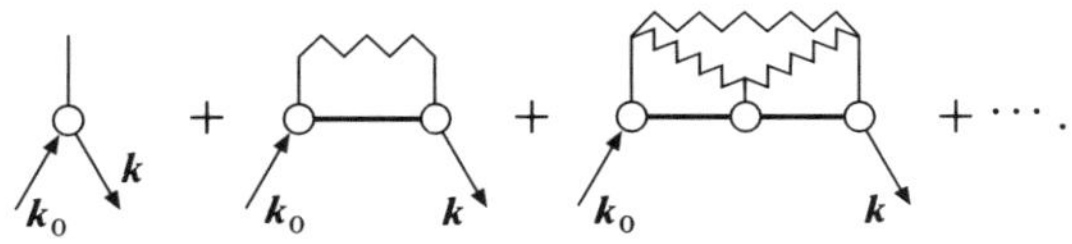

(5.4.9)

Using the above formulated rules we easily find

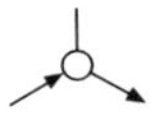

$$= -\frac{\mathrm{i}}{\pi} q \exp(-2\sigma^2 q^2)\delta(\boldsymbol{k} - \boldsymbol{k}_0)$$

so that $\bar{V} = -\exp(-2\sigma^2 q^2)$. This expression for $\bar{V}$ is obtained in the tangent plane approximation, see (5.2.25). To determine magnitudes of corrections for diagrams of the higher order and to clear up the question of what kind of small parameters the tangent plane approximation requires, we suggest the following way of calculating the diagrams.

We assume that the correlation function $\tilde{W}(\varrho)$ has some characteristic scale l (radius of correlation). Suppose that the mean square slope angle of random surface is small:

$$\delta = \sigma/l \ll 1 \ .$$

In this case, for each scattering act the vector $\boldsymbol{k}$ is bound to change little, and in integrating over $\boldsymbol{k}_n$ the area $\boldsymbol{k}_n \sim \boldsymbol{k}_0 = \boldsymbol{k}$ mainly contributes. Based on this for calculating the diagrams we use the following procedure: Let us expand in a series the product of all propagators (5.4.8) and preexponential factors in (5.4.7) at all vertexes by powers of values $(\boldsymbol{k}_n - \boldsymbol{k}_0)$ or $(\boldsymbol{k}_n - \boldsymbol{k})$. Each of these parameters can be replaced by the differentiation operator of the corresponding exponential. As a result of integrating by parts in the obtained expressions, these operators will act upon the exponentials containing correlation functions. Then integration over all $\boldsymbol{k}_n$ can be performed, which results in the expression

$$(2\pi)^{2M-2}\delta(\boldsymbol{r}_1 - \boldsymbol{r}_2)\ldots\delta(\boldsymbol{r}_{M-1} - \boldsymbol{r}_M) \ .$$

Now integrating over $\boldsymbol{r}_2, \ldots, \boldsymbol{r}_M$ reduces to calculating all derivatives of $\tilde{W}(\varrho)$ at $\varrho = 0$ and integrating over $\boldsymbol{r}_1$ gives $(2\pi)^2\delta(\boldsymbol{k} - \boldsymbol{k}_0)$. Then to obtain the final expressions the rational functions of q_n must be integrated.

Now we demonstrate the procedure described above for the first non-trivial diagram in expression for $\bar{S}$. We have

$$\text{o—o} = -\frac{4}{(2\pi)^6}\cdot\left[q_1 + \frac{\boldsymbol{k}_1\cdot(\boldsymbol{k}_1 - \boldsymbol{k}_0)}{(q_1 - q_0)}\right]\cdot\left[-q - \frac{\boldsymbol{k}\cdot(\boldsymbol{k} - \boldsymbol{k}_1)}{(q + q_1)}\right]$$
$$\times (K^2 - k_1^2 - q_1^2)^{-1}\exp[-\mathrm{i}(\boldsymbol{k}_0 - \boldsymbol{k}_1)\cdot\boldsymbol{r}_1 + \mathrm{i}(\boldsymbol{k}_1 - \boldsymbol{k})\cdot\boldsymbol{r}_2] \ . \tag{5.4.10}$$

Expand preexponential factors into a series with respect to $(\boldsymbol{k} - \boldsymbol{k}_1)$. We keep in mind that in virtue of spatial homogeneity any diagram in the general case is proportional to $\delta(\boldsymbol{k} - \boldsymbol{k}_0)$, and hence we may set $\boldsymbol{k}_0 = \boldsymbol{k}$ and $q_0 = q$. Simple calculations yield

$$\left[q_1 + \frac{\boldsymbol{k}_1\cdot(\boldsymbol{k}_1 - \boldsymbol{k})}{(q_1 - q)}\right]\cdot\left[-q - \frac{\boldsymbol{k}\cdot(\boldsymbol{k} - \boldsymbol{k}_1)}{(q + q_1)}\right]\frac{1}{(K^2 - q_1^2 - k_1^2)}$$
$$= \frac{q^2}{(q^2 - q_1^2)}\left\{\frac{q_n}{q} - \frac{2qq_1}{(q^2 - q_1^2)}\cdot\frac{\boldsymbol{k}\cdot(\boldsymbol{k}_1 - \boldsymbol{k})}{q^2} + \frac{\boldsymbol{k}\cdot(\boldsymbol{k}_1 - \boldsymbol{k})}{q^2} + \frac{q^2}{(q^2 - q_1^2)}\right.$$
$$\left.\times\frac{(\boldsymbol{k}_1 - \boldsymbol{k})^2}{q^2} + \frac{q^2}{(q^2 - q_1^2)}\cdot\left[\frac{\boldsymbol{k}\cdot(\boldsymbol{k}_1 - \boldsymbol{k})}{q^2}\right]^2 + \mathrm{O}((k - k_1)^3)\right\} \ .$$

Let us substitute the resulting expression into (5.4.10) and supress the terms $\mathrm{O}((k - k_1)^3)$:

$$\text{o—o} \approx -\frac{4}{(2\pi)^6}\cdot\frac{q^2}{(q^2 - q_1^2)}\cdot\left\{-\frac{q_1}{q} - \frac{2q_1 q}{(q^2 - q_1^2)}\cdot\frac{\boldsymbol{k}\cdot(\boldsymbol{k}_1 - \boldsymbol{k})}{q^2}\right.$$
$$\left.+\frac{\boldsymbol{k}\cdot(\boldsymbol{k}_1 - \boldsymbol{k})}{q^2} + \frac{q^2}{(q^2 - q_1^2)}\left[\frac{\boldsymbol{k}\cdot(\boldsymbol{k}_1 - \boldsymbol{k})}{q^2}\right]^2 + \frac{q^2}{(q^2 - q_1^2)}\frac{(\boldsymbol{k}_1 - \boldsymbol{k})^2}{q^2}\right\}$$
$$\cdot\exp[\mathrm{i}(\boldsymbol{k}_1 - \boldsymbol{k})\cdot\boldsymbol{r}_2 + \mathrm{i}(\boldsymbol{k}_0 - \boldsymbol{k}_1)\cdot\boldsymbol{r}_1] \ . \tag{5.4.11}$$

Equation (5.4.11) is to be integrated together with the factor

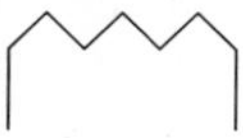

$$= \exp[-(q_0 - q_1)^2\sigma^2/2 - (q + q_1)^2\sigma^2/2 - (q_0 - q_1)(q_1 + q)$$
$$\times \tilde{W}(\boldsymbol{r}_1 - \boldsymbol{r}_2)] = \exp\{-2q^2\sigma^2 + (q^2 - q_1^2)[\sigma^2 - \tilde{W}(\boldsymbol{r}_1 - \boldsymbol{r}_2)]\} \ .$$

We replace in (5.4.11) $(\boldsymbol{k}_1 - \boldsymbol{k})$ by operator $-\mathrm{i}\nabla \boldsymbol{r}_2$ acting on $\exp[\mathrm{i}(\boldsymbol{k}_1 - \boldsymbol{k})\cdot \boldsymbol{r}_2]$ and then integrate by parts transferring its action on 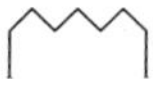. Now dependence upon $\boldsymbol{k}_1$ enters only in factor $\exp[\mathrm{i}\boldsymbol{k}_1 \cdot (\boldsymbol{r}_1 - \boldsymbol{r}_2)]$, and integrating over $\boldsymbol{k}_1$ gives $(2\pi)^2\delta(\boldsymbol{r}_1 - \boldsymbol{r}_2)$. As a result we obtain

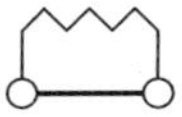

$$\approx -\frac{4}{(2\pi)^4}\exp(-2q^2\sigma^2)\int \delta(\boldsymbol{r}_1 - \boldsymbol{r}_2)\cdot \frac{q^2\,dq_1}{(q^2 - q_1^2)}$$
$$\times \left[-\frac{q_1}{q} + 2\mathrm{i}\frac{q_1}{q}\frac{1}{(q^2 - q_1^2)}(\boldsymbol{k}\cdot \nabla_{\boldsymbol{r}_2}) - \frac{\mathrm{i}(\boldsymbol{k}\cdot \nabla_{\boldsymbol{r}_2})}{q^2}\right.$$
$$\left. - \frac{1}{(q^2 - q_1^2)}\frac{(\boldsymbol{k}\cdot \nabla_{\boldsymbol{r}_2})^2}{q^2} - \frac{\nabla_{\boldsymbol{r}_2}^2}{(q^2 - q_1^2)}\right]\cdot \exp\{(q^2 - q_1^2)$$
$$\times [\sigma^2 - \tilde{W}(\boldsymbol{r}_1 - \boldsymbol{r}_2)] + \mathrm{i}(\boldsymbol{k}_0 - \boldsymbol{k})\cdot \boldsymbol{r}_1\}\, d\boldsymbol{r}_1\, d\boldsymbol{r}_2 \ .$$

Integrating over $\boldsymbol{r}_1$ yields $(2\pi)^2\delta(\boldsymbol{k}_0 - \boldsymbol{k})$. Further, the first term in brackets, $-q_1/q$, as a result of integrating over q_1, vanishes due to the integrand being odd. The second and the third terms after differentiating at the point $\boldsymbol{r}_1 = \boldsymbol{r}_2$ also become zero, since $\nabla \tilde{W}|_{\varrho=0} = 0$. As a result,

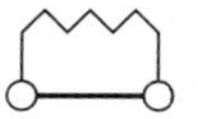

$$\approx \frac{4}{(2\pi)^2}\exp(-2q^2\sigma^2)\delta(\boldsymbol{k} - \boldsymbol{k}_0)\cdot \int \frac{q^2\,dq_1}{(q^2 - q_1^2)^2}$$
$$\times [\nabla_{\boldsymbol{r}}^2/q^2 + (\boldsymbol{k}\cdot \nabla)^2/q^2]_{\boldsymbol{r}=0}\exp\{-(q^2 - q_1^2)[\sigma^2 - \tilde{W}(\boldsymbol{r})]\} \ .$$

After the mentioned differentiations, we have

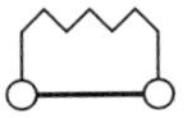

$$\approx \frac{4}{(2\pi)^2}\exp(-2q^2\sigma^2)\delta(\boldsymbol{k} - \boldsymbol{k}_0)[-q^2\nabla^2\tilde{W} - (\boldsymbol{k}\cdot \nabla)^2\tilde{W}]_{\boldsymbol{r}=0}$$
$$\times \int \frac{dq_1}{(q^2 - q_1^2)} \ .$$

The remaining integral over q_1 is equal to $-\mathrm{i}\pi/q$ [here we took into account that according to the rule ("c"), the quantity $+\mathrm{i}\varepsilon$ not written explicitly should be added to the denominator of the integrand]. Thus, the following formula is obtained for the mean reflection coefficient

$$\bar{V}_{\boldsymbol{k}} = -\exp(-2q^2\sigma^2)\cdot [1 - (\nabla^2\tilde{W})_{\boldsymbol{r}=0} - (\boldsymbol{k}\cdot \nabla/q)^2_{\boldsymbol{r}=0}\tilde{W}] \ . \tag{5.4.12}$$

Formula (5.4.12) can be also represented in the form

$$\bar{V}_{\boldsymbol{k}} = -\exp(-2q^2\sigma^2)\cdot [1 + \overline{(\nabla h)^2} + \overline{(\boldsymbol{k}\cdot \nabla h/q)^2}] \tag{5.4.13}$$

The above calculations show that for an arbitrary diagram in the general case expansion of all preexponential factors results in the power series in quantities

$$\boldsymbol{k}\cdot(\boldsymbol{k}-\boldsymbol{k}_n)/q^2 \quad \text{and} \quad (\boldsymbol{k}-\boldsymbol{k}_n)^2/q^2 \ . \tag{5.4.14}$$

The latter turns into a series in $(\boldsymbol{k}\mathscr{D})/q$ and $\mathscr{D}^2/q^2$, where $\mathscr{D}$ is a certain linear differential with respect to coordinates $\boldsymbol{r}_n$ operator with the constant coefficients, which are rational functions of q_n. It is sufficient to keep only even powers of $\mathscr{D}$ in this series due to the function $\widetilde{W}(\boldsymbol{r})$ being even and assumed as smooth. Acting of each operator in (5.4.14) on the characteristic functional $\overline{Q}$ leads alternately to arising factor $\sigma^2 q^2/l^2$ (when the exponential is differentiated) or factor l^{-2} (when the operator is applied to preexponential factors, owing to the previous differentiations). Whence it follows that parameter $\sigma^2 q^2$ enters into the expansion in a power not higher then the power of parameter l^{-2}. Thus, in the general case we have

$$\overline{V}_k = -\exp(-2q^2\sigma^2)\cdot F[\sigma^2 q^2, 1/(l^2 q^2), k^2/(l^2 q^4)]$$

and function F admits the following expansion:

$$F = 1 + \sum_{n=1}^{\infty} (\sigma^2 q^2)^n \sum_{m_1+m_2 \geqslant n} C^n_{m_1 m_2} (1/l^2 q^2)^{m_1} (K^2/l^2 q^4)^{m_2} \ . \tag{5.4.15}$$

It can be easily seen that at $n = m_1 + m_2$ the terms of this sum are independent of wavelength and, consequently, have geometric-optical origin. The rest of the terms of the second sum are proportional to $K^{2(n-m_1-m_2)}$ and are diffractional corrections vanishing in the limit $K \to \infty$. In the geometric-optical limit the solution depends only on two parameters: on the slope and on the ratio of the slope to the tangent of the grazing angle of the incident wave:

$$F \xrightarrow[K\to\infty]{} 1 + \sum_{m_1, m_2} C^{m_1+m_2}_{m_1 m_2} (\sigma^2/l^2)^{m_1} (\sigma^2/l^2 \tan^2\chi)^{m_2} \ . \tag{5.4.16}$$

Hence expression (5.4.16) takes into account in a consistent manner the effects of shadowings and multiple reflections in the statistical problem.

For the isotropic case

$$\widetilde{W}(\boldsymbol{r}) = \sigma^2\cdot[1 - r^2/2l^2 + \cdots]$$

and we easily find in virtue of (5.4.13)

$$C^1_{10} = 2 \qquad C^1_{01} = 1$$

that is

$$\overline{V}_k = -\exp(-2q^2\sigma^2)\cdot\left[1 + \sigma^2 q^2\left(\frac{2}{l^2 q^2} + \frac{k^2}{l^2 q^4} + \cdots\right) + \sigma^4 q^4\cdot(\cdots) + \cdots\right] .$$

The sufficient condition of applicability of the tangent plane approximation follows from (5.4.15). For the expansion to have sense and the correction terms to be small the following relations should be fulfilled:

$$1/l^2q^2 \ll 1 \ , \qquad \frac{k^2}{l^2q^4} \ll 1 \ , \qquad \frac{\sigma^2q^2}{l^2q^2} \ll 1 \ , \qquad \sigma^2q^2\frac{k^2}{(l^2q^4)} \ll 1 \ .$$

Let us write them down in the form

$$(\sigma^2/l^2)\cdot\max(1,\operatorname{cotan}^2\chi) \ll 1 \ , \qquad (1/l^2q^2)\cdot\max(1,\operatorname{cotan}^2\chi) \ll 1 \ . \qquad (5.4.17)$$

The first inequality meets the requirement of small mean square slope of surface as compared to the grazing angle of the incident wave (that is the absence of geometrical shadowings). The usually used applicability criteria for the tangent plane approximation result from the more precise requirements (5.4.17) in the following manner. For $\chi > \pi/4$ we have $\sigma^2/l^2 \ll 1$ and $1/l^2K^2 \ll 1$ and the geometric mean of these inequalities yields

$$K\varrho \gg 1 \ ,$$

where the value $\varrho = l^2/\sigma$ is proportional to the mean radius of curvature of the roughness. For $\chi < \pi/4$ one analogously finds

$$\sigma/l^2q\cdot\operatorname{cotan}^2\chi \ll 1$$

and since $q = K\sin\chi$ and $\operatorname{cotan}\chi \sim (\sin\chi)^{-1}$ we get

$$K\varrho\sin^3\chi \gg 1 \ . \qquad (5.4.18)$$

The requirement (5.4.18) coincides with (5.3.14). However, conditions (5.4.17) are more general than only condition (5.4.18).

The comprehensive numerical analysis of the validity of the Kirchhoff approximation which emphasise the role of the scale l is performed by *E. Thorsos* [5.9].

5.5 Two-Scale Model

It was shown in the previous sections of the present Chapter that classical approaches to the problem of wave scattering at rough surfaces – the small perturbation method and the Kirchhoff – tangent plane approximation – are applied in different situations. The former is suitable for rather small roughness at an arbitrary ratio between the characteristic horizontal scale of roughness and the incident wavelength. The tangent plane approximation is applicable for any height of roughness, but the characteristic horizontal scale must greatly exceed the radiation wavelength.

In practice roughness often does not satisfy these requirements. For example, the rough sea surface can contain both long waves with large amplitude and short waves with small amplitude for which the Kirchhoff approximation and small perturbation method can be respectively applied. In such cases scattering is estimated using the two-scale model to which the present section is dedicated. Its essense is that scattering at large-scale roughness components is considered in the tangent plane approximation and presence of small-scale components is taken into account in a perturbative manner. Firstly the two-scale model was suggested by *Kur'yanov* [5.7]. In application to the sound scattering at a rough sea surface it was developed, e.g., in the work by *McDaniel* and *Gorman* [5.10].

Thus let the boundary be represented by the superposition of two types of roughness:

$$h = h_{\mathrm{L}} + h_{\mathrm{S}} \ , \tag{5.5.1}$$

where h_{L} and h_{S} describe the large- and small-scale components. The surface consisting of large-scale components is called underlying. Suppose that the Rayleigh parameter corresponding to the small-scale components is small:

$$Kh_{\mathrm{s}} \cdot \cos\theta \ll 1 \ ,$$

where θ is the typical incidence/scattering angle. In this case the Green's function of some scalar boundary problem can be represented in the form of the power series

$$G(\boldsymbol{R}, \boldsymbol{R}_0) = G_{\mathrm{L}}(\boldsymbol{R}, \boldsymbol{R}_0) + G_{\mathrm{S}}(\boldsymbol{R}, \boldsymbol{R}_0) + \mathrm{O}(h_{\mathrm{s}}^2) \ , \tag{5.5.2}$$

where G_{L} is determined by underlying roughness and G_{S} has the first order in h_{S}. Further we shall supress corrections of the order h_{S}^2. Then for the mean Green's function we have $\langle G \rangle = \langle G_{\mathrm{L}} \rangle_{h_L}$ and

$$\Delta G = \Delta G_{\mathrm{L}} + G_{\mathrm{S}} \ .$$

Now assume the roughness h_{L} and h_{S} to be statistically independent. In this case

$$\langle |\Delta G|^2 \rangle = \langle |\Delta G_{\mathrm{L}}|^2 \rangle_{h_{\mathrm{L}}} + \langle |G_{\mathrm{S}}|^2 \rangle_{h_{\mathrm{S}}, h_{\mathrm{L}}} \ , \tag{5.5.3}$$

where the subscripts h_{L} or h_{S} indicate the statistical ensemble over which averaging is performed. To calculate averages in (5.5.3) formula (2.4.25) can be applied. In the present case in this formula we should set

$$\Psi_{\mathrm{in}}(\varrho, 0) = \frac{1}{(4\pi)^2} [(\varrho - \boldsymbol{r}_0)^2 + z_0]^{-1} \ .$$

Formula (2.4.25) for $\langle |G_{\mathrm{S}}|^2 \rangle_{h_{\mathrm{S}}}$ is obviously generalized as follows:

$$\begin{aligned} \langle |G_{\mathrm{S}}|^2 \rangle_{h_{\mathrm{S}}} = \int q^{\mathrm{loc}}(\varrho) \cdot q_0^{\mathrm{loc}}(\varrho) \cdot \sigma^{(\mathrm{S})}[\boldsymbol{k}^{\mathrm{loc}}(\varrho), \boldsymbol{k}_0^{\mathrm{loc}}(\varrho)] (4\pi)^{-2} \\ \times \{(\boldsymbol{r}_0 - \varrho)^2 + [z_0 + h_{\mathrm{L}}(\varrho)]^2\}^{-1} \\ \times \{(\varrho - \boldsymbol{r})^2 + [z + h_{\mathrm{L}}(\varrho)]^2\}^{-1} \, d\Sigma_{\mathrm{L}} \ . \end{aligned} \tag{5.5.4}$$

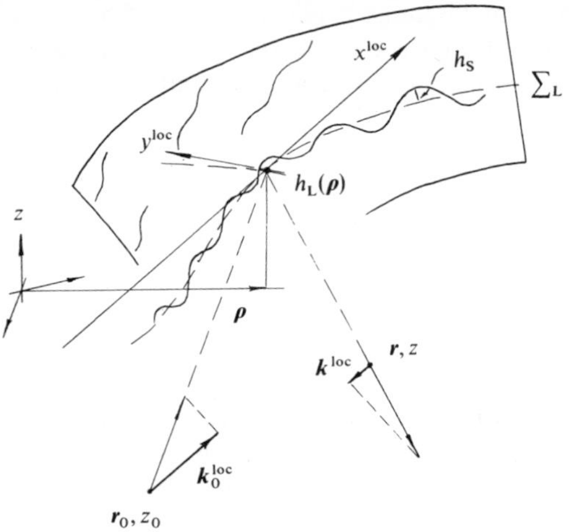

Fig. 5.5. Two-scale model: Σ_L – underlying surface, (x^{loc}, y^{loc}) – the tangent plane at the point $(\boldsymbol{\varrho}, h_L(\boldsymbol{\varrho}))$, h_s – small scale roughness. Projections of the incident and scattered wave vectors $\boldsymbol{k}_0^{loc}$ and $\boldsymbol{k}^{loc}$ on this tangent plane are also indicated

Here $\sigma^{(S)}$ is the scattering cross-section at small-scale roughness and $\boldsymbol{k}^{loc}(\varrho)$ and $\boldsymbol{k}_0^{loc}(\varrho)$ are projections of the wave vector for the incident and scattered wave on the tangent plane to the underlying surface at point $(\varrho, h_L(\varrho))$, see Fig. 5.5. For (5.5.4) to be valid it is necessary to assume that the correlation radius of small-scale roughness h_S is much less than the radius of curvature of large-scale roughness. Since as a rule $h_L \ll z, z_0$, then h_L can be omitted in the denominator of the integrand in (5.5.4). By definition of the scattering coefficient (2.4.26) we have

$$m_S(\boldsymbol{k}, \boldsymbol{k}_0) = m_S^{(L)}(\boldsymbol{k}, \boldsymbol{k}_0) + \langle q^{loc} \cdot q_0^{loc} \sigma^{(S)}(\boldsymbol{k}^{loc}, \boldsymbol{k}_0^{loc}) \rangle_{h_L} , \tag{5.5.5}$$

where $m_S^{(L)}$ is the scattering coefficient for the underlying surface.

Consider the scattering cross-section at small roughness h_S. Based on formulas (4.1.10), (4.3.8), and (4.4.23) for the general case it can be written with accuracy to the first order terms in elevations:

$$S(\boldsymbol{k}, \boldsymbol{k}_0) = V_0(k) \cdot \delta(\boldsymbol{k} - \boldsymbol{k}_0) + 2\mathrm{i}(qq_0)^{1/2} B(\boldsymbol{k}, \boldsymbol{k}_0) h(\boldsymbol{k} - \boldsymbol{k}_0) ,$$

where $h(\boldsymbol{k} - \boldsymbol{k}_0)$ is the Fourier transform of roughness shape and nondimensional function $B(\boldsymbol{k}, \boldsymbol{k}_0)$ depends on the boundary conditions. In particular, for the Dirichlet problem $B = -1$ by (4.1.10), and for the Neumann problem B is given by (4.3.9). Therefore, in the general case in the framework of the small perturbation method the scattering cross-section is given by (4.3.12):

$$\sigma(\boldsymbol{k}, \boldsymbol{k}_0) = 4qq_0 \cdot |B(\boldsymbol{k}, \boldsymbol{k}_0)|^2 W(\boldsymbol{k} - \boldsymbol{k}_0) . \tag{5.5.6}$$

As a result (5.5.5) takes the form

$$m_S(\boldsymbol{k}, \boldsymbol{k}_0) = m_S^{(L)}(\boldsymbol{k}, \boldsymbol{k}_0) + \langle 4 \cdot (1 + (\nabla h_L)^2)^{1/2} \times (q^{loc})^2 \cdot (q_0^{loc})^2 \cdot |B(\boldsymbol{k}^{loc}, \boldsymbol{k}_0^{loc})|^2 \cdot W^{(S)}(\boldsymbol{k}^{loc} - \boldsymbol{k}_0^{loc}) \rangle_{h_L} . \tag{5.5.7}$$

Here $W^{(S)}$ is the roughness spectrum h_S at the tangent plane, see Fig. 5.5. Value of elevations h_S must be taken of course along the normal to the tangent plane.

For small slopes of the underlying surface it can be approximately written:

$$\begin{aligned} \boldsymbol{k}_0^{\text{loc}} &= \boldsymbol{k}_0 + q_0\boldsymbol{a} \ , \qquad q_0^{\text{loc}} = q_0 - \boldsymbol{k}_0\cdot\boldsymbol{a} \ , \\ \boldsymbol{k}^{\text{loc}} &= \boldsymbol{k} - q\boldsymbol{a} \ , \qquad q^{\text{loc}} = q + \boldsymbol{k}\cdot\boldsymbol{a} \ , \end{aligned} \tag{5.5.8}$$

where

$$\boldsymbol{a} = \nabla h_L \ .$$

If the grazing angles of the incident and the scattered waves are comparable to the characteristic slope, then in averaging in (5.5.7) possible shadowings should be taken into account. For the single-shadowed sites we obviously have $q_0^{\text{loc}} < 0$ or $q^{\text{loc}} < 0$. Such cases should be excluded while averaging in (5.5.7). For the grazing angles greatly exceeding the slopes of the underlying surface:

$$a \ll q_0/k_0 = \operatorname{tg}\chi_0 \ , \qquad a \ll q/k = \operatorname{tg}\chi$$

we can set in (5.5.7)

$$\boldsymbol{k}_0^{\text{loc}} \approx \boldsymbol{k}_0 \ , \qquad q_0^{\text{loc}} \approx q_0 \ , \qquad \boldsymbol{k}^{\text{loc}} \approx \boldsymbol{k} \ , \qquad q^{\text{loc}} \approx q \ .$$

However in the argument of function $W^{(S)}$ it is worth keeping dependence on $\boldsymbol{k}^{\text{loc}}$ and $\boldsymbol{k}_0^{\text{loc}}$:

$$W^{(S)}(\boldsymbol{k}^{\text{loc}} - \boldsymbol{k}_0^{\text{loc}}) \approx W^{(S)}[\boldsymbol{k} - \boldsymbol{k}_0 - (q + q_0)\cdot\boldsymbol{a}] \ .$$

This is related to the spectrum $W^{(S)}$ being able to vary significantly with the changing argument by value of the order $(q + q_0)a$. In this approximation (5.5.7) reduces to the expression

$$\begin{aligned} m_S(\boldsymbol{k},\boldsymbol{k}_0) &= m_S^{(L)}(\boldsymbol{k},\boldsymbol{k}_0) + 4q^2q_0^2\cdot|B(\boldsymbol{k},\boldsymbol{k}_0)|^2 \\ &\quad \times \int W^{(S)}[\boldsymbol{k} - \boldsymbol{k}_0 - \boldsymbol{a}(q + q_0)]\cdot P(\boldsymbol{a})\,d\boldsymbol{a} \ , \end{aligned} \tag{5.5.9}$$

where $P(\boldsymbol{a})$ is the probability density of large-scale roughness slopes. Correspondingly, for the scattering cross-section by (2.4.26) we have

$$\sigma(\boldsymbol{k},\boldsymbol{k}_0) = \sigma^{(L)}(\boldsymbol{k},\boldsymbol{k}_0) + 4qq_0|B(\boldsymbol{k},\boldsymbol{k}_0)|^2 \int W^{(S)}[\boldsymbol{k} - \boldsymbol{k}_0 - \boldsymbol{a}(q + q_0)]P(\boldsymbol{a})\,d\boldsymbol{a} \ . \tag{5.5.10}$$

Formulas of the form (5.5.7, 9) consider both scattering at large-scale roughness concentrated in the narrow cone of angles in the vicinity of specular direction (the first term), and scattering at large angles determined mainly by the Bragg's scattering processes (the second term).

The shortcoming of the two-scale model is that the division of roughness into two classes by (5.5.1) is arbitrary in the known limits and the result will principally depend on the chosen parameter of scale division.

6. "Nonclassical" Approaches to Wave Scattering at Rough Surfaces

As was seen in Chap. 5, both classical approaches to our problem – the small perturbation method and the tangent plane approximation – cannot cover the whole range of practical problems. The combination of these two methods, or the two-scale model, outlined in Sect. 5.5, extends the potential of the calculations, but has a drawback: some arbitrary parameter arises which divides roughness into two classes – large- and small-scale. For the continuous spectrum of roughness this parameter cannot be uniquely determined and its choice may affect the results of the calculations.

Therefore, a method free from this drawback is desirable. There are some ways to attain this. Consider the Dirichlet problem for instance. This problem reduces to equation (5.2.2) for surface dipole density, which can be solved by iterations. It was shown in Sect. 5.2 that the first iteration corresponds to the Kirchhoff approximation. The results of Sect. 4.2, see (4.2.3, 6), reveal that to calculate SA to the 1st and the 2nd order in elevations, two and three iterations of this equation should be fulfilled, respectively. Thus, if approximation (5.4.2) is used the result will be the Kirchhoff approximation for large-scale irregularities, and the perturbation theory for small-scale irregularities (with an accuracy to the h^2 order terms). This approach was investigated by *Dunin* and *Maksimov* [6.1]. However, the resultant formulas contain quadratures of rather high order that is practically inconvenient.

In recent years some other theoretical approaches to the problem appeared which are out of scope of "classical" methods. The present chapter is dedicated to some of them.

Section 6.1 deals with the small slope approximation suggested by *Voronovich* [6.2]. This approach is based on transformation properties of SA with respect to vertical shifts. It claims that the slope of roughness is generally the only small parameter underlying this theory. Numerical calculations given in Sect. 6.2 confirm this and show high efficiency of the method.

Phase perturbation technique found by *Winebrenner* and *Ishimaru* [6.3, 4] is presented in Sect. 6.3. In a certain sense it is analogous to the Rytov method in the wave propagation theory for inhomogeneous media and is definitely related to both the Kirchhoff and the small slope approximations. Numerical experiments again confirm good accuracy of calculations by this method.

Section 6.4 proceeds the phase operator method in which instead of the scattering matrix its logarithm (phase operator) is the examined expression which is sought in the form of a power expansion in elevations [6.5]. The characteristic property of this approach is exact fulfillment of the law of conservation of energy (as well as the reciprocity theorem). Comparing the results of this approach and the exact solution show good coincidence.

The integral equation for the Dirichlet problem can be obtained not only for dipole surface source density but for monopoles as well, and the approximation due to which this equation becomes a convolution and is exactly solvable can be performed in the kernel. This approach was independently suggested by *Meecham* [6.6] and *Lysanov* [6.7]. Expressing SA in it through quadrature of rather high order complicates its practical application.

In the final Sect. 6.6 comparison of small slope approximation and Bahar's Full-wave approach is made. The full-wave approach was suggested in [6.8] and since then has been generalized for diverse situations; corresponding bibliography can be found in [6.9, 10]. This comparison is performed for the simplest case of a two-dimensional Dirichlet problem.

6.1 Small Slope Approximation

The present method has been proposed by *Voronovich* [6.2]. Its main ansatz is prompted, see (6.1.6) below, by results obtained in Chaps. 2, 5. The principal small slope approximation (SSA) formulas can be obtained under general considerations and this section presents the appropriate procedure.

This theory is based on two simple properties of SA stated in Sect. 2.4. Namely, for the boundary $z = h(\boldsymbol{r})$ shifted as a whole in horizontal directions by vector $\boldsymbol{d}$, SA transforms according to formula

$$S(\boldsymbol{k},\boldsymbol{k}_0) \to S(\boldsymbol{k},\boldsymbol{k}_0)\cdot\exp[-\mathrm{i}(\boldsymbol{k}-\boldsymbol{k}_0)\cdot\boldsymbol{d}] \tag{6.1.1}$$

and for a vertical shift by H the transformation

$$S(\boldsymbol{k},\boldsymbol{k}_0) \to S(\boldsymbol{k},\boldsymbol{k}_0)\cdot\exp[\mathrm{i}(q+q_0)H] \tag{6.1.2}$$

takes place. It is easily seen that formulas of the kind (5.1.11, 29) arising in the tangent plane approximation satisfy the requirements (6.1.1, 2) which is due to the fact that these transformation properties are of purely geometrical origin and the tangent plane approximation is based on coordinate-invariant arguments as well. The transformation properties in (5.1.11, 29) are fulfilled due to the following factor in integrands:

$$\exp[-\mathrm{i}(\boldsymbol{k}-\boldsymbol{k}_0)\cdot\boldsymbol{r} + \mathrm{i}(q+q_0)h(\boldsymbol{r})] \ . \tag{6.1.3}$$

Thus the formula for SA is naturally sought in the following form:

$$S(\boldsymbol{k},\boldsymbol{k}_0) = \int \frac{d\boldsymbol{r}}{(2\pi)^2}\exp[-\mathrm{i}(\boldsymbol{k}-\boldsymbol{k}_0)\cdot\boldsymbol{r} + \mathrm{i}(q+q_0)h(\boldsymbol{r})]\,\varphi(\boldsymbol{k},\boldsymbol{k}_0;\boldsymbol{r};[h]) \ . \tag{6.1.4}$$

Here the "Kirchhoff" factor (6.1.3) is extracted, but it should be integrated with some weight φ depending on $\boldsymbol{k}$, $\boldsymbol{k}_0$ along with $\boldsymbol{r}$. Moreover, φ is some functional of elevations h. Since the form of functional φ is not fixed yet, it is possible to seek SA according to (6.1.4).

Similar to the Taylor series any smooth functional can be expressed as an integral-power series:

$$Q[h] = Q_0 + \int Q_1(\boldsymbol{r}_1)h(\boldsymbol{r}_1)\,d\boldsymbol{r}_1 + \int Q_2(\boldsymbol{r}_1,\boldsymbol{r}_2)h(\boldsymbol{r}_1)h(\boldsymbol{r}_2)\,d\boldsymbol{r}_1\,d\boldsymbol{r}_2$$
$$+ \int Q_3(\boldsymbol{r}_1,\boldsymbol{r}_2,\boldsymbol{r}_3)h(\boldsymbol{r}_1)h(\boldsymbol{r}_2)h(\boldsymbol{r}_3)\,d\boldsymbol{r}_1\,d\boldsymbol{r}_2\,d\boldsymbol{r}_3 + \cdots \;.$$

Obviously, the coefficients Q_2, Q_3, ... can be taken as symmetric functions of $\boldsymbol{r}_1$, $\boldsymbol{r}_2$, ... (otherwise these coefficients can be symmetrized using change of integration variables of the type $\boldsymbol{r}_n \rightleftarrows \boldsymbol{r}_m$). The set of coefficient functions Q_n unambigiously determine the function Q and vice versa. In particular, for $Q = 0$ all $Q_n = 0$; the latter will not hold, if functions Q_n are not symmetric.

It is convenient to use instead of functional φ in (6.1.4) its Fourier transformation with respect to $\boldsymbol{r}$

$$\varphi = \int \exp(\mathrm{i}\xi\cdot\boldsymbol{r})\Phi(\boldsymbol{k},\boldsymbol{k}_0;\xi;[h])\,d\xi \;. \tag{6.1.5}$$

As a result (6.1.4) becomes

$$S(\boldsymbol{k},\boldsymbol{k}_0) = \int \frac{d\boldsymbol{r}}{(2\pi)^2}\exp[-\mathrm{i}(\boldsymbol{k}-\boldsymbol{k}_0)\cdot\boldsymbol{r} + \mathrm{i}(q+q_0)h(\boldsymbol{r})]\cdot\exp(\mathrm{i}\xi\cdot\boldsymbol{r})$$
$$\times\,\Phi(\boldsymbol{k},\boldsymbol{k}_0;\xi;[h])\,d\xi \;. \tag{6.1.6}$$

Functional Φ will be sought in the form of the integral-power series

$$\Phi = \delta(\xi)\Phi_0 + \int \delta(\xi-\xi_1)\Phi_1(\xi_1)h(\xi_1)\,d\xi_1$$
$$+ \int \delta(\xi-\xi_1-\xi_2)\Phi_2(\xi_1,\xi_2)h(\xi_1)h(\xi_2)\,d\xi_1\,d\xi_2 + \cdots \;. \tag{6.1.7}$$

Here $h(\xi)$ is the Fourier transform of roughness (4.1.7). Coefficient functions Φ_1, Φ_2, ... are independent of roughness; its dependency upon $\boldsymbol{k}$ and $\boldsymbol{k}_0$ is supposed. δ-functions in (6.1.7) warrant transformation property (6.1.1). For a horizontal shift of boundary by vector $\boldsymbol{d}$

$$h(\xi) \to h(\xi)\exp(-\mathrm{i}\xi\cdot\boldsymbol{d})$$

and thus (6.1.1) is fulfilled in (6.1.6).

The vertical shift by H means

$$h(\xi) \to h(\xi) + H\delta(\xi) \;. \tag{6.1.8}$$

The exponential factor in the integrand of (6.1.6) assures fulfillment of transformation property (6.1.2). Hence, it is natural to assume that functional (6.1.7) remains invariant with respect to this transformation. This assumption is rather plausible, but it cannot be proved in the general case and should be verified in each concrete physical situation.

This invariancy appeared to occur for the Dirichlet and Neumann problems and for scattering of electromagnetic waves at the perfectly conducting boundary.

If invariance of functional (6.1.7) in transforming (6.1.8) takes place then it directly follows that the coefficient functions Φ_n, $n \geqslant 1$ vanish when one of the arguments becomes zero. In fact, as h is arbitrary, this invariance holds independently for all terms in (6.1.7). Consider the first order term. Since

$$\int \delta(\xi - \xi_1)\Phi_1(\xi_1)h(\xi_1)\,d\xi_1 = \Phi_1(\xi)h(\xi)$$

then in transformation (6.1.8) we have

$$\Phi_1(\xi)h(\xi) \rightarrow \Phi_1(\xi)h(\xi) + H\Phi_1(0)\delta(\xi)$$

and consequently $\Phi_1(0) = 0$. Whence it follows that factor ξ in function Φ_1 can be extracted:

$$\Phi_1(\xi) = \sum_{\alpha=1,2} \xi^{(\alpha)}\tilde{\Phi}_1^{(\alpha)}(\xi) \ , \qquad (6.1.9)$$

where $\tilde{\Phi}_1$ is a regular function of ξ.

Proceed to the second order term in (6.1.7). Performing (6.1.8) and taking the H order term yields

$$\int \delta(\xi-\xi_2)\Phi_2(0,\xi_2)h(\xi_2)\,d\xi_2 + \int \delta(\xi-\xi_1)\Phi_2(\xi_1,0)h(\xi_1)\,d\xi_1 = 2\Phi_2(0,\xi)h(\xi)$$
$$= 0 \ .$$

Due to the arbitrariness of $h(\xi)$ we obtain $\Phi_2(0,\xi) = 0$. Therefore, it is possible to extract again in Φ_2 factor ξ_1. As function Φ_2 is symmetric, we obtain

$$\Phi_2(\xi_2,\xi_2) = \sum_{\alpha_1,\alpha_2=1,2} \xi_1^{(\alpha_1)}\xi_2^{(\alpha_2)}\tilde{\Phi}_2^{(\alpha_1,\alpha_2)}(\xi_1,\xi_2) \ ,$$

where function $\tilde{\Phi}_2$ is regular. In a quite similar manner we can write for the nth order term

$$\Phi_n = \sum_{\alpha_1,\dots\alpha_n=1,2} \xi_1^{(\alpha_1)}\dots\xi_n^{(\alpha_n)}\cdot\tilde{\Phi}_2^{(\alpha_1,\alpha_2,\dots\alpha_n)}(\xi_1,\xi_2,\dots,\xi_n) \ , \qquad (6.1.10)$$

where $\tilde{\Phi}_n$ are regular functions without singularities at $\xi_i = 0$. As a result, (6.1.7) becomes

$$\Phi = \delta(\xi)\tilde{\Phi}_0 + \int \delta(\xi - \xi_1)\tilde{\Phi}_1\xi_1 h(\xi_1)\,d\xi_1$$
$$+ \int \delta(\xi - \xi_1 - \xi_2)\tilde{\Phi}_2\xi_1 h(\xi_1)\xi_2 h(\xi_2)\,d\xi_1\,d\xi_2 + \cdots \ , \qquad (6.1.11)$$

where $\tilde{\Phi}_n = \tilde{\Phi}_n(\boldsymbol{k}, \boldsymbol{k}_0; \xi_1, \ldots, \xi_n)$ [the superscripts α_i in (6.1.11) are omitted for simplicity].

When substituting (6.1.11) into (6.1.5) the nth order term can be estimated as follows:

$$|\int \tilde{\Phi}_n(\boldsymbol{k}, \boldsymbol{k}_0; \xi_1, \ldots, \xi_n) \xi_1 h(\xi_1) \exp(\mathrm{i}\xi_1 \cdot \boldsymbol{r}) \cdot \ldots \cdot \xi_n h(\xi_n) \exp(\mathrm{i}\xi_n \cdot \boldsymbol{r}) \, d\xi_1 \ldots d\xi_n|$$
$$\sim |\tilde{\Phi}_n| \cdot |\nabla h|^n \, , \tag{6.1.12}$$

where $|\tilde{\Phi}_n|$ is some typical value of the coefficient function (note that $\tilde{\Phi}_n$ is nondimensional). When $\tilde{\Phi}_n$ is bounded at all $\xi_1, \xi_2, \ldots$ then for

$$|\nabla h| \ll 1 \tag{6.1.13}$$

series (6.1.11) can be considered as a small slope expansion.

Various methods are available to calculate the kernel functions $\tilde{\Phi}_n$. One of them is presented in Sect. 6.2. In particular, these kernels can be determined using a transition to the small perturbation theory, if the common integral-power expansion of SA is known:

$$S(\boldsymbol{k}, \boldsymbol{k}_0) = V_0(k)\delta(\boldsymbol{k} - \boldsymbol{k}_0) + 2\mathrm{i}(qq_0)^{1/2} B(\boldsymbol{k}, \boldsymbol{k}_0) h(\boldsymbol{k} - \boldsymbol{k}_0)$$
$$+ (qq_0)^{1/2} \sum_{n=2}^{\infty} \int B_n(\boldsymbol{k}, \boldsymbol{k}_0; \boldsymbol{k}_1, \ldots, \boldsymbol{k}_{n-1})$$
$$\times h(\boldsymbol{k} - \boldsymbol{k}_1) \ldots h(\boldsymbol{k}_{n-1} - \boldsymbol{k}_0) \, d\boldsymbol{k}_1 \ldots d\boldsymbol{k}_{n-1} \, . \tag{6.1.14}$$

Equation (6.1.14) agrees with relations (4.1.9, 10), (4.3.8), and (4.4.23) obtained in the problems discussed previously. The structure of arguments of function h in (6.1.14) assures the transformation property (6.1.1). Functions B, B_2, … are independent of elevations and related to the boundary conditions.

The kernels $\tilde{\Phi}_0$, $\tilde{\Phi}_1$, … can be determined after substituting (6.1.11) into (6.1.6), expanding the exponent in (6.1.6) in h and comparing the result to (6.1.14). However, an extremely significant circumstance should be taken into account here. That is, the expression for Φ in (6.1.6) is not determined unambigously and permits some gauge arbitrariness, since $\tilde{\Phi}_n$ can be added with an arbitrary function vanishing at $\boldsymbol{k} - \boldsymbol{k}_0 = \xi_1 + \cdots + \xi_n$ with simultaneous introduction of some compensation term into $\tilde{\Phi}_{n+1}$. In fact, with replacement

$$\tilde{\Phi}_n \to \tilde{\Phi}_n + (\boldsymbol{k} - \boldsymbol{k}_0 - \xi_1 - \cdots - \xi_n) g \, , \qquad \tilde{\Phi}_{n+1} \to \tilde{\Phi}_{n+1} - \mathrm{i}(q + q_0) g \, , \tag{6.1.15}$$

where g is an arbitrary function, the value of SA in (6.1.6) does not change. This statement results from the following calculation:

$$\int \frac{d\boldsymbol{r}}{(2\pi)^2} \exp[-\mathrm{i}(\boldsymbol{k} - \boldsymbol{k}_0) \cdot \boldsymbol{r} + \mathrm{i}(q + q_0) h(\boldsymbol{r})] \cdot \exp(\mathrm{i}\xi \cdot \boldsymbol{r})$$
$$\times \int (\boldsymbol{k} - \boldsymbol{k}_0 - \xi_1 - \cdots - \xi_n) g \cdot \xi_1 h(\xi_1) \ldots \xi_n h(\xi_n) \, d\xi_1 \ldots d\xi_n$$

$$= \int \frac{d\boldsymbol{r}}{(2\pi)^2} \exp[\mathrm{i}(q + q_0)h(\boldsymbol{r})] \cdot \mathrm{i}\nabla\{\exp[-\mathrm{i}(\boldsymbol{k} - \boldsymbol{k}_0 - \xi)\cdot\boldsymbol{r}]\}$$

$$\times \int \delta(\xi - \xi_1 - \cdots - \xi_n) g \cdot \xi_1 h(\xi_1) \ldots \xi_n h(\xi_n)\, d\xi_1 \ldots d\xi_n$$

$$= \int \frac{d\boldsymbol{r}}{(2\pi)^2} (q + q_0)\nabla h \exp[\mathrm{i}(q + q_0)h(\boldsymbol{r}) - \mathrm{i}(\boldsymbol{k} - \boldsymbol{k}_0 - \xi)\cdot\boldsymbol{r}]$$

$$\times \int \delta(\xi - \xi_1 - \cdots - \xi_n) g \cdot \xi_1 h(\xi_1) \ldots \xi_n h(\xi_n)\, d\xi_1 \ldots d\xi_n$$

$$= \mathrm{i}(q + q_0) \int \frac{d\boldsymbol{r}}{(2\pi)^2} \exp[-\mathrm{i}(\boldsymbol{k} - \boldsymbol{k}_0)\cdot\boldsymbol{r} + \mathrm{i}(q + q_0)h(\boldsymbol{r})] \cdot \exp(\mathrm{i}\xi\cdot\boldsymbol{r})\, d\xi$$

$$\times \int \delta(\xi - \xi_1 - \cdots - \xi_{n+1}) \cdot g \cdot \xi_1 h(\xi_1) \ldots \xi_{n+1} h(\xi_{n+1})\, d\xi_1 \ldots d\xi_{n+1} \,. \tag{6.1.16}$$

The last expression in this chain is precisely compensated by the addition which, according to (6.1.15), is introduced to the $(n + 1)$ order term. In other words, when extracting in $\tilde{\Phi}_n$ some portion vanishing at manifold

$$\boldsymbol{k} - \boldsymbol{k}_0 = \xi_1 + \cdots + \xi_n \tag{6.1.17}$$

we can "pass" this portion to the $(n + 1)$ order term as a result of integrating by parts.

This ambiguity in choice of function $\tilde{\Phi}_n$ is due to the fact that the same scattered field can be represented with the help of various sets of dipole and monopole sources at the boundary. The gauge transformations (6.1.15) and (3.1.38) are probably related to each other.

We set

$$\tilde{\Phi}_n = (\tilde{\Phi}_n - \tilde{\Phi}_n|_{\xi_n = \boldsymbol{k} - \boldsymbol{k}_0 - \xi_1 - \cdots - \xi_{n-1}}) + \tilde{\Phi}_n|_{\xi_n = \boldsymbol{k} - \boldsymbol{k}_0 - \xi_1 - \cdots - \xi_{n-1}} \,. \tag{6.1.18}$$

It is clear that the difference in the parentheses vanishes over manifold (6.1.17) and hence can be transformed into the $(n + 1)$th order term. The last term in (6.1.18) after the indicated substitution becomes independent of ξ_n, which allows us to make the following transformation in (6.1.6):

$$\int \frac{d\boldsymbol{r}}{(2\pi)^2} \exp[-\mathrm{i}(\boldsymbol{k} - \boldsymbol{k}_0)\cdot\boldsymbol{r} + \mathrm{i}(q + q_0)h(\boldsymbol{r})] \exp(\mathrm{i}\xi\cdot\boldsymbol{r}) \int \delta(\xi - \xi_1 - \cdots - \xi_n)$$

$$\times \tilde{\Phi}_n(\boldsymbol{k}, \boldsymbol{k}_0; \xi_1, \ldots, \xi_{n-1}, \boldsymbol{k} - \boldsymbol{k}_0 - \xi_1 - \cdots - \xi_{n-1})$$

$$\times \xi_1 h(\xi_1) \ldots \xi_n h(\xi_n)\, d\xi_1 \ldots d\xi_n\, d\xi$$

$$= \int \frac{d\boldsymbol{r}}{(2\pi)^2} \exp[-\mathrm{i}(\boldsymbol{k} - \boldsymbol{k}_0)\cdot\boldsymbol{r} + \mathrm{i}(q + q_0)h(\boldsymbol{r})] \cdot \exp[\mathrm{i}(\xi_1 + \cdots + \xi_{n-1})\cdot\boldsymbol{r}]$$

$$\times (-\mathrm{i}\nabla h(\boldsymbol{r}))\tilde{\Phi}_n(\boldsymbol{k},\boldsymbol{k}_0;\xi_1,\ldots,\xi_{n-1},\boldsymbol{k}-\boldsymbol{k}_0-\xi_1-\cdots-\xi_{n-1})$$

$$\times\ \xi_1 h(\xi_1)\ldots\xi_{n-1}h(\xi_{n-1})\,d\xi_1\ldots d\xi_{n-1}$$

$$= -\mathrm{i}\int\frac{d\boldsymbol{r}}{(2\pi)^2}\,\frac{\boldsymbol{k}-\boldsymbol{k}_0-\xi_1-\cdots-\xi_{n-1}}{(q+q_0)}$$

$$\times \exp[-\mathrm{i}(\boldsymbol{k}-\boldsymbol{k}_0-\xi_1-\cdots-\xi_{n-1})\cdot\boldsymbol{r}+\mathrm{i}(q+q_0)h(\boldsymbol{r})]\tilde{\Phi}_n$$

$$\times\ \xi_1 h(\xi_1)\ldots\xi_{n-1}h(\xi_{n-1})\,d\xi_1\ldots d\xi_{n-1}$$

$$= -\frac{\mathrm{i}}{(q+q_0)}\int\frac{d\boldsymbol{r}}{(2\pi)^2}\exp[-\mathrm{i}(\boldsymbol{k}-\boldsymbol{k}_0)\cdot\boldsymbol{r}+\mathrm{i}(q+q_0)h(\boldsymbol{r})]$$

$$\times \exp(\mathrm{i}\xi\cdot\boldsymbol{r})\cdot\int\delta(\xi-\xi_1-\cdots-\xi_{n-1})$$

$$\times\ \Phi_n(\boldsymbol{k},\boldsymbol{k}_0;\xi_1,\ldots,\xi_{n-1},\boldsymbol{k}-\boldsymbol{k}_0-\xi_1-\cdots-\xi_{n-1})$$

$$\times\ h(\xi_1)h(\xi_2)\ldots h(\xi_{n-1})\,d\xi_1\ldots d\xi_{n-1}\ . \tag{6.1.19}$$

It is easily seen that the resulting expression has the structure of the $(n-1)$th order term in series (6.1.6).

Thus, using transformations (6.1.18, 19) allows us to exclude any given term from the series (6.1.7). This implies also that functions Φ_n can be chosen so that n terms of series (6.1.7) ensure the accuracy of calculating SA within the $(n+1)$ order in parameter ∇h.

Now we proceed to determining the coefficient functions and calculate in representation (6.1.6) the terms up to the first order in h, assuming that h is small and the term Φ_1 is excluded by the above procedure. Then

$$S(\boldsymbol{k},\boldsymbol{k}_0)=\Phi_0\cdot\delta(\boldsymbol{k}-\boldsymbol{k}_0)+\mathrm{i}(q+q_0)\Phi_0\cdot h(\boldsymbol{k}-\boldsymbol{k}_0)+\mathrm{O}(h^2)\ . \tag{6.1.20}$$

Comparing this formula to (6.1.14) we find

$$\Phi_0(\boldsymbol{k},\boldsymbol{k}_0)=\frac{2(qq_0)^{1/2}}{(q+q_0)}\cdot B(\boldsymbol{k},\boldsymbol{k}_0)\ . \tag{6.1.21}$$

According to (5.2.15)

$$V_0(k)=B(\boldsymbol{k},\boldsymbol{k})=\Phi_0(\boldsymbol{k},\boldsymbol{k})$$

thus the 0th order terms in series (6.1.14) and (6.1.20) coincide also.

Thus, after calculating the Bragg's scattering described by function $B(\boldsymbol{k},\boldsymbol{k}_0)$, we pass to the small slope approximation of the first order with the simple replacement of the factor $h(\boldsymbol{k}-\boldsymbol{k}_0)$ in (6.1.14) by an integral of the kind (6.1.6):

$$S(\boldsymbol{k},\boldsymbol{k}_0) = B(\boldsymbol{k},\boldsymbol{k}_0)\frac{2(qq_0)^{1/2}}{(q+q_0)}\int \exp[-\mathrm{i}(\boldsymbol{k}-\boldsymbol{k}_0)\cdot\boldsymbol{r} + \mathrm{i}(q+q_0)h(\boldsymbol{r})]\,d\boldsymbol{r}/(2\pi)^2 \ . \tag{6.1.22}$$

This relation is valid under the assumption that $|\nabla h| \ll 1$ within an accuracy of the first order in this parameter (the applicability of this approximation is specified in Sect. 6.2).

Comparing (5.2.17) obtained in the Kirchhoff-tangent plane approximation to (6.1.22) shows that the only difference is in the factor $g(\boldsymbol{k},\boldsymbol{k}_0)$ whose role was discussed in Sect. 5.2.

By (4.1.10 and 14) we have $B = -1$ for the Dirichlet problem, and for the Neumann problem the kernel B is given by (4.3.9). It is easily seen that we have nowhere used the scalar character of the problem and hence SA S can be taken as matrix-valued, describing transitions of waves of various polarizations into each other Sect. 2.2. According to this, in the general case Φ, Φ_n, and B are matrices. For instance, for electromagnetic waves scattering from the perfectly conducting surface, matrix $\hat{B}$ is given by (4.4.21).

Now let us determine Φ_2. For this we first write down the expression for SA within an accuracy of the second order values in h:

$$\begin{aligned} S(\boldsymbol{k},\boldsymbol{k}_0) = {} & V_0(k)\delta(\boldsymbol{k}-\boldsymbol{k}_0) + 2\mathrm{i}(qq_0)^{1/2}B(\boldsymbol{k},\boldsymbol{k}_0)h(\boldsymbol{k}-\boldsymbol{k}_0) \\ & + \frac{(qq_0)^{1/2}}{2}\int [B_2(\boldsymbol{k},\boldsymbol{k}_0;\boldsymbol{k}-\xi_1) + B_2(\boldsymbol{k},\boldsymbol{k}_0;\boldsymbol{k}-\xi_2)] \\ & \times h(\xi_1)h(\xi_2)\delta(\boldsymbol{k}-\boldsymbol{k}_0-\xi_1-\xi_2)\,d\xi_1\,d\xi_2 \ . \end{aligned}$$

In the last terms of this equality we perform symmetrization over the integration variables ξ_1, ξ_2. When defining in (6.1.6, 7) the terms to the second order in h and assuming that the term with $n = 1$ is excluded, i.e., $\Phi_1 = 0$, we have

$$\begin{aligned} & \int [-\tfrac{1}{2}(q+q_0)^2\cdot\Phi_0 + \Phi_2]h(\xi_1)h(\xi_2)\delta(\boldsymbol{k}-\boldsymbol{k}_0-\xi_1-\xi_2)\,d\xi_1\,d\xi_2 \\ & = \frac{(qq_0)^{1/2}}{2}\int [B_2(\boldsymbol{k},\boldsymbol{k}_0;\boldsymbol{k}-\xi_1) + B_2(\boldsymbol{k},\boldsymbol{k}_0;\boldsymbol{k}-\xi_2)]h(\xi_1)h(\xi_2) \\ & \quad \times \delta(\boldsymbol{k}-\boldsymbol{k}_0-\xi_1-\xi_2)\,d\xi_1\,d\xi_2 \ . \end{aligned}$$

Due to the arbitrariness of h this relation determines the symmetric function Φ_2 with an accuracy to an arbitrary term vanishing at $\boldsymbol{k}-\boldsymbol{k}_0 = \xi_1 + \xi_2$. It was seen, however, that this term can be transformed by (6.1.16) into the third order term in ∇h and included in Φ_3. Thus, taking into account (6.1.21) also, one can set

$$\begin{aligned} \Phi_2(\boldsymbol{k},\boldsymbol{k}_0;\xi_1,\xi_2) = {} & \frac{(qq_0)^{1/2}}{2}[B_2(\boldsymbol{k},\boldsymbol{k}_0;\boldsymbol{k}-\xi_1) + B_2(\boldsymbol{k},\boldsymbol{k}_0;\boldsymbol{k}-\xi_2) \\ & + 2(q+q_0)B(\boldsymbol{k},\boldsymbol{k}_0)] \ . \end{aligned}$$

After the kernel Φ_2 is thus determined we can transform in (6.1.6, 7) this just calculated term with the help of (6.1.18, 16, 19). According to (6.1.19) the contribution of this terms to Φ_1 is as follows:

$$\Phi_1(\boldsymbol{k},\boldsymbol{k}_0;\xi) = -\frac{\mathrm{i}}{q+q_0}\cdot \Phi_2(\boldsymbol{k},\boldsymbol{k}_0;\xi,\boldsymbol{k}-\boldsymbol{k}_0-\xi)$$

$$= -\frac{\mathrm{i}}{2}\frac{(qq_0)^{1/2}}{q+q_0}\cdot[B_2(\boldsymbol{k},\boldsymbol{k}_0;\boldsymbol{k}-\xi)+B_2(\boldsymbol{k},\boldsymbol{k}_0;\boldsymbol{k}_0+\xi)$$

$$+ 2(q+q_0)B(\boldsymbol{k},\boldsymbol{k}_0)] \ . \tag{6.1.23}$$

The remaining part of the term Φ_2 is transformed by (6.1.16) and is included into the term Φ_3.

Thus, we obtained the small slope expansion within the accuracy of $(\nabla h)^2$ order values in the following form:

$$S(\boldsymbol{k},\boldsymbol{k}_0) = \frac{(qq_0)^{1/2}}{(q+q_0)}\int \exp[-\mathrm{i}(\boldsymbol{k}-\boldsymbol{k}_0-\xi)\cdot\boldsymbol{r}+\mathrm{i}(q+q_0)h(\boldsymbol{r})]$$

$$\times\Bigg\{2B(\boldsymbol{k},\boldsymbol{k}_0)\delta(\xi)-\frac{\mathrm{i}}{2}[B_2(\boldsymbol{k},\boldsymbol{k}_0;\boldsymbol{k}-\xi)+B_2(\boldsymbol{k},\boldsymbol{k}_0;\boldsymbol{k}_0+\xi)$$

$$+2\cdot(q+q_0)B(\boldsymbol{k},\boldsymbol{k}_0)]\cdot h(\xi)\Bigg\}\frac{d\boldsymbol{r}\,d\xi}{(2\pi)^2} \ . \tag{6.1.24}$$

We can go on with this procedure and calculate functions Φ_n of any order. That is, defining in representation (6.1.6, 7) without the term $n = 2$ the h^3 order values and comparing the result with the known power series, we find Φ_3, then transform it according to (6.1.19) into the term with $n = 2$ and include the rest into Φ_4, and so on.

Comparison of series (4.1.10) and (6.1.14) shows that for the Dirichlet problem

$$B_2(\boldsymbol{k},\boldsymbol{k}_0;\boldsymbol{k}') = 2q_{k'} \ . \tag{6.1.25a}$$

Similarly, for the Neumann problem we find from (4.3.7, 9) and (6.1.14) that

$$B_2(\boldsymbol{k},\boldsymbol{k}_0;\boldsymbol{k}') = -\frac{2}{(qq_0)^{1/2}}(K^2-\boldsymbol{k}\boldsymbol{k}')(K^2-\boldsymbol{k}'\boldsymbol{k}_0)/q_{k'} \ ,$$

$$= -2q_{k'}B(\boldsymbol{k},\boldsymbol{k}')B(\boldsymbol{k}',\boldsymbol{k}_0) \ . \tag{6.1.25b}$$

For the problem of electromagnetic wave scattering from the perfectly conducting surface, according to (4.4.22),

$$B_2(\boldsymbol{k},\boldsymbol{k}_0;\boldsymbol{k}') = -2q_{k'}\hat{B}(\boldsymbol{k},\boldsymbol{k}')\hat{V}_0\hat{B}(\boldsymbol{k}',\boldsymbol{k}_0) \ . \tag{6.1.26}$$

It is obvious that (6.1.25) can be also represented in the form (6.1.26); besides that, in virtue of (5.2.15) one can write down

$$B_2(\boldsymbol{k},\boldsymbol{k}_0;\boldsymbol{k}') = -2q_{k'}B(\boldsymbol{k},\boldsymbol{k}')\cdot B(\boldsymbol{k}',\boldsymbol{k}')B(\boldsymbol{k}',\boldsymbol{k}_0)\ . \tag{6.1.27}$$

For the time being it is not clear why B_2 can be expressed through the Bragg's kernel B; however, there is the possibility that relation (6.1.26) occurs for other scattering problems also. In this case, the expression for SA, accurate within the second order values in ∇h, is expressed in the form

$$\begin{aligned} S(\boldsymbol{k},\boldsymbol{k}_0) = \frac{(qq_0)^{1/2}}{q+q_0}\int &\exp[-\mathrm{i}(\boldsymbol{k}-\boldsymbol{k}_0-\boldsymbol{\xi})\cdot\boldsymbol{r}+\mathrm{i}(q+q_0)h(\boldsymbol{r})] \\ &\times\{2B(\boldsymbol{k},\boldsymbol{k}_0)\delta(\boldsymbol{\xi})+\mathrm{i}[q(\boldsymbol{k}-\boldsymbol{\xi})\cdot B(\boldsymbol{k},\boldsymbol{k}-\boldsymbol{\xi})V_0B(\boldsymbol{k}-\boldsymbol{\xi},\boldsymbol{k}_0) \\ &+q(\boldsymbol{k}_0+\boldsymbol{\xi})B(\boldsymbol{k},\boldsymbol{k}_0+\boldsymbol{\xi})V_0B(\boldsymbol{k}_0+\boldsymbol{\xi},\boldsymbol{k}_0)-(q+q_0)B(\boldsymbol{k},\boldsymbol{k}_0)] \\ &\times h(\boldsymbol{\xi})\}\,d\boldsymbol{\xi}\,d\boldsymbol{r}/(2\pi)^2\ . \end{aligned} \tag{6.1.28}$$

The surface scattering properties (i.e., the form of boundary condition) enters this formula only through the kernel B. For all three previously presented cases $V_0^2=1$. Whence it readily follows that the expression in brackets in the second term vanishes at $\boldsymbol{\xi}=0$ and $\boldsymbol{\xi}=\boldsymbol{k}-\boldsymbol{k}_0$. Hence it can be divided by $\boldsymbol{\xi}\cdot(\boldsymbol{k}-\boldsymbol{k}_0-\boldsymbol{\xi})$ and it is immediately clear that the second term in (6.1.28) transforms into the second order term in ∇h in (6.1.6, 7) with regular function $\tilde{\Phi}_2$.

Consider (6.1.28) in application to the case of an acoustically soft surface with $V_0 = B = -1$:

$$\begin{aligned} S(\boldsymbol{k},\boldsymbol{k}_0) = -\frac{(qq_0)^{1/2}}{(q+q_0)}\int &\exp[-\mathrm{i}(\boldsymbol{k}-\boldsymbol{k}_0-\boldsymbol{\xi})\cdot\boldsymbol{r}+\mathrm{i}(q+q_0)h(\boldsymbol{r})] \\ &\times[2\delta(\boldsymbol{\xi})+\mathrm{i}(q_{\boldsymbol{k}-\boldsymbol{\xi}}+q_{\boldsymbol{k}_0+\boldsymbol{\xi}}-q_{\boldsymbol{k}}-q_{\boldsymbol{k}_0})h(\boldsymbol{\xi})]\cdot d\boldsymbol{r}\,d\boldsymbol{\xi}/(2\pi)^2\ . \end{aligned} \tag{6.1.29}$$

We point out that the reciprocity theorem $S(\boldsymbol{k},\boldsymbol{k}_0)=S(-\boldsymbol{k}_0,-\boldsymbol{k})$ is fulfilled due to the evenness of function q_k.

Now we pass to the high-frequency limit in (6.1.29). The area of integration over $\boldsymbol{\xi}$ related to the second term in brackets of this formula, is determined by the width of the irregularities spectrum $h(\boldsymbol{\xi})$. Hence for $K\to\infty$ we can take that $|\boldsymbol{\xi}|<k,\ k_0$ and accordingly set

$$q_{\boldsymbol{k}-\boldsymbol{\xi}}+q_{\boldsymbol{k}_0+\boldsymbol{\xi}}-q_{\boldsymbol{k}}-q_{\boldsymbol{k}_0}\approx(\boldsymbol{k}/q_k-\boldsymbol{k}_0/q_{k_0})\cdot\boldsymbol{\xi}\ .$$

Using this approximation in (6.1.29) and integrating the result over $\boldsymbol{\xi}$ yields

$$\begin{aligned} S(\boldsymbol{k},\boldsymbol{k}_0) = -\frac{(qq_0)^{1/2}}{(q+q_0)}\int &\exp[-\mathrm{i}(\boldsymbol{k}-\boldsymbol{k}_0)\cdot\boldsymbol{r}+\mathrm{i}(q+q_0)h(\boldsymbol{r})] \\ &\times[2+(\boldsymbol{k}/q-\boldsymbol{k}_0/q_0)\cdot\nabla h]\frac{d\boldsymbol{r}}{(2\pi)^2}\ . \end{aligned}$$

Then integrating by parts in virtue of relation

$$\exp[i(q+q_0)h(\boldsymbol{r})]\cdot\nabla h = -\frac{i}{(q+q_0)}\nabla\{\exp[i(q+q_0)h(\boldsymbol{r})]\} ,$$

see (5.1.10), and taking into account that

$$2+\frac{\boldsymbol{k}-\boldsymbol{k}_0}{(q+q_0)}\cdot\left(\frac{\boldsymbol{k}}{q}-\frac{\boldsymbol{k}_0}{q_0}\right)=\frac{K^2+qq_0-\boldsymbol{k}\cdot\boldsymbol{k}_0}{qq_0}$$

leads readily to the expression for SA of the form (5.1.11). Thus, the formula for $S(\boldsymbol{k},\boldsymbol{k}_0)$ calculated by the small slope approximation with the accuracy to the $(\nabla h)^2$ order values in a high-frequency limit transforms exactly into the expression following from the tangent plane approximation. The same takes place for scattering on an acoustically rigid surface.

The way of constructing (6.1.29) shows that when proceeding to the limit $h\to 0$ and making a series expansion of $\exp[i(q+q_0)h]$, (6.1.29) transforms into (4.1.10) with an accuracy to the h^2 order term.

Now we proceed to the statistical case and consider the spatially homogeneous Gaussian ensemble of elevations. For averaging appropriate relations we use (5.2.24) and write down

$$\begin{aligned}\overline{\exp[A\cdot h(\boldsymbol{r})]\cdot h(\boldsymbol{r}')} &= \frac{\partial}{\partial\alpha}\overline{\exp[A\cdot h(\boldsymbol{r})+\alpha h(\boldsymbol{r}')]}_{\alpha=0}\\ &= A\exp[A^2\tilde{W}(0)/2]\,\tilde{W}(\boldsymbol{r}-\boldsymbol{r}') ,\end{aligned}\tag{6.1.30}$$

where $\tilde{W}$ is a correlation function, see (4.1.24):

$$\tilde{W}(\boldsymbol{r})=\overline{h(\boldsymbol{r}+\boldsymbol{\varrho})h(\boldsymbol{\varrho})}=\int\exp(i\boldsymbol{k}\cdot\boldsymbol{r})W(\boldsymbol{k})\,d\boldsymbol{k} .\tag{6.1.31}$$

Assume as well

$$\sigma^2=\tilde{W}(0)=\int W(\xi)\,d\xi .\tag{6.1.32}$$

Using (6.1.30) we easily find

$$\begin{aligned}&\exp(i\xi\cdot\boldsymbol{r})\cdot\langle\exp[i(q+q_0)h(\boldsymbol{r})]\cdot\int h(\boldsymbol{r}')\exp(-i\xi\cdot\boldsymbol{r}')\,d\boldsymbol{r}'/(2\pi)^2\rangle\\ &\quad = i(q+q_0)\cdot\exp[-(q+q_0)^2\sigma^2/2]\cdot W(\xi) .\end{aligned}$$

Then the formula for mean reflection coefficient $\bar{V}$ readily results from (6.1.24)

$$\begin{aligned}\bar{V}(\boldsymbol{k}) &= \exp(-2q^2\sigma^2)\cdot[B(\boldsymbol{k},\boldsymbol{k})\\ &\quad + q_k\int(B_2(\boldsymbol{k},\boldsymbol{k};\boldsymbol{k}+\xi)+2q_kB(\boldsymbol{k},\boldsymbol{k}))W(\xi)\,d\xi] .\end{aligned}\tag{6.1.33}$$

For the particular case of the Dirichlet problem we find

$$\bar{V}(\boldsymbol{k}) = -\exp(-2q^2\sigma^2)[1 - 2q_k \int (q_{\boldsymbol{k}+\xi} - q_k) W(\xi)\, d\xi] \ . \tag{6.1.34}$$

Thus, for the acoustically soft boundary (6.1.34) can be directly used in the problem of wave scattering from Gaussian roughness with small slopes and an arbitrary Rayleigh parameter. For instance it is applicable to calculations of the mean reflection coefficient for underwater sound scattered at the rough sea surface.

In the next section we will specify the applicability of our theory which is not simply in small roughness slopes compared to unity, but in small roughness slopes compared to the grazing angle χ_0 and the scattering grazing angle χ:

$$|\nabla h| \ll \chi_0, \chi \ . \tag{6.1.35}$$

Thus situations with shadowings cannot be considered in the framework of the small slope approximation.

When proceeding to the small perturbation method (6.1.34) gives the relation

$$\bar{V}(\boldsymbol{k}) = -1 + 2q_k \int q_{\boldsymbol{k}+\xi} W(\xi)\, d\xi + \mathrm{O}(W) \tag{6.1.36}$$

which coincides with (4.1.22). Now we examine the case $kl \gg 1$ where l is the radius of correlation of roughness. This time integration in (6.1.34) is concentrated at small ξ: $|\xi| \lesssim l^{-1}$. For simplicity we consider the one-dimensional problem. Two cases can occur: with nongrazing angle χ: $\chi \gg (kl)^{-1/2}$, and with grazing angles $\chi \ll (kl)^{-1/2}$. For nongrazing angles we have

$$q_{k+\xi} = q_k - \frac{k}{q} \cdot \xi - \frac{K^2}{2q^3} \xi^2$$

and

$$\Delta = 2q_k \cdot \int (q_{k+\xi} - q_k) W(\xi)\, d\xi \approx -\frac{K^2}{q^2} \int \xi^2 W(\xi)\, d\xi \ ,$$

$$= (\sigma^2/l^2) \cdot \sin^{-2}\chi \ll 1 \ .$$

The last inequality is due to the condition of the approximation applicability – small slopes of inhomogeneities σ/l as compared to the grazing angle. For grazing angles we have $q_k \sim K \cdot \chi$, $q_{k+\xi} \sim (K\xi)^{1/2}$ and

$$\Delta \sim K \cdot \chi \cdot (K/l)^{1/2} \cdot \sigma^2 = \chi \cdot (Kl)^{1/2} \cdot (\sigma^2/l^2) \cdot (Kl) \ll 1$$

since

$$\frac{\sigma}{l} \ll \chi \ll (Kl)^{-1/2} \ll 1$$

and both $\chi \cdot (Kl)^{1/2} \ll 1$ and $(\sigma^2/l^2) \cdot (Kl) \ll 1$ as well. At $Kl \ll 1$ estimation $q_{k+\xi} \sim l^{-1}$ takes place and

$$\Delta \sim K \cdot \sin\chi \cdot (\sigma^2/l) = Kl \sin\chi \cdot (\sigma^2/l^2) \ll 1 \ .$$

Thus, for all cases the condition

$$\sigma^2/l^2 \ll \sin^2\chi \tag{6.1.37}$$

warrants a small correction: $\Delta \ll 1$ in (6.1.34).

It is true that when the Rayleigh parameters are not very small $q\sigma \gtrsim 1$ the exponential factor plays the main role in (6.1.34) and the mean reflection coefficient is close to its value $\overline{V} = -\exp(-2q^2\sigma^2)$ originating in the tangent plane approximation, see (5.2.25).

To calculate $\overline{V}$ using (6.1.34) is fairly convenient, since it requires, in fact, performing the same quadrature as in the small perturbation method and can be written as

$$\overline{V}^{(\mathrm{SSA})} = \exp(-2q^2\sigma^2)[\overline{V}^{(\mathrm{MSP})} - 2q^2\sigma^2] \ , \tag{6.1.38}$$

where $\overline{V}^{(\mathrm{MSP})}$ is given by (6.1.36). Note that for large Rayleigh parameters $\overline{V}^{(\mathrm{MSP})}$ itself has no physical sense (for example, it can exceed unity). Thus, transition from the small perturbation method to the small slope approximation when calculating the mean reflection coefficient reduces to the elementary transformation (6.1.38) which eliminates the essential restriction of smallness of the Rayleigh parameter. We emphasize that this simple relation between $\overline{V}^{(\mathrm{SSA})}$ and $\overline{V}^{(\mathrm{MSP})}$ originates under the assumption that the statistical ensemble of elevations is Gaussian.

Calculating the quadrature participating in (6.1.36) with respect to sound scattering at sea surface was discussed in [6.11].

To study the second moments of SA we limit ourselves to expression (6.1.22). Averaging this formula is quite analogous to the Kirchhoff approximation case, see (5.2.27). As a result, we obtain

$$\begin{aligned}\sigma(\boldsymbol{k},\boldsymbol{k}_0) = {} & B(\boldsymbol{k},\boldsymbol{k}_0) \otimes B^+(\boldsymbol{k},\boldsymbol{k}_0) \cdot 4qq_0 \cdot (q+q_0)^{-2} \int \exp[-\mathrm{i}(\boldsymbol{k}-\boldsymbol{k}_0)\cdot\boldsymbol{r}] \\ & \times \{\exp[-\tfrac{1}{2}(q+q_0)^2 D(\boldsymbol{r})] - \exp[-\tfrac{1}{2}(q+q_0)^2 D(\infty)]\}\, d\boldsymbol{r}/(2\pi)^2 \ ,\end{aligned} \tag{6.1.39}$$

where

$$D(\boldsymbol{r}) = 2\cdot[\sigma^2 - \tilde{W}(\boldsymbol{r})] = \overline{[h(\boldsymbol{r}+\boldsymbol{\varrho}) - h(\boldsymbol{\varrho})]^2}$$

is the structure function of elevations. The form of (6.1.39) reminds us that in the general case the kernel B is a matrix, see Sect. 2.4. In the small perturbation method $\sigma(\boldsymbol{k},\boldsymbol{k}_0)$ is given, according to (6.1.14) and (2.4.9, 14) by formula

$$\sigma(\boldsymbol{k},\boldsymbol{k}_0) = B(\boldsymbol{k},\boldsymbol{k}_0) \otimes B^+(\boldsymbol{k},\boldsymbol{k}_0) \cdot 4qq_0 \int \exp[-\mathrm{i}(\boldsymbol{k}-\boldsymbol{k}_0)\cdot\boldsymbol{r}]\,\tilde{W}(\boldsymbol{r}) \frac{d\boldsymbol{r}}{(2\pi)^2} \ . \tag{6.1.40}$$

Comparing this relation to (6.1.39) shows that the transition of the small perturbation method to the small slope approximation reduces to redefining the correlation function by formula

$$\tilde{W}(\boldsymbol{r}) \to \tilde{W}_M(\boldsymbol{r}) = (q + q_0)^{-2}(\exp\{-(q + q_0)^2[\sigma^2 - \tilde{W}(r)]\} - \exp[-(q + q_0)^2\sigma^2])$$

after which relation (6.1.40), following from ordinary perturbation theory, should be applied. It is worth pointing out that the correlation function is redefined differently for different waves, but the reference transformation depends on parameter $(q + q_0)$ only. This receipt is suitable only for the first approximation in the small slopes expansion when geometrical effects related to large-scale roughness and Bragg's scattering at roughness with horizontal scales of the wavelength order factorizes.

The integral in (6.1.39) with statistical characteristics of elevations has the same form as in the tangent plane approximation, see (5.1.15), and has essentially pure geometrical origin (it is independent of the boundary conditions). The scattering properties of the surface enter in SA in the first order of small slope approximation in the form of a factor having the same structure as when using (6.1.40) for the Bragg's scattering.

We will discuss now the relationship of (6.1.39) with the two-scale model considered in Sect. 5.5. For simplicity take the scalar problem. The formula for the cross-section is

$$\sigma(\boldsymbol{k}, \boldsymbol{k}_0) = 4qq_0 \cdot (q + q_0)^{-2} \cdot |B(\boldsymbol{k}, \boldsymbol{k}_0)|^2 \cdot \int \exp[-\mathrm{i}(\boldsymbol{k} - \boldsymbol{k}_0) \cdot \boldsymbol{r}] \times (\exp\{-(q + q_0)^2[\sigma^2 - \tilde{W}(\boldsymbol{r})]\} - \exp[-(q + q_0)^2\sigma^2])\, d\boldsymbol{r}/(2\pi)^2 \ . \tag{6.1.41}$$

Let the roughness spectrum consist of the part W_S describing the short-wave components with $|\boldsymbol{k}| > \boldsymbol{k}_*$, and the part corresponding to the large-scale components $|\boldsymbol{k}| < \boldsymbol{k}_*$:

$$W = W_\mathrm{L} + W_\mathrm{S} \ .$$

Here k_* is some parameter separating scales. The correlation function, respectively, decomposes into two terms

$$\tilde{W}_\mathrm{L}(\boldsymbol{r}) = \int W_\mathrm{L}(\boldsymbol{k}) \exp(\mathrm{i}\boldsymbol{k} \cdot \boldsymbol{r})\, d\boldsymbol{k} \ ; \qquad \tilde{W}_\mathrm{S}(\boldsymbol{r}) = \int W_\mathrm{S}(\boldsymbol{k}) \exp(\mathrm{i}\boldsymbol{k} \cdot \boldsymbol{r})\, d\boldsymbol{k} \ .$$

Parameter k_* is chosen so that condition

$$(q + q_0)^2 \tilde{W}_\mathrm{S}(0) \ll 1$$

is fulfilled and high-frequency components of roughness can be taken into account as perturbations. Then

$$\exp[(q+q_0)^2\tilde{W}(\boldsymbol{r})] \approx \exp[(q+q_0)^2\tilde{W}_\mathrm{L}(\boldsymbol{r})]\cdot[1+(q+q_0)^2 W_\mathrm{S}(\boldsymbol{r})] \ .$$

We expand $\tilde{W}_\mathrm{L}$ into a series and limit ourselves to the first terms

$$\tilde{W}_\mathrm{L}(\boldsymbol{r}) \approx \tilde{W}_\mathrm{L}(0) - \tfrac{1}{2}\delta_x^2\cdot x^2 - \tfrac{1}{2}\delta_y^2\cdot y^2 \ ,$$

where $\delta_{x,y}^2 = \overline{(\delta_{x,y}h_\mathrm{L})^2}$ is the mean square slopes of long wave components of roughness. Substituting this approximation for $\tilde{W}_\mathrm{L}$ into the exponent (which is valid for sound of rather high frequency, see (5.2.28), we obtain

$$\sigma(\boldsymbol{k},\boldsymbol{k}_0) = 4qq_0\cdot(q+q_0)^{-2}|B(\boldsymbol{k},\boldsymbol{k}_0)|^2\cdot\int \exp[-\mathrm{i}(\boldsymbol{k}-\boldsymbol{k}_0)\cdot\boldsymbol{r}$$

$$-(q+q_0)^2(\delta_x^2\cdot x^2/2+\delta_y^2\cdot y^2/2)]\cdot[1+(q+q_0)^2\tilde{W}_\mathrm{S}(\boldsymbol{r})]\frac{d\boldsymbol{r}}{(2\pi)^2} \ .$$

In the second term, when going over from the correlation function $\tilde{W}_\mathrm{S}$ to its spectrum and calculating the Gaussian quadrature, we find

$$\sigma(\boldsymbol{k},\boldsymbol{k}_0) = 4qq_0\cdot(q+q_0)^{-4}|B(\boldsymbol{k},\boldsymbol{k}_0)|^2\cdot P\left(\frac{\boldsymbol{k}-\boldsymbol{k}_0}{q+q_0}\right) + 4qq_0\cdot(q+q_0)^{-2}$$

$$\times|B(\boldsymbol{k},\boldsymbol{k}_0)|^2\int P\left(\frac{\boldsymbol{k}-\boldsymbol{k}_0-\boldsymbol{k}'}{q+q_0}\right)W_\mathrm{S}(\boldsymbol{k}')\,d\boldsymbol{k}' \ , \qquad (6.1.42)$$

where

$$P(\boldsymbol{a}) = (2\pi\delta_x\delta_y)^{-1}\exp(-a_x^2/2\delta_x^2 - a_y^2/2\delta_y^2)$$

is the probability density of the large-scale surface slopes. The first term in (6.1.42) describes scattering by long wave components of roughness, which under the conditon of small slopes is concentrated in the narrow cone of angles around the specular direction. It coincides with the result following from the tangent plane approximation with an accuracy to g – factor discussed in Sect. 5.2, see (5.2.27, 18). The second term in (6.1.42) describes scattering at small-scale components and also takes into account that these components are situated at the underlying inclined surface. Really, setting

$$\boldsymbol{a} = (\boldsymbol{k}-\boldsymbol{k}_0-\boldsymbol{k}')/(q+q_0)$$

we can represent the contribution of the small-scale component in (6.1.42) as

$$\sigma^{(\mathrm{S})}(\boldsymbol{k},\boldsymbol{k}_0) = \int P(\boldsymbol{a})\cdot 4q_0q\cdot|B(\boldsymbol{k},\boldsymbol{k}_0)|^2 W_\mathrm{S}[\boldsymbol{k}-\boldsymbol{k}_0-\boldsymbol{a}(q+q_0)]\,d\boldsymbol{a} \ . \qquad (6.1.43)$$

Since projection of the difference of the scattered and incident wave vectors onto the plane with a slope $\nabla h = \boldsymbol{a}$ is

$$[\Delta k]_h = \boldsymbol{k}-\boldsymbol{k}_0-\boldsymbol{a}(q+q_0)$$

then (6.1.43) corresponds to the common Bragg's scattering averaged over slopes of the underlying surface. Thus (6.1.42) transforms into the usual two-scale model expression (5.5.10).

It is important to stress that the two-scale model was used here only as a method of calculating the integral (6.1.41). The scale-dividing parameter significant for the two-scale model, strictly speaking, does not arise at all when using (6.1.41). This fact can be considered as an advantage of the small slope approximation. But this does not mean that the proposed procedure of calculating the interal (6.1.41) should not be used. Quite the reverse, as real spectra of roughness are rather complicated, analytic calculation of (6.1.41) seems to be hopeless. Taking into account the "tail" of spectrum in the perturbative manner can be considered as the computational method related with the correct physical picture underlying the two-scale model. However, the integral (6.1.41) can be evaluated also as such without using the mode referred to above; appropriate investigation was performed by *Dashen* et al. [6.12].

Finally we consider the first order small slope approximation for electromagnetic wave scattering at the interface between two dielectrics. Derivation of (6.1.22) directly corresponds to this case. Transformation properties (6.1.1, 2) are responsible for the integral factors and the pre-integral ones could be determined when passing to the limit $h \to 0$ and comparing the result with perturbational formula (4.6.21):

$$\hat{S}^{NN_0}(\boldsymbol{k},\boldsymbol{k}_0) = \hat{V}_0^{NN_0}(\boldsymbol{k})\delta(\boldsymbol{k}-\boldsymbol{k}_0) + 2\mathrm{i}q_{\boldsymbol{k}}^{(N)1/2}\cdot q_{\boldsymbol{k}_0}^{(N_0)1/2}\hat{B}^{NN_0}(\boldsymbol{k},\boldsymbol{k}_0)h(\boldsymbol{k}-\boldsymbol{k}_0)$$
$$+ \mathrm{O}(h^2)$$

($\hat{S}^{NN_0}$, $\hat{B}^{NN_0}$ and $\hat{V}_0^{NN_0}$ at fixed N and N_0 are the 2×2 matrices). Taking into account representation (4.6.26) for the reflection coefficient, we can write down the following formulas:

$$\hat{S}^{11}(\boldsymbol{k},\boldsymbol{k}_0) = \hat{B}^{11}(\boldsymbol{k},\boldsymbol{k}_0)\cdot\frac{2(q^{(1)}q_0^{(1)})^{1/2}}{(q^{(1)}+q_0^{(1)})}$$
$$\times \int \exp[-\mathrm{i}(\boldsymbol{k}-\boldsymbol{k}_0)\cdot\boldsymbol{r} + \mathrm{i}(q^{(1)}+q_0^{(1)})h(\boldsymbol{r})]\frac{d\boldsymbol{r}}{(2\pi)^2}\ ,$$

$$\hat{S}^{12}(\boldsymbol{k},\boldsymbol{k}_0) = \hat{B}^{12}(\boldsymbol{k},\boldsymbol{k}_0)\cdot\frac{2(q^{(1)}q_0^{(2)})^{1/2}}{(q^{(1)}-q_0^{(2)})}$$
$$\times \int \exp[-\mathrm{i}(\boldsymbol{k}-\boldsymbol{k}_0)\cdot\boldsymbol{r} + \mathrm{i}(q^{(1)}-q_0^{(2)})h(\boldsymbol{r})]\frac{d\boldsymbol{r}}{(2\pi)^2}\ ,$$

$$\hat{S}^{21}(\boldsymbol{k},\boldsymbol{k}_0) = \hat{B}^{21}(\boldsymbol{k},\boldsymbol{k}_0)\cdot\frac{2(q^{(2)}q_0^{(1)})^{1/2}}{(-q^{(2)}+q_0^{(1)})}$$
$$\times \int \exp[-\mathrm{i}(\boldsymbol{k}-\boldsymbol{k}_0)\cdot\boldsymbol{r} + \mathrm{i}(-q^{(2)}+q_0^{(2)})h(\boldsymbol{r})]\frac{d\boldsymbol{r}}{(2\pi)^2}\ ,$$

$$\hat{S}^{22}(\boldsymbol{k},\boldsymbol{k}_0) = \hat{B}^{22}(\boldsymbol{k},\boldsymbol{k}_0)\cdot\frac{2(q^{(2)}q_0^{(2)})^{1/2}}{(-q^{(2)}-q_0^{(2)})}$$

$$\times\int \exp[-\mathrm{i}(\boldsymbol{k}-\boldsymbol{k}_0)\cdot\boldsymbol{r} - \mathrm{i}(q^{(2)}+q_0^{(2)})h(\boldsymbol{r})]\frac{d\boldsymbol{r}}{(2\pi)^2}\ . \tag{6.1.44}$$

Relations (6.1.44) are, in fact, of general character and valid for waves of any nature (under the condition that wave propagation velocity is independent of their polarization – otherwise vertical wavenumbers $q^{(1,2)}$ in these formulas will depend on polarization).

6.2 Small Slope Approximation and Rayleigh Equation. Numerical Experiments

Because of the abstract character of the considerations used in Sect. 6.1 in deriving the expansion in small slopes and calculating the coefficient functions it is desirable to obtain this expansion by other more direct ways. The present section shows how this can be done for the simplest Dirichlet scalar problem.

Suppose that the shape of roughness $h(\boldsymbol{r})$ is an analytic function of coordinates and its slopes are so small that the Rayleigh hypothesis is fulfilled (it was shown in Sect. 3.2 that this can be achieved when considering the family of roughness of the form $\varepsilon h(\boldsymbol{r})$ and choosing a small enough ε). In fact we use the scheme of considerations as in Chap. 4.

The coefficient functions in (6.1.7) are roughness independent and can be calculated using any (maybe a rather narrow) class of roughness. Assume that the convergence of the series (6.1.7) is of no interest and this series will be considered as an asymptotic expansion. Then, after determining functions Φ_n, we may forget the way they were obtained. The applicability conditions for (6.1.7) as an asymptotic expansion follow from the requirement of subsequently decreasing the correction terms.

Thus, in the reference case we can directly write down (4.1.2) for SA (which we call the Rayleigh equation)

$$\int S(\boldsymbol{k}',\boldsymbol{k}_0)q_{k'}^{-1/2}\exp[\mathrm{i}\boldsymbol{k}'\cdot\boldsymbol{r} - \mathrm{i}q_{k'}h(\boldsymbol{r})]\,d\boldsymbol{k}' = -q_0^{-1/2}\exp[\mathrm{i}\boldsymbol{k}_0\cdot\boldsymbol{r} + \mathrm{i}q_0h(\boldsymbol{r})]\ . \tag{6.2.1}$$

Multiply (6.2.1) by $q_k^{-1/2}\exp[-\mathrm{i}\boldsymbol{k}\cdot\boldsymbol{r} + \mathrm{i}q_kh(\boldsymbol{r})]\,d\boldsymbol{r}/(2\pi)^2$, integrate over $\boldsymbol{r}$ and reduce the result to the form

$$S(\boldsymbol{k},\boldsymbol{k}_0) = -\left(\frac{q}{q_0}\right)^{1/2} \int \exp[-\mathrm{i}(\boldsymbol{k}-\boldsymbol{k}_0)\cdot\boldsymbol{r} + \mathrm{i}(q+q_0)h(\boldsymbol{r})]\frac{d\boldsymbol{r}}{(2\pi)^2}$$
$$- \int S(\boldsymbol{k}',\boldsymbol{k}_0)(q/q')^{1/2}\exp[-\mathrm{i}(\boldsymbol{k}-\boldsymbol{k}')\cdot\boldsymbol{r}]$$
$$\times\{\exp[\mathrm{i}(q-q')h(\boldsymbol{r})]-1\}\frac{d\boldsymbol{r}\,d\boldsymbol{k}'}{(2\pi)^2}\ . \tag{6.2.2}$$

We have the identity

$$\int \exp[-\mathrm{i}(\boldsymbol{k}-\boldsymbol{k}')\cdot\boldsymbol{r}]\cdot\{\exp[\mathrm{i}(q-q')h(\boldsymbol{r})]-1\}\,d\boldsymbol{r}/(2\pi)^2$$
$$= -\frac{(\boldsymbol{k}+\boldsymbol{k}')}{(q+q')}\int \exp[-\mathrm{i}(\boldsymbol{k}-\boldsymbol{k}')\cdot\boldsymbol{r} + \mathrm{i}(q-q')h(\boldsymbol{r})]\nabla h\cdot d\boldsymbol{r}/(2\pi)^2\ . \tag{6.2.3}$$

In fact, since

$$\exp[\mathrm{i}(q-q')h(\boldsymbol{r})]\nabla h = -\frac{\mathrm{i}}{q-q'}\cdot\nabla\{\exp[\mathrm{i}(q-q')h(\boldsymbol{r})]-1\}$$

then integration by parts gives the required equality. Using (6.2.3) in (6.2.2) reduces this equation to the form

$$S(\boldsymbol{k},\boldsymbol{k}_0) = -(q/q_0)^{1/2}\int \exp[-\mathrm{i}(\boldsymbol{k}-\boldsymbol{k}_0)\cdot\boldsymbol{r} + \mathrm{i}(q+q_0)h(\boldsymbol{r})]\frac{d\boldsymbol{r}}{(2\pi)^2}$$
$$+ \int S(\boldsymbol{k}',\boldsymbol{k}_0)\cdot\left(\frac{q}{q'}\right)^{1/2}\left(\frac{\boldsymbol{k}+\boldsymbol{k}'}{q+q'}\right)$$
$$\times\int \exp[-\mathrm{i}(\boldsymbol{k}-\boldsymbol{k}')\cdot\boldsymbol{r} + \mathrm{i}(q-q')h(\boldsymbol{r})]\nabla h\frac{d\boldsymbol{r}\,d\boldsymbol{k}'}{(2\pi)^2}\ . \tag{6.2.4}$$

The given integral equation contains a slope as the required small parameter and can be solved by iterations:

$$S(\boldsymbol{k},\boldsymbol{k}_0) = -(q/q_0)^{1/2}\int \exp[-\mathrm{i}(\boldsymbol{k}-\boldsymbol{k}_0)\cdot\boldsymbol{r} + \mathrm{i}(q+q_0)h(\boldsymbol{r})]\frac{d\boldsymbol{r}}{(2\pi)^2}$$
$$\times\left\{1+\int\left(\frac{\boldsymbol{k}+\boldsymbol{k}'}{q+q'}\right)\cdot\exp[\mathrm{i}(\boldsymbol{k}-\boldsymbol{k}')\cdot(\boldsymbol{r}-\boldsymbol{r}')\right.$$
$$\left.-\,\mathrm{i}(q-q')(h_{\boldsymbol{r}}-h_{\boldsymbol{r}'})]\nabla h_{\boldsymbol{r}'}\cdot\frac{d\boldsymbol{r}'\,d\boldsymbol{k}'}{(2\pi)^2}+\cdots\right\}\ . \tag{6.2.5}$$

To reduce (6.2.5) to the form (6.1.6), the expression in curly brackets should be Fourier transformed, and as a result we get

$$\Phi(\boldsymbol{k}, \boldsymbol{k}_0; \xi; [h]) = -(q/q_0)^{1/2} \sum_{\pi=0}^{\infty} \int \exp(-\mathrm{i}\xi \cdot \boldsymbol{r})\, d\boldsymbol{r}/(2\pi)^2 \prod_{m=1}^{n} \left(\frac{\boldsymbol{k}_{m+1} + \boldsymbol{k}_m}{q_{m+1} + q_m} \right)$$
$$\times \nabla h_m \exp[\mathrm{i}(\boldsymbol{k}_{m+1} - \boldsymbol{k}_m) \cdot (\boldsymbol{r} - \boldsymbol{r}_m)$$
$$- \mathrm{i}(q_{m+1} - q_m)(h_{\boldsymbol{r}} - h_m)] \frac{d\boldsymbol{r}_m\, d\boldsymbol{k}_m}{(2\pi)^2} \,. \tag{6.2.6}$$

Here $h_m = h(\boldsymbol{r}_m)$ and it is set by definition $\boldsymbol{k}_{n+1} = \boldsymbol{k}$, $q_{n+1} = q$, and $\prod_{m=1}^{0} = 1$. It is obvious that functional Φ is invariant with respect to transformation $h \to h + H$ with $H = \text{const}$, and therefore the coefficient functions of its expansion in the integral-power series (6.1.7) are proportional to ξ_i when $\xi_i \to 0$, as was noted previously.

To illustrate the general procedure suggested in the previous section consider representation (6.2.5). With an accuracy to the first order values in ∇h in (6.2.5) we can take

$$S_1(\boldsymbol{k}, \boldsymbol{k}_0) = -(q/q_0)^{1/2} \int \frac{d\boldsymbol{r}}{(2\pi)^2} \exp[-\mathrm{i}(\boldsymbol{k} - \boldsymbol{k}_0) \cdot \boldsymbol{r} + \mathrm{i}(q + q_0) h(\boldsymbol{r})]$$
$$\times \left[1 + \int \frac{(\boldsymbol{k} + \boldsymbol{k}')}{(q + q')} \exp[\mathrm{i}(\boldsymbol{k} - \boldsymbol{k}') \cdot (\boldsymbol{r} - \boldsymbol{r}')] \nabla h_{\boldsymbol{r}'} \frac{d\boldsymbol{r}'\, d\boldsymbol{k}'}{(2\pi)^2} \right] . \tag{6.2.7}$$

In fact, since $\exp[-\mathrm{i}(q - q')(h_{\boldsymbol{r}} - h_{\boldsymbol{r}'})] - 1$ does not change at shifts $h \to h + H$, then the integral-power expansion of the value is in ∇h and begins, obviously, with the first order terms. Therefore, the second exponential in (6.2.5) can be replaced by unity with the first order accuracy in ∇h. Relation (6.2.7) is expressed through the Fourier components as

$$S_1(\boldsymbol{k}, \boldsymbol{k}_0) = -(q/q_0)^{1/2} \int \frac{d\boldsymbol{r}}{(2\pi)^2} \exp[-\mathrm{i}(\boldsymbol{k} - \boldsymbol{k}_0) \cdot \boldsymbol{r} + \mathrm{i}(q + q_0) h(\boldsymbol{r})]$$
$$\times \left[1 + \int \frac{\mathrm{i}\xi \cdot (2\boldsymbol{k} - \xi)}{(q_k + q_{\boldsymbol{k}-\xi})} \cdot \exp(\mathrm{i}\xi \cdot \boldsymbol{r}) h(\xi)\, d\xi \right] . \tag{6.2.8}$$

Looking at (6.2.8) we see that it differs greatly from (6.1.22) with $B = -1$. However, the ambiguousness of the coefficients in expansion (6.1.7) noted in Sect. 6.1 manifests itself here. In fact, (6.2.8) and (6.1.22) coincide to an accuracy of the ∇h order terms which can be demonstrated by representing (6.2.8) in the form

$$S_1(\boldsymbol{k}, \boldsymbol{k}_0) = -(q/q_0)^{1/2} \int \frac{d\boldsymbol{r}}{(2\pi)^2} \cdot \exp[-\mathrm{i}(\boldsymbol{k} - \boldsymbol{k}_0) \cdot \boldsymbol{r} + \mathrm{i}(q + q_0) h(\boldsymbol{r})]$$
$$\times \left(1 + \frac{(\boldsymbol{k} + \boldsymbol{k}_0) \cdot \nabla h}{(q + q_0)} \right) + T \,, \tag{6.2.9}$$

where

$$T = -(q/q_0)^{1/2} \int \frac{d\boldsymbol{r}}{(2\pi)^2} \exp[-\mathrm{i}(\boldsymbol{k} - \boldsymbol{k}_0 - \xi)\cdot \boldsymbol{r} + \mathrm{i}(q + q_0)h_{\boldsymbol{r}}]$$

$$\times \mathrm{i}\cdot(\boldsymbol{k} - \boldsymbol{k}_0 - \xi)\left[\frac{1}{(q + q_{\boldsymbol{k}-\xi})}\right.$$

$$\left. + \frac{(\boldsymbol{k} + \boldsymbol{k}_0 - \xi)\cdot(\boldsymbol{k} + \boldsymbol{k}_0)}{(q + q_{\boldsymbol{k}-\xi})(q_0 + q_{\boldsymbol{k}-\xi})(q_0 + q_\xi)}\right]\xi h(\xi)\, d\xi \ . \tag{6.2.10}$$

Integration by parts in the first term in (6.2.9) is equivalent to replacement (5.1.10) and therefore it can be written down as

$$S_1(\boldsymbol{k}, \boldsymbol{k}_0) = -\frac{2(qq_0)^{1/2}}{(q + q_0)} \int \exp[-\mathrm{i}(\boldsymbol{k} - \boldsymbol{k}_0)\cdot \boldsymbol{r} + \mathrm{i}(q + q_0)h(\boldsymbol{r})] \frac{d\boldsymbol{r}}{(2\pi)^2} + T \ . \tag{6.2.11}$$

Here the first term perfectly coincides with (6.1.22) and quantity T has an order of $(\nabla h)^2$: we show this replacing

$$\mathrm{i}(\boldsymbol{k} - \boldsymbol{k}_0 - \xi)\exp[-\mathrm{i}(\boldsymbol{k} - \boldsymbol{k}_0 - \xi)\cdot \boldsymbol{r}] = -\nabla \exp[-\mathrm{i}(\boldsymbol{k} - \boldsymbol{k}_0 - \xi)\cdot \boldsymbol{r}]$$

and integrating by parts which gives factor $\mathrm{i}(q + q_0)\nabla h$. Quantity $\mathrm{i}\xi h(\xi)$, being the Fourier transform of ∇h, also contains the required small parameter which proves the above statement. Formula (6.2.10) for T is easily reduced to the form of (6.1.11):

$$T = \int \frac{d\boldsymbol{r}}{(2\pi)^2} \cdot \exp[-\mathrm{i}(\boldsymbol{k} - \boldsymbol{k}_0)\cdot \boldsymbol{r} + \mathrm{i}(q + q_0)h(\boldsymbol{r})] \cdot \exp(\mathrm{i}\xi\cdot \boldsymbol{r})\delta(\xi - \xi_1 - \xi_2)$$

$$\times\, t_2^{\alpha\beta}(\boldsymbol{k}, \boldsymbol{k}_0; \xi_1, \xi_2)\cdot \xi_1^{(\alpha)} h(\xi_1)\cdot \xi_2^{(\beta)} h(\xi_2)\, d\xi_1\, d\xi_2\, d\xi \ ; \qquad (\alpha, \beta = 1, 2)$$

(summation over α, β is assumed) where

$$t_2^{\alpha\beta}(\boldsymbol{k}, \boldsymbol{k}_0; \xi_1, \xi_2) = (q/q_0)^{1/2}\cdot 2^{-1}\cdot \left\{(q + q_0)\cdot\left[\frac{1}{(q + q_{\boldsymbol{k}-\xi_1})}\right.\right.$$

$$\left. + \frac{1}{(q + q_{\boldsymbol{k}-\xi_2})}\right]\delta_{\alpha\beta} + \frac{(k + k_0)^{(\alpha)}(k + k_0 - \xi_1)^{(\beta)}}{(q + q_{\boldsymbol{k}-\xi_1})(q_0 + q_{\boldsymbol{k}-\xi_1})}$$

$$\left. + \frac{(k + k_0)^{(\beta)}(k + k_0 - \xi_2)^{(\alpha)}}{(q + q_{\boldsymbol{k}-\xi_2})(q_0 + q_{\boldsymbol{k}-\xi_2})}\right] \ .$$

Nondimensional quantity t_2 is a part of $\tilde{\Phi}_2(\boldsymbol{k}, \boldsymbol{k}_0; \xi_1, \xi_2)$ and is bounded for all ξ_1, ξ_2, including $|\xi_1|, |\xi_2| \to \infty$.

Now consider applicability conditions for the small slope approximation. This problem was partly discussed in Sect. 6.1 where the statistical case was considered. It is apriori clear that we need $|\nabla h| \ll 1$. To specify this condition we consider the correction term in (6.1.29) and write

$$q_{\boldsymbol{k}-\xi} + q_{\boldsymbol{k}+\xi} - q - q_0 = 2\cdot(q+q_0)^{-1}\varphi^{\alpha\beta}(\boldsymbol{k},\boldsymbol{k}_0;\xi)\xi^{(\alpha)}(k-\boldsymbol{k}_0-\xi)^{(\beta)} , \tag{6.2.12}$$

where

$$\begin{aligned}\varphi^{\alpha\beta}(\boldsymbol{k},\boldsymbol{k}_0;\xi) = \frac{(q+q_0)}{2}\Bigg\{&\left[\frac{1}{(q+q_{\boldsymbol{k}-\xi})}+\frac{1}{(q_0+q_{\boldsymbol{k}_0+\xi})}\right]\delta^{\alpha\beta} \\ &+\frac{(k+k_0)^{(\alpha)}}{(q+q_{\boldsymbol{k}-\xi})(q_0+q_{\boldsymbol{k}_0+\xi})} \\ &\times\left[\frac{(k+k_0+\xi)^{(\beta)}}{(q_{\boldsymbol{k}_0+\xi}+q)}+\frac{(k+k_0-\xi)^{(\beta)}}{(q_0+q_{\boldsymbol{k}-\xi})}\right]\Bigg\} .\end{aligned} \tag{6.2.13}$$

Using our usual integration by parts, we write

$$\begin{aligned}S(\boldsymbol{k},\boldsymbol{k}_0) = -\frac{2(qq_0)^{1/2}}{q+q_0}\int\frac{d\boldsymbol{r}}{(2\pi)^2}&\exp[-\mathrm{i}(\boldsymbol{k}-\boldsymbol{k}_0)\cdot\boldsymbol{r}+\mathrm{i}(q+q_0)h(\boldsymbol{r})] \\ &\times\left[1+(\nabla h)^{(\alpha)}\cdot\int\varphi^{\alpha\beta}.\mathrm{i}\xi^{(\beta)}\exp(\mathrm{i}\xi\cdot\boldsymbol{r})h(\xi)\,d\xi\right] .\end{aligned} \tag{6.2.14}$$

Since the integral in square brackets is $(\nabla h)^{(\beta)}$ for the constant φ-factor, then the problem reduces to estimation of the nondimensional kernel φ. It is seen from (6.2.13) that this value is bounded for any ξ and at $|\xi|\to\infty$ is of $|\xi|^{-1}$ order. If the incident and scattered waves propagate in a nongrazing direction: $q_0\sim K$ and $q\sim K$, then $\varphi\sim 1$ and the correction has an order $(\nabla h)^2$. For one of these directions being grazing, the maximum φ is related to the minimum of denominators when $q_{\boldsymbol{k}-\xi}=0$ or $q_{\boldsymbol{k}_0+\xi}=0$. Proceeding from these considerations the following estimation for φ can be done:

$$\varphi\sim(K/q+K/q_0)^2 .$$

According to (6.2.14) correction Δ is estimated as

$$\Delta\sim(K/q+K/q_0)^2\cdot(\nabla h)^2$$

whence follows the applicability criterion for the small slope approximation

$$|\nabla h|\ll\min(\sin\chi_0,\sin\chi) , \tag{6.2.15}$$

where χ_0, χ are the grazing angles of the incident and scattered waves. Consider the case of large-scale roughness when spectrum $h(\xi)$ is concentrated at $|\xi|\ll K$ and $S(\boldsymbol{k},\boldsymbol{k}_0)$ is not negligible only at $\boldsymbol{k}\approx\boldsymbol{k}_0$. It is clear that

$$\varphi^{\alpha\beta}(\boldsymbol{k},\boldsymbol{k};0)=\delta^{\alpha\beta}+k^{(\alpha)}k^{(\beta)}/q^2 .$$

Therefore, in this case (6.2.14) approximately transforms into

$$S(\boldsymbol{k},\boldsymbol{k}_0) \approx -\int \exp[-\mathrm{i}(\boldsymbol{k}-\boldsymbol{k}_0)\cdot\boldsymbol{r} + \mathrm{i}(q+q_0)h(\boldsymbol{r})]$$

$$\times \{1 + (\nabla h)^2 + [(\boldsymbol{k}\cdot\nabla)h/q]^2\} \frac{d\boldsymbol{r}}{(2\pi)^2} . \quad (6.2.16)$$

The condition of small corrections following from this formula also reduces to (6.2.15). This criterion implies a geometrical absence of shadowings with respect to the incident and scattered waves and should be fulfilled independently of boundary conditions. Finally, it is due to the structure of denominators in expansion (6.2.6) which is common to all boundary problems. However, in other scattering problems additional applicability conditions for small slope expansion may arise owing to the structure of the kernel $B(\boldsymbol{k},\boldsymbol{k}_0)$. In particular, for a rigid surface, B has singularity at $q_k \to 0$, see (4.3.9). Therefore, the second order correction in (6.1.28) can turn to infinity if one of the virtually generated spectra turns to be grazing. This example shows that criterion (6.2.15) must, generally speaking, be fulfilled with respect to all intermediate waves significantly generated due to interaction.

Let us average (6.2.16) for the Gaussian ensemble of roughness. Since $\overline{h\cdot\nabla h} = 0$, then elevations and slopes at the same point are statistically independent and we easily obtain

$$\overline{V}_k = -\exp(-2q^2\sigma^2)\cdot[1 + \overline{(\nabla h)^2} + \overline{(\boldsymbol{k}\cdot\nabla h/q)^2}] .$$

It is clear that this formula perfectly coincides with expression (5.4.13) obtained in Sect. 5.4 when analyzing the statistical problem by quite another technique.

Hence the small slope approximation was considered in detail from the theoretical point of view. The question naturally arises on experimental verification of the theory. Comparing scattering cross-sections calculated by theoretical formula and experimental data, we should take into account a lot of additional factors accompanying the experiment. For example, for scattering of undersea acoustic signals at the rough sea surface, the presence of air bubbles should, under certain conditions, be taken into account. Besides that, statistical properties and the roughness spectrum itself can be poorly known. For instance, direct measuring of the spatial spectrum of roughness for the sea surface is a complicated experimental problem. In any case, thorough control of external conditions is necessary in real experiments. However, even in laboratory measurements an all-side account of experimental conditions is sometimes difficult.

But if we rely on the applicability of the wave equation and on the form of boundary conditions for a specific problem then numerical simulation should be considered as most adequate for the investigation of diffractional effects in wave scattering at rough surfaces.

Appropriate calculations, especially for two-dimensional (plane) problems can be performed by modern computers. There are some approaches to the numerical solution of such scattering problems, but numerical algorithms are beyond the scope of the present book. We will refer to the results of appropriate computations as the "exact" solutions of the problem.

Now consider the case of periodic roughness with the period L. We restrict ourselves to the plane problem. In this case integration over transverse coordinate y in corresponding formulas will lead to the trivial δ-functions of the transverse component of the wave vector which will be omitted. Respectively, k, k_0, etc. denote the x-component of the corresponding wave vector.

For periodic roughness $h(x) = h(x + L)$ (4.1.7) takes the form

$$h(k) = \int_{-\infty}^{+\infty} \exp(-\mathrm{i}kx)h(x)\frac{dx}{(2\pi)}$$

$$= \int_{-L/2}^{L/2} \exp(-\mathrm{i}kx)h(x)\frac{dx}{(2\pi)} \cdot \sum_{n=-\infty}^{+\infty} \exp(\mathrm{i}\cdot kL\cdot n) \ . \qquad (6.2.17)$$

The last sum is the Fourier series for 2π periodic sequence of δ-functions:

$$\frac{1}{2\pi}\sum_{n=-\infty}^{+\infty} \exp(\mathrm{i}nx) = \sum_{n=-\infty}^{+\infty} \delta(x - 2\pi n) \ .$$

Whence we easily find

$$h(k) = \sum_{n=-\infty}^{+\infty} h_n\delta(k - pn) \ ,$$

where $p = 2\pi/L$ and

$$h_n = L^{-1}\int_{-L/2}^{L/2} \exp(-\mathrm{i}pnx)h(x)\,dx \ . \qquad (6.2.18)$$

Using the same approach as in (6.2.17), (6.1.24) reduces to

$$S(\boldsymbol{k},\boldsymbol{k}_0) = \sum_{N=-\infty}^{+\infty} S_N(\boldsymbol{k}_0)\cdot\delta(k - k_0 - pN) \ , \qquad (6.2.19)$$

where

$$S_N = (q_N q_0)^{1/2}\cdot(q_N + q_0)^{-1}\sum_{n=-\infty}^{+\infty}\left\{2\cdot B(k_N,k_0)\delta_{n0} - \frac{\mathrm{i}}{2}[B_2(k_N,k_0;k_{N-n}) + B_2(k_N,k_0;k_n) + 2(q_N + q_0)B(k_N,k_0)]h_n\right\}Q_{N,n} \qquad (6.2.20)$$

and the following notation is introduced:

$$k_n = k_0 + p\cdot n \ , \qquad q_n = (K^2 - k_n^2)^{1/2} \qquad (\mathrm{Re}\{q_n\},\mathrm{Im}\{q_n\} \geqslant 0)$$

and

$$Q_{N,n} = L^{-1}\int_{-L/2}^{L/2} \exp[-\mathrm{i}(N - n)px + \mathrm{i}(q_0 + q_N)h(x)]\,dx \ . \qquad (6.2.21)$$

For the Dirichlet problem (6.2.20) reduces to

$$S_N = -(q_N q_0)^{1/2} \cdot (q_N + q_0)^{-1} \times \sum_{n=-\infty}^{+\infty} [2\delta_{n0} + \mathrm{i}(q_{N-n} + q_n - q_N - q_0)h_n] Q_{N,n} \; . \tag{6.2.22}$$

Here the second term in the parenthesis is a correction to $2\delta_{n0}$ which has an order of $(\nabla h)^2$.

If the tangent plane approximation is applied, the formula for S_N becomes

$$S_N = -\frac{2(q_N q_0)^{1/2}}{(q_N + q_0)^{-1}} \cdot g_N \cdot Q_{N,0} \; , \tag{6.2.23}$$

where

$$g_N = 1 + \frac{(pN)^2 + (q_0 - q_N)^2}{4 q_N q_0} \; .$$

The first order small slopes approximation corresponding to the account of only term $2\delta_{n0}$ in the sum (6.2.22) and formula (6.2.23) differ only in factor g_N. It becomes unity at $N = 0$ and is essential for the high order spectra.

In the works devoted to calculation of scattering at periodic undulations the reflection coefficients $\mathscr{R}_N$ are usually introduced related to the field representation in the form

$$\Psi = \exp(\mathrm{i}k_0 x + \mathrm{i}q_0 z) + \sum_{N=-\infty}^{+\infty} \mathscr{R}_N \cdot \exp(\mathrm{i}k_n x - \mathrm{i}q_n z) \; .$$

It is obvious that

$$\mathscr{R}_N = (q_0/q_N)^{1/2} S_N \; .$$

Consider scattering at a periodic echelette grating surface:

$$h(x) = -a \cdot |x| \; .$$

This case is of special interest since at $a \ll 1$ roughness has small slopes but infinite values of the second derivative. Values h_n and $Q_{N,n}$ participating in (6.2.22) are easily determined in this case. The slope parameter was taken equal to $a = 0.2$ and the angle of incidence was 45°. The exact numerical solution of the problem was obtained using the algorithm described in [6.13].

The results are shown in Figs. 6.1, 2. Here deviations of the reflection coefficients modules $|\mathscr{R}_N|$ from their exact values are demonstrated for various spectra orders N. It is clear that the exact solution and the first approximation of this approach (dashed line) sufficiently agree. With account of the second order corrections (circles) agreement became nearly complete. The tangent plane approximation correctly describing reconstruction of the phase front of the wave

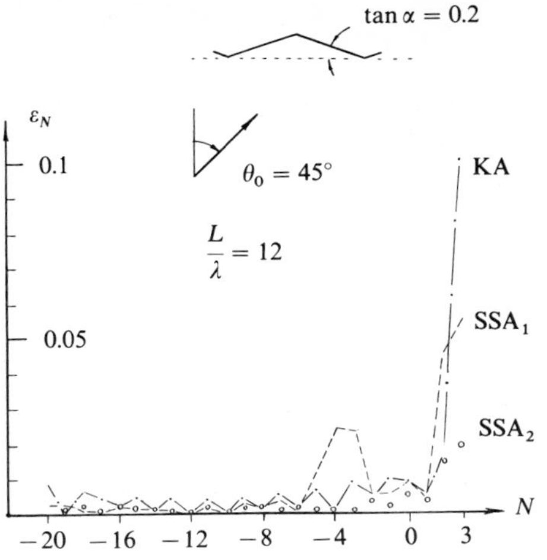

Fig. 6.1. Deviations of the reflection coefficient modules from exact values versus the number of homogeneous spectra for scattering at echelette grating. The angle of incidence $\theta_0 = 45°$, roughness slope $a = 0.2$, and the ratio of grating period to the wave length $L/\lambda = 12$

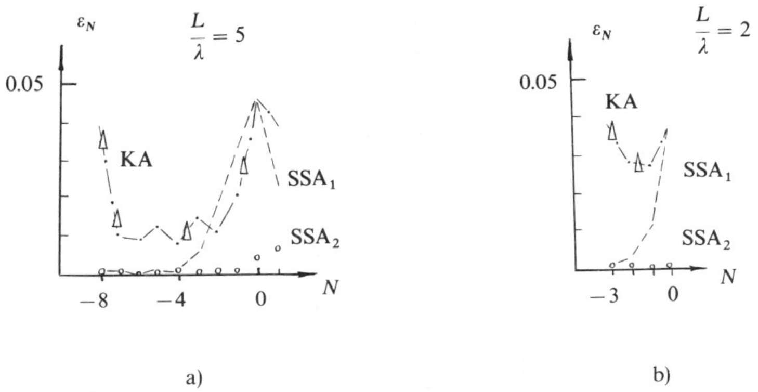

Fig. 6.2a,b. The same as in Fig. 5.1 but for other value of parameter L/λ

describes scattering well also; errors are about 10% for the grazing spectra only, but their amplitudes are rather small.

Thus the results of numerical computations confirm that the small slope approximation becomes applicable independently of the wavelength and for spectra of all orders.

Applicability of the small slope approximation was studied also in [6.14]. A periodic surface was constructed out of ten Fourier harmonics with amplitudes corresponding to the Pierson-Moscovitz spectrum of sea surface roughness (the shape of undulationsis is demonstrated in Fig. 6.3a). Reflection coefficients for all homogeneous spectra were calculated with the help of a special numerical algorithm. The accuracy of the approximate solution with respect to homogeneous spectra was estimated according to the formula

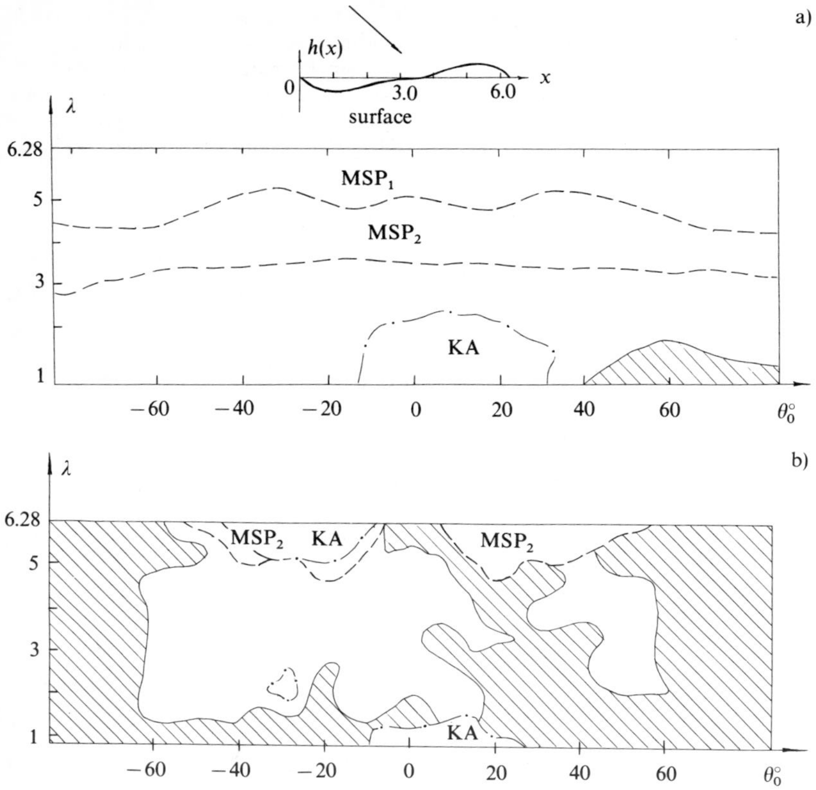

Fig. 6.3a,b. Comparison of two approximation method for 2π – periodic roughness with the shape shown above [max $h(x) = 0.4$]: (**a**) Dirichlet problem, (**b**) Neumann problem. Dashed lines separate areas of 10% accuracy for the first and second order small perturbation method; dot-dash line – the same for Kirchhoff approximation. The second order small slope approximation ensures 10% accuracy everywhere except the shaded area

$$\delta = \frac{\max\limits_n q_n \cdot \left| |\mathscr{R}_n^{(\mathrm{Ap})}| - |\mathscr{R}_n^{(\mathrm{Ex})}| \right|}{\max\limits_n q_n \cdot |\mathscr{R}_n^{(\mathrm{Ex})}|} .$$

In Fig. 6.3 areas in which a 10% accuracy of different approaches is achieved are indicated. The method of small perturbation of the first and second order, the second order Kirchoff, and the small slope approximations are considered. Figures 6.3a, b correspond to the Dirichlet and Neumann problems, respectively. In the latter case small slope approximation doesn't work so well as in the former. That is natural since for the Neumann problem excitation of virtual spectra propagating at small grazing angles can distort the solution, cf. (6.1.25a, b) in the case $q_{k'} \to 0$. *Berman* [6.15] compared the small slope approximation and the small perturbation method for the case of a boundary separating two media. The results obtained also confirmed good efficiency of the small slope approach for this case.

6.3 Phase Perturbation Technique

Phase perturbation technique was suggested in papers by *Winebrenner* and *Ishimaru* [6.3, 4], although its principal ansatz was encountered in the work by *Shen* and *Maradudin* [6.16]. Later this approach was developed further in several works (inter alia [6.17, 18]).

Consider the Dirichlet problem and search for the solution in the form

$$\Psi(\boldsymbol{R}) = \Psi_{\text{in}} + \Psi_{\text{sc}} = q_0^{-1/2} \exp(\mathrm{i}\boldsymbol{k}_0 \cdot \boldsymbol{r} + \mathrm{i}q_0 z) + \int \mu(\boldsymbol{r}') \cdot G_0(\boldsymbol{R}' - \boldsymbol{R})\, d\boldsymbol{r}' \; ;$$
$$\boldsymbol{R}' = (\boldsymbol{r}', h(\boldsymbol{r}')) \in \Sigma \; . \tag{6.3.1}$$

Substituting expression (2.1.24) for G_0 and comparing the result with (2.2.10) we obtain the following representation for SA:

$$S(\boldsymbol{k}, \boldsymbol{k}_0) = -\frac{\mathrm{i}}{2q^{1/2}} \int \mu(\boldsymbol{r}) \cdot \exp[-\mathrm{i}\boldsymbol{k} \cdot \boldsymbol{r} + \mathrm{i}qh(\boldsymbol{r})] \frac{d\boldsymbol{r}}{(2\pi)^2} \; . \tag{6.3.2}$$

The density μ depends on the incident wave parameters, certainly. For the trivial case of the horizontal plane $h = 0$ we have

$$S(\boldsymbol{k}, \boldsymbol{k}_0) = -\delta(\boldsymbol{k} - \boldsymbol{k}_0)$$

and correspondingly,

$$\mu(\boldsymbol{r}) = -2\mathrm{i}q_0^{1/2} \exp(\mathrm{i}\boldsymbol{k}_0 \cdot \boldsymbol{r}) \; .$$

In the general case we shall seek μ in the form

$$\mu(\boldsymbol{r}) = -2\mathrm{i}q_0^{1/2} \exp[\mathrm{i}\boldsymbol{k}_0 \cdot \boldsymbol{r} + \mathrm{i}\varphi(\boldsymbol{k}_0, \boldsymbol{r})] \; . \tag{6.3.3}$$

This ansatz is analogous to the representation for the field in the Rytov method [6.19]. Then (6.3.2) reduces to formula

$$S(\boldsymbol{k}, \boldsymbol{k}_0) = -(q_0/q)^{1/2} \int \exp[-\mathrm{i}(\boldsymbol{k} - \boldsymbol{k}_0) \cdot \boldsymbol{r} + \mathrm{i}qh(\boldsymbol{r}) + \mathrm{i}\varphi(\boldsymbol{k}_0, \boldsymbol{r})] \frac{d\boldsymbol{r}}{(2\pi)^2} \; . \tag{6.3.4}$$

The unknown phase function will be sought in the form of an expansion in powers of elevations

$$\varphi = \varphi_1 + \varphi_2 + \cdots \; . \tag{6.3.5}$$

Substitute (6.3.5) into (6.3.4) and expand the exponent into a series. We have

$$S(\boldsymbol{k},\boldsymbol{k}_0) = -(q_0/q)^{1/2}\int \exp[-\mathrm{i}(\boldsymbol{k}-\boldsymbol{k}_0)\cdot\boldsymbol{r}]\cdot\{1+\mathrm{i}qh(\boldsymbol{r})+\mathrm{i}\varphi_1(\boldsymbol{k}_0,\boldsymbol{r})$$

$$+\mathrm{i}\varphi_2(\boldsymbol{k}_0,\boldsymbol{r})-[q\cdot h(\boldsymbol{r})+\varphi_1]^2/2+\mathrm{O}(h^3)\}\frac{d\boldsymbol{r}}{(2\pi)^2}\ . \tag{6.3.6}$$

Comparing the first order terms in h in (6.3.6) and expansion (4.1.10) we find

$$\int \exp[-\mathrm{i}(\boldsymbol{k}-\boldsymbol{k}_0)\cdot\boldsymbol{r}]\varphi_1(\boldsymbol{k}_0,\boldsymbol{r})\frac{d\boldsymbol{r}}{(2\pi)^2} = q_k h(\boldsymbol{k}-\boldsymbol{k}_0)\ ,$$

where $h(\boldsymbol{k}-\boldsymbol{k}_0)$ is the Fourier transform of the roughness shape, see (4.1.7). Whence we obtain

$$\varphi_1(\boldsymbol{k}_0,\boldsymbol{r}) = \int q_{\boldsymbol{k}_0+\xi}\exp(\mathrm{i}\xi\cdot\boldsymbol{r})h(\xi)\,d\xi\ . \tag{6.3.7}$$

Suppose that roughness is smooth, so that $h(\xi)$ is concentrated in the small ξ. Assume that the incident wave is running along the x-axis: $(\boldsymbol{k}_0)_y = 0$. Then for small ξ it can be approximately written as

$$q_{\boldsymbol{k}_0+\xi} \approx q_0 - \frac{k_0\xi_x}{q_0} - \frac{K^2\xi_x^2}{2q_0^3} - \frac{\xi_y^2}{2q_0}\ .$$

Substituting this expansion into (6.3.7) we find

$$\varphi_1(\boldsymbol{k}_0,\boldsymbol{r}) = q_0 h(\boldsymbol{r}) + \frac{\mathrm{i}\boldsymbol{k}_0}{q_0}\cdot\frac{\partial h}{\partial x} + \frac{K^2}{2q_0^3}\frac{\partial^2 h}{\partial x^2} + \frac{1}{2q_0}\frac{\partial^2 h}{\partial y^2}\ .$$

Compare this formula and the result obtained in Sect. 5.3 when calculating the correction to the Kirchhoff approximation. Taking into account (5.3.7), we have

$$\varphi_1(\boldsymbol{k}_0,0) \approx (2K\varrho_1\cdot\sin^3\chi_0)^{-1} + (2K\varrho_2\sin\chi_0)^{-1}\ ,$$

where χ_0 is the grazing angle of the incident wave. It is easily seen that this formula coincides with (5.3.15), which was pointed out by *Rodriguez* [6.20].

Substituting (6.3.7) into (6.3.4) we obtain

$$S(\boldsymbol{k},\boldsymbol{k}_0) = -(q_0/q)^{1/2}\int \exp[-\mathrm{i}(\boldsymbol{k}-\boldsymbol{k}_0)\cdot\boldsymbol{r}+\mathrm{i}\int(q+q_{\boldsymbol{k}_0+\xi})\exp(\mathrm{i}\xi\cdot\boldsymbol{r})$$

$$\times\ h(\xi)\,d\xi]\frac{d\boldsymbol{r}}{(2\pi)^2}\ . \tag{6.3.8}$$

The last formula can also be represented in the form

$$S(\boldsymbol{k},\boldsymbol{k}_0) = -(q_0/q)^{1/2}\int \exp\Bigg[-\mathrm{i}(\boldsymbol{k}-\boldsymbol{k}_0)\cdot\boldsymbol{r} + \mathrm{i}(q+q_0)h(\boldsymbol{r})$$
$$+\,\mathrm{i}\int (q_{\boldsymbol{k}_0+\xi} - q_{k_0})\exp(\mathrm{i}\xi\cdot\boldsymbol{r})h(\xi)\,d\xi\Bigg]\frac{d\boldsymbol{r}}{(2\pi)^2}\ . \qquad (6.3.9)$$

Pay attention to the fact that (6.3.9) satisfies transformation properties (6.1.1, 2), see also (6.1.8). As is seen, the reciprocity theorem (2.3.6) is not satisfied, since

$$S(-\boldsymbol{k}_0,-\boldsymbol{k}) = -(q/q_0)^{1/2}\int \exp\Bigg[-\mathrm{i}(\boldsymbol{k}-\boldsymbol{k}_0)\cdot\boldsymbol{r} + \mathrm{i}(q+q_0)h(\boldsymbol{r})$$
$$+\,\mathrm{i}\int (q_{\boldsymbol{k}-\xi} - q_k)\exp(\mathrm{i}\xi\cdot\boldsymbol{r})h(\xi)\,d\xi\Bigg]\frac{d\boldsymbol{r}}{(2\pi)^2} \qquad (6.3.10)$$

(here we used eveness of function $q_{\boldsymbol{k}} = q_{-\boldsymbol{k}}$). Now suppose that roughness has small slopes

$$\nabla h = \int \mathrm{i}\xi h(\xi)\exp(\mathrm{i}\xi\cdot\boldsymbol{r})\,d\xi \ll 1\ .$$

In virtue of the obvious transformation we have

$$\int (q_{\boldsymbol{k}_0+\xi} - q_{k_0})\exp(\mathrm{i}\xi\cdot\boldsymbol{r})h(\xi)\,d\xi = -\int \frac{2\boldsymbol{k}_0+\xi}{(q_{\boldsymbol{k}_0+\xi}+q_0)}\cdot\xi h(\xi)\exp(\mathrm{i}\xi\cdot\boldsymbol{r})\,d\xi\ . \qquad (6.3.11)$$

Since the first factor in the integrand (6.3.11) is limited at any ξ (including $|\xi|\to\infty$), then the integral term in the exponent of (6.3.9) is of the roughness slope order. Hence it can be approximately set as

$$\exp[\mathrm{i}\int (q_{\boldsymbol{k}_0+\xi} - q_0)\exp(\mathrm{i}\xi\cdot\boldsymbol{r})h(\xi)\,d\xi]$$
$$\approx 1 + \mathrm{i}\int (q_{\boldsymbol{k}_0+\xi} - q_0)\exp(\mathrm{i}\xi\cdot\boldsymbol{r})h(\xi)\,d\xi\ . \qquad (6.3.12)$$

Let us consider the symmetric combination of SA given by (6.3.9, 10):

$$S^{(\mathrm{Sm})}(\boldsymbol{k},\boldsymbol{k}_0) = (q+q_0)^{-1}[q\cdot S(\boldsymbol{k},\boldsymbol{k}_0) + q_0 S(-\boldsymbol{k}_0,-\boldsymbol{k})]\ .$$

It is clear that $S^{(\mathrm{Sm})}$ satisfies the reciprocity theorem

$$S^{(\mathrm{Sm})}(\boldsymbol{k},\boldsymbol{k}_0) = S^{(\mathrm{Sm})}(-\boldsymbol{k}_0,-\boldsymbol{k})\ .$$

With account of (6.3.12) we find from (6.3.9, 10)

$$S^{(\mathrm{Sm})}(\boldsymbol{k},\boldsymbol{k}_0) = -\frac{(qq_0)^{1/2}}{(q+q_0)}\int \exp[-\mathrm{i}(\boldsymbol{k}-\boldsymbol{k}_0-\xi)\cdot\boldsymbol{r} + \mathrm{i}(q+q_0)h(\boldsymbol{r})]$$
$$\times\left[2\delta(\xi) + \mathrm{i}\int (q_{\boldsymbol{k}-\xi} + q_{\boldsymbol{k}_0+\xi} - q - q_0)h(\xi)\,d\xi\right]\frac{d\boldsymbol{r}\,d\xi}{(2\pi)^2}\ . \qquad (6.3.13)$$

This formula coincides with (6.1.29) which is the result of the second order small slope approximation for the Dirichlet problem.

The reference relation between phase perturbation techniques and small slope approximation can also be applied as follows. Write the formula for SA in the second order approximation of small slopes as

$$S(\boldsymbol{k},\boldsymbol{k}_0) = -\frac{2(qq_0)^{1/2}}{(q+q_0)}\int \exp[-\mathrm{i}(\boldsymbol{k}-\boldsymbol{k}_0)\boldsymbol{r} + \mathrm{i}(q+q_0)h(\boldsymbol{r})]$$
$$\times\left[1 + \frac{\mathrm{i}}{2}\int (q_{\boldsymbol{k}-\xi} + q_{\boldsymbol{k}_0+\xi} - q - q_0)h(\xi)\exp(\mathrm{i}\xi\cdot\boldsymbol{r})\,d\xi\right]\frac{d\boldsymbol{r}}{(2\pi)^2}\ . \tag{6.3.14}$$

It was emphasized in Sect. 6.1 that for roughness with small slopes the integral term in the square bracket is small compared to unity. Therefore, the corresponding expression in the integrand in (6.3.14) can be taken as the two first terms of exponential expansion. Based on this the following approximated representation for SA can be written:

$$S(\boldsymbol{k},\boldsymbol{k}_0) = -\frac{2(qq_0)^{1/2}}{q+q_0}\int \exp\left[-\mathrm{i}(\boldsymbol{k}-\boldsymbol{k}_0)\cdot\boldsymbol{r} + \frac{\mathrm{i}}{2}\int (q_{\boldsymbol{k}-\xi} + q_{\boldsymbol{k}_0+\xi}\right.$$
$$\left. + q + q_0)h(\xi)\exp(\mathrm{i}\xi\cdot\boldsymbol{r})\,d\xi\right]\frac{d\boldsymbol{r}}{(2\pi)^2}\ . \tag{6.3.15}$$

This formula for SA was suggested in [6.21]. Its structure is very close to that of (6.3.4), but in contrast to it, (6.3.15) satisfies the reciprocity theorem (2.3.6).

Now examine the correction φ_2. We compare the h^2 order terms in (6.3.6) and (4.1.10) and have

$$\int \exp[-\mathrm{i}(\boldsymbol{k}-\boldsymbol{k}_0)\cdot\boldsymbol{r}]\varphi_2(\boldsymbol{k}_0,\boldsymbol{r})\frac{d\boldsymbol{r}}{(2\pi)^2} = \int \exp[-\mathrm{i}(\boldsymbol{k}-\boldsymbol{k}_0)\cdot\boldsymbol{r}]$$
$$\times\,[\varphi_1(\boldsymbol{k}_0,\boldsymbol{r}) + qh(\boldsymbol{r})]^2\frac{d\boldsymbol{r}}{(2\pi)^2}$$
$$+\,2\mathrm{i}q\int q_{\boldsymbol{k}_1}\cdot h(\boldsymbol{k}-\boldsymbol{k}_1)h(\boldsymbol{k}_1-\boldsymbol{k}_0)\,d\boldsymbol{k}_1\ .$$

Following after simple transformations

$$\varphi_2(\boldsymbol{k}_0,\boldsymbol{r}) = \frac{\mathrm{i}}{2}\int (q_{\boldsymbol{k}_0+\xi_1+\xi_2}\cdot q_{\boldsymbol{k}_0+\xi_2} + q_{\boldsymbol{k}_0+\xi_1+\xi_2}\cdot q_{\boldsymbol{k}_0+\xi_1} - q^2_{\boldsymbol{k}_0+\xi_1+\xi_2}$$
$$- q_{\boldsymbol{k}_0+\xi_1}\cdot q_{\boldsymbol{k}_0+\xi_2})\exp[\mathrm{i}(\xi_1+\xi_2)\cdot\boldsymbol{r}]h(\xi_1)h(\xi_2)\,d\xi_1\,d\xi_2\ . \tag{6.3.16}$$

Now consider the statistical characteristics of SA assuming the ensemble of elevations to be Gaussian. To calculate $\overline{V}$ we should average the exponential in the integrand, according to (6.3.4). The result of this averaging is also represented in exponential form

$$\overline{\exp[\mathrm{i}q_k h(\boldsymbol{r}) + \mathrm{i}\varphi(\boldsymbol{k}_0, \boldsymbol{r})]} = \exp[-N(\boldsymbol{k}, \boldsymbol{k}_0)] \ . \tag{6.3.17}$$

Then value N is sought in the form of a series in σ^2:

$$N = N_2 + N_4 + \cdots , \tag{6.3.18}$$

where N_{2n} is proportional to σ^{2n}. Expanding into a series the exponent on the left-hand side of (6.3.17), see (6.3.6), we obtain

$$N_2 = -\mathrm{i}\overline{\varphi_2} + \tfrac{1}{2}\overline{\varphi_1^2} + q \cdot \overline{h \cdot \varphi_1} + \tfrac{1}{2}q^2\sigma^2 \ .$$

Using (6.3.7, 16) we easily find

$$N_2(\boldsymbol{k}, \boldsymbol{k}_0) = \tfrac{1}{2}(q_k^2 - q_0^2)\sigma^2 + (q_k q_0) \int q_{\boldsymbol{k}_0+\xi} W(\xi)\, d\xi \ . \tag{6.3.19}$$

Whence

$$\overline{V}_{\boldsymbol{k}} = -\exp[-N_2(\boldsymbol{k}, \boldsymbol{k})] = -\exp\left[-2q_k \int q_{\boldsymbol{k}+\xi} W(\xi)\, d\xi\right] \ . \tag{6.3.20}$$

It is true that this formula can be directly obtained from (4.1.22) if the mean reflection coefficient is written as

$$\overline{V}_k = -\exp(-N)$$

and N is sought in the form of the power expansion (6.3.18). Suppose now that irregularities are smooth, so that their spectrum is concentrated in small ξ. Then we can approximately set in (6.3.20) $q_{\boldsymbol{k}+\xi} \approx q_k$, and this formula takes the form

$$\overline{V}_k = -\exp(-2q^2\sigma^2) \ .$$

This result was obtained in the Kirchhoff approximation, see (5.2.25). Thus despite our permanent use of expansions in elevation powers, the result (6.3.20) turned out to be applicable for the large Rayleigh parameter as well. This is related to the fact that (6.3.8) gives an exact result for the horizontal plane, note the comments following (5.2.21).

A similar situation takes place concerning the scattering cross-section $\sigma(\boldsymbol{k}, \boldsymbol{k}_0)$. This value is directly calculated by (2.4.9, 12). It is obvious that in this case it is necessary to calculate the following correlator:

$$\begin{aligned} &\langle \{\exp[\mathrm{i}q_k h(\boldsymbol{r}_1) + \mathrm{i}\varphi(\boldsymbol{k}_0, \boldsymbol{r}_1)] - \exp[-N(\boldsymbol{k}, \boldsymbol{k}_0)]\} \\ &\quad \times \{\exp[-\mathrm{i}q_k^* h(\boldsymbol{r}_2) - \mathrm{i}\varphi^*(\boldsymbol{k}_0, \boldsymbol{r}_2)] - \exp[-N^*(\boldsymbol{k}, \boldsymbol{k}_0)]\}\rangle \\ &= \exp[-N(\boldsymbol{k}, \boldsymbol{k}_0) - N^*(\boldsymbol{k}, \boldsymbol{k}_0)] \cdot \{\exp[M(\boldsymbol{k}, \boldsymbol{k}_0; \boldsymbol{r}_1 - \boldsymbol{r}_2)] - 1\} \ . \end{aligned} \tag{6.3.21}$$

This equality determines, in fact, the value M which will be sought in the form of a series in powers σ^2:

$$M = M_2 = M_4 + \cdots \ .$$

Expanding both parts of (6.3.21) and averaging yields:

$$M_2(\boldsymbol{k}, \boldsymbol{k}_0; \boldsymbol{r}) = \int |q_k + q_{\boldsymbol{k}_0+\xi}|^2 \exp(\mathrm{i}\xi \cdot \boldsymbol{r}) W(\xi)\, d\xi \ . \tag{6.3.22}$$

As a result, the formula for the scattering cross-section takes the form

$$\sigma(\boldsymbol{k}, \boldsymbol{k}_0) = (q_0/q) \cdot \exp[-2\,\mathrm{Re}\{N_2(\boldsymbol{k}, \boldsymbol{k}_0)\}] \cdot \int \exp[-\mathrm{i}(\boldsymbol{k} - \boldsymbol{k}_0) \cdot \boldsymbol{r}] \times \{\exp[M_2(\boldsymbol{k}, \boldsymbol{k}_0; \boldsymbol{r})] - 1\}\, d\boldsymbol{r}/(2\pi)^2 \ . \tag{6.3.23}$$

Assume again that irregularities are smooth and set in (6.3.19, 22) $q_{\boldsymbol{k}_0+\xi} \approx q_{k_0}$. Then,

$$N_2(\boldsymbol{k}, \boldsymbol{k}_0) \approx \sigma^2 (q + q_0)^2/2 \ ,$$

$$M_2(\boldsymbol{k}, \boldsymbol{k}_0; \boldsymbol{r}) \approx (q + q_0)^2 \tilde{W}(\boldsymbol{r}) \ ,$$

and (6.3.23) transforms into

$$\sigma(\boldsymbol{k}, \boldsymbol{k}_0) = \frac{q_0}{q} \int \exp[-\mathrm{i}(\boldsymbol{k} - \boldsymbol{k}_0) \cdot \boldsymbol{r}] \cdot (\exp\{-(q + q_0)^2 [\sigma^2 - \tilde{W}(\boldsymbol{r})]\} - \exp[-(q + q_0)^2 \sigma^2]) \frac{d\boldsymbol{r}}{(2\pi)^2} \ . \tag{6.3.24}$$

This expression for the scattering cross-section with an accuracy to the pre-exponential factor coincides with (5.2.27) obtained in the Kirchhoff approximation.

Numerical experiments whose results were published in particular in [6.17, 18] show the high efficiency of the phase perturbation method. Here we present only a few results from [6.17]. Calculations were carried out for the two-dimensional roughness by (6.3.23). The correlation function of elevations was chosen as Gaussian:

$$\tilde{W}(x) = \sigma^2 \exp(-x^2/l^2) \ .$$

In Fig. 6.4a the bold line shows the dependence of the scattering strength, see (2.4.28), versus the scattering angle θ for the following parameters of the problem: the angle of incidence $\theta_0 = 45°$, $K\sigma = 0.67$ and $Kl = 2.80$. The root-mean-square angle of roughness γ: $\tan\gamma = 2^{1/2}\sigma/l$ was $\gamma = 18.6°$. In addition to dependences representing the field perturbation technique (i.e., the common method of small perturbations) and the Kirchhoff approximation, Figs. 6.4, 5 demonstrate the results obtained with the help of the exact solution of the Dirichlet

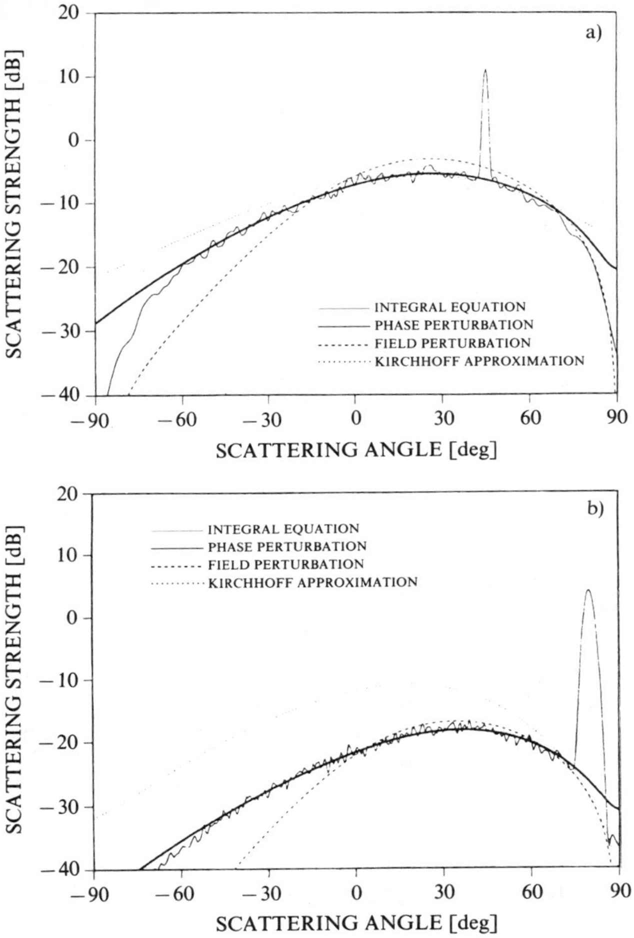

Fig. 6.4a,b. Comparison of phase perturbation technique with exact (numerical) solution and "classical" approximations: (**a**) $\theta_0 = 45°$ and (**b**) $\theta_0 = 80°$ (grazing incidence)

boundary problem. The boundary problem was solved numerically for each realization of rough surface which formed the statistical ensemble, and then the results were averaged. Since the coherent part of SA (2.4.6) was not excluded, in the specular direction the solution has a pike.

The plots in Fig. 6.4 show that the phase perturbation technique gives results which are very close to an exact solution. The dependence given in Fig. 6.4b coincides with the same parameters of the problem as in Fig. 6.4a, however, here the grazing incidence occurs: $\theta_0 = 80°$. Nevertheless, there is a good agreement with the exact solution for scattering directions far from grazing ones. Numerical experiments show that the range of scattering angles for which the

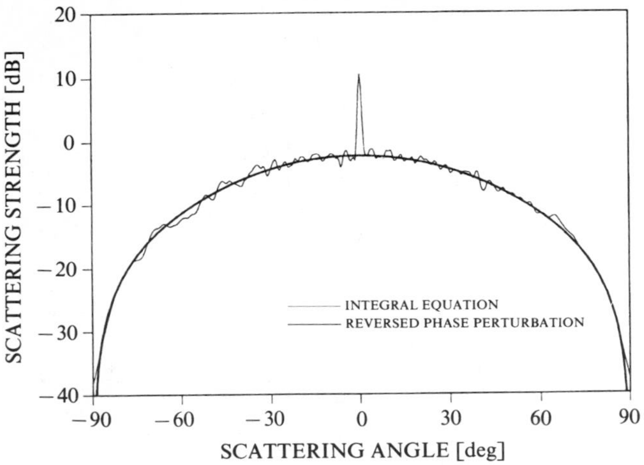

Fig. 6.5. Comparison of reversed phase perturbation technique with exact solution for normal incidence

phase perturbation technique is applicable is estimated as $|\theta| \lesssim 90^\circ - 2\gamma$. To calculate scattering in grazing direction one can also use the reciprocity theorem (2.4.16) and replace the directions of incident and scatterig waves. This approach is referred to as the *reversed* phase perturbation technique. Fig. 6.5 shows the dependence of the scattering strength calculated by such an approach for normal incidence: $\theta_0 = 0$. In this case agreement with the exact solution is very good for all ranges of scattering angles. The reversed phase perturbation technique is developed further in [6.18].

6.4 Phase Operator Method for Dirichlet Problem

The approach described in the present section reminds us of the Winebrenner-Ishimaru method presented in Sect. 6.3. Here the series expansion is also obtained, not for SA, but for some "phase", which is the operator in this case. This method was suggested in the work [6.5].

As was noted in Sect. 2.2 SA $S(\boldsymbol{k}, \boldsymbol{k}_0)$ can be considered as a matrix element of some operator $\hat{S}$ (caps denote matrix or operator values), the first argument $\boldsymbol{k}$ corresponds to the line's indices, and the second $\boldsymbol{k}_0$ corresponds to the column's indices. In this case relation (2.3.10)

$$\int\limits_{|\boldsymbol{k}'|<K} S(\boldsymbol{k}', \boldsymbol{k}_1) \cdot S^*(\boldsymbol{k}', \boldsymbol{k}_2)\, d\boldsymbol{k}' = \delta(\boldsymbol{k}_1 - \boldsymbol{k}_2) \; ; \qquad (|\boldsymbol{k}_1|, |\boldsymbol{k}_2| < K) \tag{6.4.1}$$

is the unitarity condition of the block of S-matrix which is selected by conditions $|\boldsymbol{k}|, |\boldsymbol{k}_0| < K$. This block describes processes of mutual scattering of homogeneous waves. We take $\hat{S}_h$ for this block and then have in matrix notation

$$\hat{S}_{\mathrm{h}}^{+} \cdot \hat{S}_{\mathrm{h}} = \hat{1} \ , \tag{6.4.2}$$

where operator $\hat{S}_{\mathrm{h}}^{+}$ is Hermitian conjugated to $\hat{S}_{\mathrm{h}}$, i.e.,

$$S_{\mathrm{h}}^{+}(\boldsymbol{k}_1, \boldsymbol{k}_2) = S_{\mathrm{h}}^{*}(\boldsymbol{k}_2, \boldsymbol{k}_1) \ .$$

We recall some principal facts concerning calculating the function of matrices. Let $f(x)$ be an analytic function

$$f(x) = f_0 + f_1 \cdot x + f_2 \cdot x^2 + \cdots \ .$$

Then the function of matrix $f(\hat{A})$ can be determined using the same series

$$f(\hat{A}) = f_0 + f_1 \cdot \hat{A} + f_2 \cdot \hat{A}^2 + \cdots \ . \tag{6.4.3}$$

All matrices we shall deal with (and generally most of matrices in physics) can be represented in the following form:

$$\hat{A} = \hat{T}\hat{\Lambda}\hat{T}^{-1} \ , \tag{6.4.4}$$

where $\hat{\Lambda} = \mathrm{diag}(\lambda_1, \ldots, \lambda_n)$ is a diagonal matrix consisting of eigenvalues of matrix $\hat{A}$, and $\hat{T}$ is a matrix whose columns are the correspondent eigenvectors. Transformation of an arbitrary matrix according to the rule

$$\hat{B}' = \hat{T}^{-1}\hat{B}\hat{T}$$

is called transition into a diagonal representation of matrix $\hat{A}$ (since it is apparent that $\hat{A}' = \hat{\Lambda}$). Substituting representation (6.4.4) into (6.4.3) we obtain

$$f(\hat{A}) = f_0 + f_1 \cdot \hat{T} \cdot \hat{\Lambda}\hat{T}^{-1} + f_2\hat{T} \cdot \hat{\Lambda}^2\hat{T}^{-1} + \cdots = \hat{T} \cdot f(\hat{\Lambda})\hat{T}^{-1} \ , \tag{6.4.5}$$

where the function of diagonal matrix $\hat{\Lambda}$ obviously is

$$f(\hat{\Lambda}) = \mathrm{diag}[f(\lambda_1), \ldots, f(\lambda_n)] \ .$$

It can be easily verified that when matrix $\hat{S}_{\mathrm{h}}$ is unitary, then all its eigenvectors are orthogonal. Therefore, matrix $\hat{T}$ for $\hat{S}_{\mathrm{h}}$ is also some unitary matrix $\hat{U}$ and (6.4.4) acquires the form

$$\hat{S}_{\mathrm{h}} = \hat{U}\hat{\Lambda}\hat{U}^{+} \ . \tag{6.4.6}$$

Substituting (6.4.6) into (6.4.2), we find that $|\lambda_i|^2 = 1$. Hence, we can write down

$$\hat{\Lambda} = \mathrm{diag}[\exp(\mathrm{i}\varphi_1), \ldots, \exp(\mathrm{i}\varphi_n)] = \exp(\mathrm{i}\hat{\varphi}) \ ,$$

where $\hat{\varphi} = \mathrm{diag}(\varphi_1, \ldots, \varphi_n)$ and all φ_i are real. Whence

$$\hat{S}_{\mathrm{h}} = \hat{U} \cdot \exp(\mathrm{i}\hat{\varphi})\hat{U}^{+} = \exp(\mathrm{i}\hat{U}\hat{\varphi}\hat{U}^{+}) = \exp(\mathrm{i}\hat{F}) \ , \tag{6.4.7}$$

where, in virtue of real φ_i matrix $\hat{F} = \hat{U}\hat{\varphi}\hat{U}^{+}$ is Hermitian: $\hat{F}^{+} = \hat{F}$. As a result, we obtain that unitary matrix $\hat{S}_{\mathrm{h}}$ can be represented as

$$\hat{S}_{\mathrm{h}} = -\exp(2\mathrm{i}\hat{H}) \ , \tag{6.4.8}$$

where $\hat{H} = \hat{H}^{+}$ is some Hermitian matrix. Minus in this formula is related to the Dirichlet problem (see below). In what follows we call matrix $\hat{H}$ a phase operator.

Proceeding in (6.4.8) to the matrix elements and expanding the exponential in a series, we obtain the following representation for SA:

$$-S(\boldsymbol{k},\boldsymbol{k}_0) = \delta(\boldsymbol{k}-\boldsymbol{k}_0) + 2\mathrm{i}H(\boldsymbol{k},\boldsymbol{k}_0) + \frac{(2\mathrm{i})^2}{2!}\int H(\boldsymbol{k},\boldsymbol{k}_1)H(\boldsymbol{k}_1,\boldsymbol{k}_0)\,d\boldsymbol{k}_1$$

$$+\frac{(2\mathrm{i})^3}{3!}\int\cdots d\boldsymbol{k}_1\,d\boldsymbol{k}_2 + \cdots \tag{6.4.9}$$

with integrating being over areas $|\boldsymbol{k}_n| < K$. So, the suggestion is, in fact, to seek SA in the form (6.4.9). If the conditions

$$H(\boldsymbol{k}_1,\boldsymbol{k}_2) = H^*(\boldsymbol{k}_2,\boldsymbol{k}_1) \ , \tag{6.4.10a}$$

$$H(\boldsymbol{k}_1,\boldsymbol{k}_2) = H(-\boldsymbol{k}_2,-\boldsymbol{k}_1) \tag{6.4.10b}$$

are fulfilled, then the reciprocity theorem (2.3.6) and the unitarity condition (6.4.1) will be fulfilled automatically.

The main idea, on which the phase operator approach is based, is to seek in the form of a power series in elevations h, not S, as it is usually done in the ordinary theory of perturbations, see Sect. 4.1, but matrix $\hat{H}$:

$$\hat{H} = \hat{H}_1 + \hat{H}_2 + \hat{H}_3 + \cdots \ . \tag{6.4.11}$$

Substituting (6.4.11) into (6.4.9) and comparing the result with expansion (4.1.10), we easily find

$$H_1(\boldsymbol{k},\boldsymbol{k}_0) = (qq_0)^{1/2}h(\boldsymbol{k}-\boldsymbol{k}_0) \ , \tag{6.4.12}$$

$$H_2(\boldsymbol{k},\boldsymbol{k}_0) = \mathrm{i}(qq_0)^{1/2}\int_{|\boldsymbol{k}_1|>K} q(k_1)h(\boldsymbol{k}-\boldsymbol{k}_1)h(\boldsymbol{k}_1-\boldsymbol{k}_0)\,d\boldsymbol{k}_1 \ . \tag{6.4.13}$$

It is clear that relations (6.4.10) are fulfilled and hermiticity of $\hat{H}_{1,2}$ results from the fact that q_k is real for $|\boldsymbol{k}| < K$, and if function $h(\boldsymbol{r})$ is real, then for the Fourier transform we have $h(\boldsymbol{k}) = h^*(-\boldsymbol{k})$:

$$H_1(\boldsymbol{k},\boldsymbol{k}_0) = (qq_0)^{1/2}h(\boldsymbol{k}-\boldsymbol{k}_0) = [(q_0q)^{1/2}h(\boldsymbol{k}_0-\boldsymbol{k})]^* = H_1^*(\boldsymbol{k}_0,\boldsymbol{k}) \ .$$

To obtain more accurate results one needs to take into account the subsequent terms in the expansion of the phase operator. As is seen from (6.4.13) $\hat{H}_2$ is related to the excitation of inhomogeneous waves. Formally, this term arises due to differences in areas of integration over $\boldsymbol{k}_1$ in (4.1.10) and (6.4.9): in the first case integrations are applied to all $\boldsymbol{k}_1$, and in the second case only to $|\boldsymbol{k}_1| < K$.

If, in the second case integration is applied to all $\boldsymbol{k}_1$, then we would have $\hat{H}_2 = 0$. Really,

$$\exp[2\mathrm{i}(\hat{H}_1 + \hat{H}_2)] = \hat{1} + 2\mathrm{i}\hat{H}_1 + 2\mathrm{i}\hat{H}_2 - 2\cdot\hat{H}_1^2 + \mathrm{O}(h^3) = -\hat{S}$$

and for matrix elements of the order h^2, we have

$$2\mathrm{i}H_2(\boldsymbol{k},\boldsymbol{k}_0) = 2\int H_1(\boldsymbol{k},\boldsymbol{k}')H_1(\boldsymbol{k}',\boldsymbol{k}_0)\,d\boldsymbol{k}' - S_2(\boldsymbol{k},\boldsymbol{k}_0) = 0$$

in virtue of (4.1.10) and (6.4.12). This comment prompts us to modify our approach somewhat.

That is, we get rid of the limitation to search in the form of (6.4.3) the only "homogeneous" block of the scattering matrix and will use exponential representation for the total S-matrix

$$\hat{S} = -\exp(2\mathrm{i}\hat{H}) \ . \tag{6.4.14}$$

Representation (6.4.14) implies that in (6.4.9) $\boldsymbol{k}$ and $\boldsymbol{k}_0$ take any values and, correspondingly, integrations are applied to all $\boldsymbol{k}_n$ and the phase operator is sought in the form of expansion (6.4.11), as previously. Here (6.4.12) remains valid for $\hat{H}_1$, and as was mentioned above, $\hat{H}_2 = 0$. However, in this version of the method the unitarity condition (6.4.1) is fulfilled only approximately. Thus, when introducing representation (6.4.14) for the S-matrix, the arguments based on the energy conservation are valid only for the case when inhomogeneous waves are excited weakly.

However (6.4.14) can be proved also from other considerations. Consider the case of the horizontal plane situated at the level $z = h_0 = \mathrm{const}$. We have

$$h(\boldsymbol{k} - \boldsymbol{k}_0) = h_0\delta(\boldsymbol{k} - \boldsymbol{k}_0)$$

and according to (6.4.12), formula (6.4.14) with $\hat{H} = \hat{H}_1$ gives the exact result. Indeed, in this case the operator is diagonal and calculating the operator's exponent yields

$$-\exp(2\mathrm{i}\hat{H}_1)(\boldsymbol{k},\boldsymbol{k}_0) = -\exp(2\mathrm{i}q_k h_0)\cdot\delta(\boldsymbol{k} - \boldsymbol{k}_0) \ .$$

This elementary example shows that the small parameter of expansion (6.4.11) is not Rayleigh's and should be related to the deviation of surface from the horizontal plane.

The case of slope plane $z = ax$, and $a = \tan\alpha$ gives another less trivial example. We assume a wave to be incident in the plane (x, z) and omit further the trivial dependences on transverse coordinates. We have

$$h(k) = \int \exp(-\mathrm{i}kx)\cdot ax\cdot dx/(2\pi) = \mathrm{i}a\delta'(k)$$

and formal application of the perturbation theory (4.1.6) to this situation yields, according to (2.2.14), increasing with coordinates "secular" terms

$$\Psi_{sc} = -2i \int (q_k q_{k_0})^{1/2} ia\delta'(k - k_0) q_k^{-1/2} \exp(ikx - iq_k z)\, dk + O(a^2)$$

$$= -2iaq_0^{1/2}(x + k_0 z/q_0) \exp(ik_0 x - iq_0 z) \ .$$

The result has no physical meaning and is caused by the Rayleigh parameter not being small here.

Let us clear up what the first approximation of the phase operator method gives in this case. By virtue of (6.4.12), we have

$$H_1(k, k_0) = ia(qq_0)^{1/2} \delta'(k - k_0) \ .$$

The operator exponent in (6.4.14) is calculated in diagonal representation of operator $\hat{H}_1$. It can be easily verified that eigenfunctions of operator $\hat{H}_1$ are

$$y_\mu(k) = \cos^{-1/2}\theta \cdot \exp(i\mu\theta) \ ,$$

where μ is a parameter and $\theta = \sin^{-1}(k/K)$ is the incidence angle. Indeed,

$$\hat{H}_1 y_\mu = iaq^{1/2} \int q_0^{1/2} \delta'(k - k_0) y_\mu(k_0)\, dk_0 \ ,$$

$$= iaq^{1/2} \left. \frac{d}{dk_0} \right|_{k_0 = k} [q_0^{1/2} y_\mu(k_0)] = -a\mu y_\mu \ . \qquad (6.4.15)$$

It can be easily shown that expansion of $\delta(k - k_0)$ in eigenfunctions y_μ is

$$\delta(k - k_0) = q_0^{-1} \delta(\theta - \theta_0) = \frac{1}{2\pi K} \int_{-\infty}^{+\infty} (\cos\theta_0)^{-1/2} \cdot \exp(-i\mu\theta_0) y_\mu(\theta)\, d\mu \ .$$

Then, according to (6.4.15)

$$\exp(2i\hat{H}_1) y_\mu(\theta_0) = \exp(-2ia\mu) y_\mu(\theta)$$

and

$$S(k, k_0) = -\frac{1}{2\pi K} \int (\cos\theta_0)^{-1/2} \exp(-i\mu\theta_0) \cdot \exp(-2ia\mu) \exp(i\mu\theta)\, d\mu$$

$$= (q_0 q)^{-1/2} \delta(\theta - \theta_0 - 2a) \ . \qquad (6.4.16)$$

Let the wave incident onto the slope plane have unit amplitude. Substituting (6.4.16) into (2.2.14) shows that the scattered field becomes the plane wave with amplitude equal to -1 which propagates at the angle $\theta = \theta_0 + 2a$ to the vertical:

$$\Psi_{sc} = q_0^{1/2} \int S(k, k_0) q_k^{-1/2} \exp(ikx - iq_k z)\, dk \ ,$$

$$= -\int \delta(\theta - \theta_0 - 2a) \cdot \exp(ikx - iq_k z)\, dk/q_k \ ,$$

$$= -\exp(iKx \sin\theta_r - iKz \cos\theta_r) \ ,$$

where $\theta_r = \theta_0 + 2a$. Since an accurate result is certainly $\theta_r = \theta_0 + 2\alpha$, then the error in propagation direction is of an order of α^3. Thus using the perturbation theory in the phase operator method allows us to examine not only situations with small Rayleigh parameter but also the case of large undulation as well, provided their slopes are small.

To find correction $\hat{H}_3$ in the phase operator compare the third order terms by h in equation

$$-\exp[2i(\hat{H}_1 + \hat{H}_3)] = \hat{S}_1 + \hat{S}_2 + \hat{S}_3 + \cdots$$

that yields

$$\hat{H}_3 = \frac{i}{2}\hat{S}_3 + \frac{2}{3}\hat{H}_1^3 .$$

Whence according to (4.1.9b) and (6.4.12) we obtain

$$H_3(k, k_0) = (qq_0)^{1/2} \int \left[-\frac{1}{3}q_{k_1}q_{k_2} + \frac{1}{4}(q_{k_1}^2 + q_{k_2}^2) - \frac{(q_k^2 + q_0^2)}{12} \right] \times h(k - k_1)h(k_1 - k_2)h(k_2 - k_0)\, dk_1\, dk_2 . \tag{6.4.17}$$

Now return to the problem of the slope plane and test that function $y_\mu(\theta)$ is an eigenfunction for operator $\hat{H}_3$, similar to (6.4.15). Indeed,

$$\hat{H}_3 y_\mu(\theta) = (qK)^{1/2}(ia)^3 \left.\frac{d}{dk_1}\right|_{k_1=k} \left.\frac{d}{dk_2}\right|_{k_2=k_1} \left.\frac{d}{dk_0}\right|_{k_0=k_2} \times \left[-\frac{q_{k_1}q_{k_2}}{3} + \frac{(q_{k_1}^2 + q_{k_2}^2)}{4} - \frac{(q_k^2 + q_0^2)}{12} \right] \exp(i\mu\theta_0) .$$

Direct calculation show that the last expression reduces to

$$\hat{H}_3 y_\mu(\theta) = \tfrac{1}{3}\cdot a^3 \mu y_\mu(\theta)$$

and $y_\mu(\theta)$ is, in fact, an eigenfunction of both operators $\hat{H}_1$ and $\hat{H}_3$. Respectively

$$(\hat{H}_1 + \hat{H}_3)y_\mu(\theta) = \left(-a\mu + \frac{a^3\mu}{3} \right) y_\mu = -\left(a - \frac{a^3}{3} \right) \cdot \mu \cdot y_\mu(\theta) .$$

Hence, after replacement $a \to a - a^3/3$ calculations for the slope plane with $\hat{H} = \hat{H}_1 + \hat{H}_3$ coincide to the case $\hat{H} = \hat{H}_1$. Now the reflected field is again a plane wave with amplitude -1 which, however, propagates at an angle to vertical equal to

$$\theta_r = \theta_0 + 2\tan\alpha - \tfrac{2}{3}\tan^3\alpha = \theta_0 + 2\alpha + O(\alpha^5) . \tag{6.4.18}$$

Thus, using the third approximation yields an error in the direction of the reflected wave of the order of α^5. It is natural to suppose that an account of the higher expansion terms in the phase operator will lead to the further specification of the reflection direction and (6.4.18) reduces generally to the series for $\tan^{-1}$

$$\theta_r = \theta_0 + 2\tan\alpha - \tfrac{2}{3}\tan^3\alpha + \tfrac{3}{40}\tan^5\alpha + \cdots ,$$
$$= \theta_0 + 2\tan^{-1}(\tan\alpha) = \theta_0 + 2\alpha . \qquad (6.4.19)$$

The example of the slope plane indicates that the necessary condition for applicability of the phase operator method is a small slope of irregularities.

The absence of terms with even powers in (6.4.19) suggests that operator $\hat{H}$ possesses the same property, and the series (6.4.11) consists, in fact, only of odd terms. This appears to be the case.

This fact can be formally established in the following way. Let us consider the Rayleigh equation (4.1.2):

$$q_0^{-1/2}\exp[\mathrm{i}\boldsymbol{k}_0\cdot\boldsymbol{r} + \mathrm{i}q_0 h(\boldsymbol{r})] + \int q_{k'}^{-1/2}\exp[\mathrm{i}\boldsymbol{k}'\cdot\boldsymbol{r} - \mathrm{i}q_{k'}h(\boldsymbol{r})]S(\boldsymbol{k}',\boldsymbol{k}_0;[h])\,d\boldsymbol{k}'$$
$$= 0 . \qquad (6.4.20)$$

The dependence of SA on the roughness shape h is clearly indicated as a functional argument in S. Introduce reverse matrix S^{-1} and take $\hat{S}^{-1}(\boldsymbol{k},\boldsymbol{k}_0;[h])$ for its matrix elements. Then we multiply (6.4.20) by $S^{-1}(\boldsymbol{k}_0,\boldsymbol{k};[h])$ and integrate over $\boldsymbol{k}_0$. We have

$$q_k^{-1/2}\exp(\mathrm{i}\boldsymbol{k}\cdot\boldsymbol{r} - \mathrm{i}q_k h) + \int q_0^{-1/2}\exp(\mathrm{i}\boldsymbol{k}_0\cdot\boldsymbol{r} + \mathrm{i}q_0\cdot h)S^{-1}(\boldsymbol{k}_0,\boldsymbol{k};[h])\,d\boldsymbol{k}_0 = 0 . \qquad (6.4.21)$$

After replacing $\boldsymbol{k}_0 \to \boldsymbol{k}'$ in (6.4.21) and setting $\boldsymbol{k} = \boldsymbol{k}_0$ (6.4.21) transfers into (6.4.20) with a changed sign of elevations: $h \to -h$. Therefore,

$$\hat{S}[h] = \hat{S}^{-1}[-h] .$$

If exponential representation is used for (6.4.14), then we get

$$\exp(2\mathrm{i}\hat{H}[h]) = \exp(-2\mathrm{i}\hat{H}[-h])$$

and consequently, all the even order terms in expansion (6.4.11) vanish.

Now we consider in more detail the first approximation of the phase operator method, that is that the S-matrix will be calculated by (6.4.14) with $\hat{H} = \hat{H}_1$. For SA we have representation

$$-S(\boldsymbol{k},\boldsymbol{k}_0) = \delta(\boldsymbol{k}-\boldsymbol{k}_0) + 2\mathrm{i}(qq_0)^{1/2}h(\boldsymbol{k}-\boldsymbol{k}_0)$$
$$+ \frac{(2\mathrm{i})^2}{2!}(qq_0)^{1/2}\int q_{k_1}h(\boldsymbol{k}-\boldsymbol{k}_1)h(\boldsymbol{k}_1-\boldsymbol{k}_0)\,d\boldsymbol{k}_1 + \cdots . \qquad (6.4.22)$$

Factors $q_{k_1}, q_{k_2}, \ldots$ in integrands hinder summing up this series. Let us apply the following representations:

$$q_{k_1} = \tfrac{1}{2}(q_{k_1} - q_{k_0}) + \tfrac{1}{2}(q_{k_1} - q_k) + \tfrac{1}{2}(q_0 + q)$$

in the first integral term;

$$q_{k_1} = \tfrac{1}{2}(q_{k_1} - q_k) - \tfrac{1}{2}(q_{k_2} - q_{k_1}) - \tfrac{1}{2}(q_0 - q_{k_2}) + \tfrac{1}{2}(q_0 + q) \ ,$$

$$q_{k_2} = \tfrac{1}{2}(q_{k_2} - q_{k_1}) + \tfrac{1}{2}(q_{k_1} - q_k) - \tfrac{1}{2}(q_0 - q_{k_2}) + \tfrac{1}{2}(q_0 + q) \ ,$$

in the second integral term; and so on. The values

$$(q_{k_i} - q_{k_{i+1}})h(\boldsymbol{k}_i - \boldsymbol{k}_{i+1}) = -\left(\frac{\boldsymbol{k}_i + \boldsymbol{k}_{i+1}}{q_{k_i} + q_{k_{i+1}}}\right)\cdot(\boldsymbol{k}_i - \boldsymbol{k}_{i+1})h(\boldsymbol{k}_i - \boldsymbol{k}_{i+1})$$

are proportional to the slope of undulations, since

$$\nabla h = \mathrm{i}\int \boldsymbol{k}h(\boldsymbol{k})\cdot\exp(\mathrm{i}\boldsymbol{k}\cdot\boldsymbol{r})\,d\boldsymbol{k} \ .$$

Thus, for a small slope of irregularities, one can approximately set in (6.4.22)

$$q_{k_i} \approx (q_0 + q_k)/2 \ . \tag{6.4.23}$$

Since in this approximation the kernels q_{k_i} in integrands are replaced by constant values, then series (6.4.22) is readily summed up using the convolution theorem (the irregularities spectra can be expressed by (4.1.7) and then integration in $\boldsymbol{k}_i$ by (2.1.20) can be directly made). As a result, we find

$$S(\boldsymbol{k}, \boldsymbol{k}_0) = -\frac{2(qq_0)^{1/2}}{(q + q_0)}\int \exp[-\mathrm{i}(\boldsymbol{k} - \boldsymbol{k}_0)\cdot\boldsymbol{r} + \mathrm{i}(q + q_0)h(\boldsymbol{r})]\frac{d\boldsymbol{r}}{(2\pi)^2} \ . \tag{6.4.24}$$

As is seen, expression (6.4.24) entirely coincides with (6.1.22) for $B = -1$. This is the way in which this formula has been obtained primarily in [6.5].

Now we treat the algorithm of calculating scattering for the case of two-dimensional periodical irregularities. We shall omit, as in Sect. 6.2, trivial dependences upon transverse coordinates in all formulas. Hence let irregularities be described by function $h(x)$ with period L and be represented in the form of the Fourier transformation

$$h(x) = \sum_{n=-\infty}^{+\infty} A_n\cdot\exp(\mathrm{i}pnx) \ ,$$

where $p = 2\pi/L$ and $A_n = A^*_{-n}$ in virtue of real $h(x)$.

It was seen in Sect. 2.3 that SA's are not equal to zero only for some discrete directions, see (4.1.15):

$$S(\boldsymbol{k}, \boldsymbol{k}_0) = \sum S_n\delta(k - k_0 - pn) \ .$$

First we examine the "unitary" variant of our approach related to representation (6.4.8). For $z \to -\infty$ with only homogeneous waves taken into account, the field of incident wave with unit amplitude is represented as

$$\Psi = \exp(\mathrm{i}k_0 x + \mathrm{i}q_0 z) + \sum_{-N_1}^{N_2} \mathscr{R}_n \exp(\mathrm{i}k_n x - \mathrm{i}q_n z) \ , \tag{6.4.25}$$

where $k_n = k_0 + pn$, $q_n = q(k_n)$, and $N_1 = [(K + k_0)/p]$ and $N_2 = [(K - k_0)/p]$ are the marginal numbers for the homogeneous diffraction spectra, and $\mathscr{R}_n$ are the reflection coefficients introduced in Sect. 6.2:

$$\mathscr{R}_n = (q_0/q_n)^{1/2} S_n \ . \tag{6.4.26}$$

Calculation of reflection coefficients using the first approximation of the phase operator $\hat{H} = \hat{H}_1$ will be as follows: Introduce the diagonal matrix

$$\hat{q}^{1/2} = \mathrm{diag}(q_{-N_1}^{1/2}, \dots, q_{N_2}^{1/2})$$

and Hermitian matrix

$$\hat{A} = \begin{bmatrix} A_0 & A_1 & & \dots & A_{N_1+N_2} \\ A_{-1} & A_0 & A_1 & \dots & \\ A_{-N_1-N_2} & \dots & & \dots & A_0 \end{bmatrix}$$

composed of the Fourier coefficients of elevations; lines and columns are numerated by indices varying from $-N_1$ to N_2. Then

$$\hat{S} = -\exp(2\mathrm{i}\hat{H}) \ ,$$

where

$$\hat{H} = \hat{q}^{1/2} \cdot \hat{A} \cdot \hat{q}^{1/2} \ .$$

SA S_n are given by the matrix $\hat{S}$ column $S_n = (\hat{S})_{n0}$ corresponding to the 0th index. Then the reflection coefficients are calculated by (6.4.26). In the general case it is reasonable not to use the series expansion (6.4.3) for calculating the matrix exponent, but to take advantage of the diagonal representation of the phase operator. Namely, the Hermitian matrix $\hat{H}$ should be represented in the form $\hat{H} = \hat{U}\hat{\Lambda}\hat{U}^+$, where $\hat{U}$ is the unitary matrix consisting of the eigenvectors of matrix $\hat{H}$, and $\hat{\Lambda} = \mathrm{diag}(\lambda_{-N_1}, \dots, \lambda_{N_2})$ is the diagonal matrix composed of the appropriate real eigenvalues. Then

$$\exp(2\mathrm{i}\hat{H}) = \hat{U}\,\mathrm{diag}[\exp(2\mathrm{i}\lambda_{-N_1}), \dots, \exp(2\mathrm{i}\lambda_{N_2})]\,\hat{U}^+ \ .$$

Only determining the eigenvalues and eigenvectors of matrix $\hat{H}$ is nontrivial computationally in the procedure described.

In the case of periodical surfaces, the transfer to such a variant of this approach which applies the exponential representation (6.4.14) for the total S-matrix reduces to widening matrix $\hat{H}$ through the marginal values of indices $-N_1$ and N_2 to infinity, saving the structure of the formulas for $\hat{q}^{1/2}$, $\hat{A}$, and $\hat{H}_1$, written previously. However, matrix $\hat{q}^{1/2}$ stops being real, and matrix $\hat{H}_1$ stops being Hermitian.

To test the phase operator method efficiency we made some numerical calculations for sinusoidal inhomogeneities $h = a \cos px$. The results of calculations of scattering for this case with the help of different methods are available in literature.

We start with the results of the numerical experiment for the "unitary" version of the approach related to representation (6.4.8) for $\hat{H} = \hat{H}_1$. For this case, only Fourier coefficients $A_{\pm 1} = a/2$ are not zero, and matrix $\hat{H}_1$ is two-diagonal. However, we failed to find its spectrum analytically, and all calculations we made numerically, according to the scheme described above. Figures 6.6, 7 show the parameter Ka dependence of the reflection coefficients for various values of slope ap and various diffraction spectra. In Fig. 6.6a–c $ap = 0.46$ and the spectra numbers are 0 and -1. The incident angle θ_0 is 40° for Fig. 6.6a, b, and $\theta_0 = 0°$ (normal incidence) for Fig. 6.6c. The crosses on the plots taken from [6.22] indicate our results compared to the data of other authors based on other theories. The curves represent calculations by the small pertur-

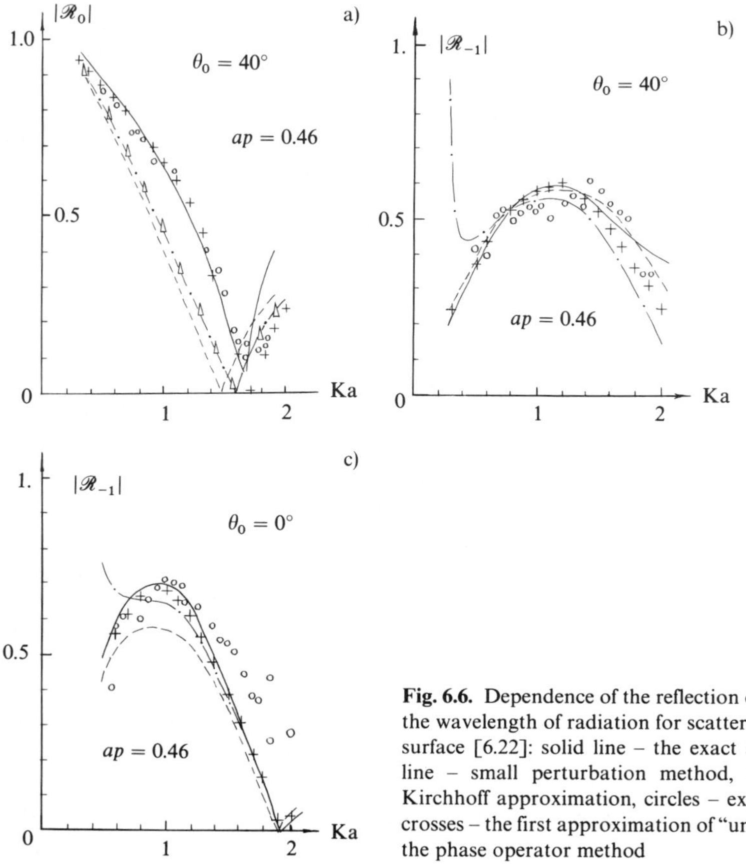

Fig. 6.6. Dependence of the reflection coefficients upon the wavelength of radiation for scattering at sinusoidal surface [6.22]: solid line – the exact solution, dashed line – small perturbation method, dot-dash line – Kirchhoff approximation, circles – experiment [6.24], crosses – the first approximation of "unitary" version of the phase operator method

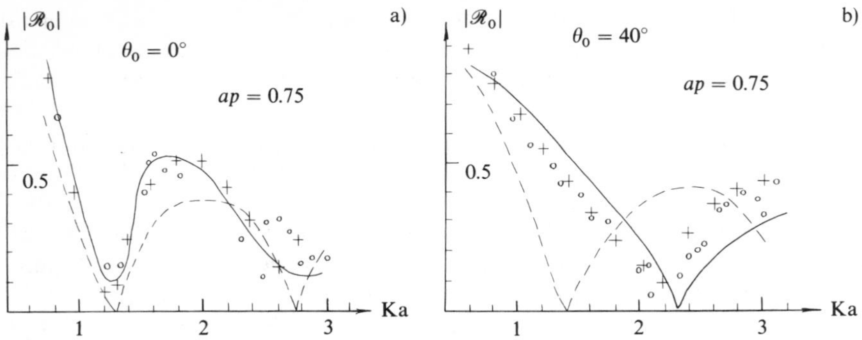

Fig. 6.7a,b. The same as for Fig. 6.6. Numerical results are taken from [6.23]

bation and the tangent plane approaches, as well as the exact numerical solution obtained by *Uretsky* [6.23]. The circles show the experimental results of *Lacasce* and *Tamarkin* [6.24]. It is clear that results obtained by our technique have the best correspondence with the exact solution. Figure 6.7a, b gives the case of a relatively large slope and the plots are taken from the work [6.23]. The coincidence with the exact solution and the experimental data is satisfactory. It is curious that in Fig. 6.7b our computation is closer to experimental data than to the exact solution and this is probably an accidental fact.

Now we pass to the second variant of our approach with the phase operator described by the infinite matrix. The exponent of this matrix was also determined numerically by (6.4.5) with matrix $\hat{H}$ cut at rather large indices when reduction weakly influences the answer. The results are given in Figs. 6.8, 9. Figure 6.8a, b shows the reflection coefficient for $n = -3, 7$ as a function of slope ap at the normal incidence $\theta_0 = 0$ and various values of parameter K/p. The solid line corresponds to the exact solution obtained using the numerical *Vainstein* and *Sukhov* algorithm [6.13]. The dashed line corresponds to the first approximation of this method and crosses show the results obtained according to the third approximation $\hat{H} = \hat{H}_1 + \hat{H}_3$. As is seen in the figures the results obtained coincide satisfactorily to the exact solution including the slopes less than unity. Figure 6.9 represents the case of grazing incidence ($\theta_0 = 80°$) when geometric shadowings occur. Apparently, the transfer from the first to the third approximation extends the applicability region for this case till $ap \lesssim 0.6$.

For the given slope and small values of parameter $K/p \to 0$ the phase operator method automatically transforms into the perturbation method (the operator exponent can be replaced by the first terms of the power series). The method works well also for intermediate values of K/p. Table 6.1 lists the reflection coefficients for various parameters of the problem which demonstrate the accuracy of the results obtained.

It is obvious that the phase operator method is readily generalized for the case of the Neumann problem (in this case minus in exponential representations

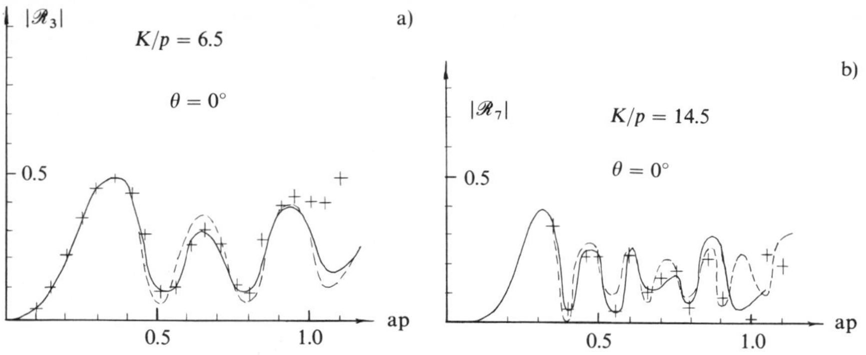

Fig. 6.8a,b. The reflection coefficients of the 3rd and 7th spectra versus the roughness slope (normal incidence). Solid line – exact solution, dashed line and crosses – the first and third approximations of the phase operator method

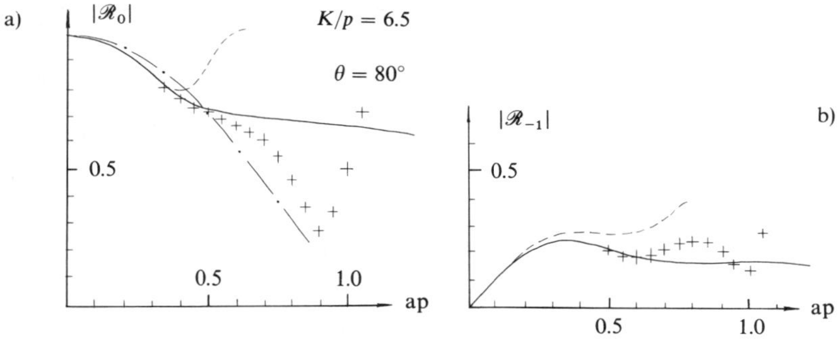

Fig. 6.9a,b. The reflection coefficients of the 0th and 1st spectra versus the roughness slope for the case of grazing incidence (notation as in Fig. 6.8); the dot-dash line in Fig. 6.9a corresponds to the Kirchhoff approximation

(6.4.8, 14) should be omitted). But the appropriate study has not been carried out until now.

The phase operator method is not purely analytical, since to obtain the numerical result the matrix exponent (6.4.12 or 6.4.14) should be calculated. However, for this operation a lot of efficient algorithms were developed which could be easily applied numerically.

6.5 Meecham's–Lysanov's Approach

The present approach was independently suggested in the works by *Meecham* [6.6] and *Lysanov* [6.7] for analyzing the Dirichlet problem.

Table 6.1. Numerical comparison of different approaches for $h(x) = a\cos(px)$; $\theta_0 = 0$ and $ap = 0.5$

		Exact solution	Phase operator method	Kirchoff approximation	Method of small perturbations
$K/p = 1.1$	R_0	0.885	0.886	0.719	1.
	R_1	0.508	0.507	0.959	0.550
$K/p = 1.2$	R_0	0.821	0.821	0.671	1.
	R_1	0.542	0.542	0.742	0.600
$K/p = 1.3$	R_0	0.758	0.758	0.620	1.
	R_1	0.576	0.576	0.792	0.650
$K/p = 1.4$	R_0	0.694	0.694	0.567	1.
	R_1	0.608	0.608	0.697	0.700

It was shown in Sect. 3.1 that for the Dirichlet problem the scattered field can be sought in the form

$$\Psi_{sc}(\boldsymbol{R}_*) = -\int \frac{\partial \Psi}{\partial \boldsymbol{n}} G_0(\boldsymbol{R} - \boldsymbol{R}_*)\, d\Sigma \tag{6.5.1}$$

see (3.1.32) and the unknown monopole density at the boundary equal to $\partial\psi/\partial\boldsymbol{n}$ satisfies equation

$$\int \frac{\partial \Psi}{\partial \boldsymbol{n}_{\boldsymbol{R}'}} G_0(\boldsymbol{R} - \boldsymbol{R}')\, d\Sigma_{\boldsymbol{R}'} = \Psi_{in}(\boldsymbol{R}) \; ; \qquad \boldsymbol{R}, \boldsymbol{R}' \in \Sigma \; . \tag{6.5.2}$$

We assume the incident field to be a plane wave of the kind (2.2.7), and pass in (6.5.2) to integration over the horizontal plane $z = 0$ and set

$$\mu(\boldsymbol{r}) = \frac{\partial \Psi}{\partial \boldsymbol{n}_{\boldsymbol{R}}} [1 + (\nabla h)^2]^{1/2} \; .$$

Equation (6.5.2) takes the following form:

$$\int \mu(\boldsymbol{r}') G_0(\boldsymbol{R} - \boldsymbol{R}')\, d\boldsymbol{r}' = q_0^{-1/2} \exp[\mathrm{i}\boldsymbol{k}_0 \cdot \boldsymbol{r} + \mathrm{i} q_0 h(\boldsymbol{r})] \; . \tag{6.5.3}$$

The Green's function G_0 is given by (2.1.21)

$$G_0(\boldsymbol{R} - \boldsymbol{R}') = -(4\pi|\boldsymbol{R} - \boldsymbol{R}'|)^{-1} \cdot \exp(\mathrm{i}K|\boldsymbol{R} - \boldsymbol{R}'|)$$

and in this case

$$|\boldsymbol{R} - \boldsymbol{R}'| = \{|\boldsymbol{r} - \boldsymbol{r}'|^2 + [h(\boldsymbol{r}) - h(\boldsymbol{r}')]^2\}^{1/2} \; . \tag{6.5.4}$$

For small roughness slopes one can set in (6.5.4)

$$|\boldsymbol{R} - \boldsymbol{R}'| \approx |\boldsymbol{r} - \boldsymbol{r}'| \ . \tag{6.5.5}$$

Approximation (6.5.5) is valid under condition:

$$K \cdot \frac{(h(\boldsymbol{r}) - h(\boldsymbol{r}'))^2}{|\boldsymbol{r} - \boldsymbol{r}'|} = K \cdot [h(\boldsymbol{r}) - h(\boldsymbol{r}')] \cdot \frac{h(\boldsymbol{r}) - h(\boldsymbol{r}')}{|\boldsymbol{r} - \boldsymbol{r}'|} \ll 1 \ . \tag{6.5.6}$$

The left-hand side of (6.5.6) can be estimated as a product of the Rayleigh parameter and the characteristic slope. Hence, in this case the Rayleigh parameter can be large (at sufficiently small roughness slopes).

Equation (6.5.3) with account of (6.5.5) takes the form of convolution

$$\int \mu(\boldsymbol{r}') G_0(\boldsymbol{r} - \boldsymbol{r}') \, d\boldsymbol{r}' = q_0^{-1/2} \exp[\mathrm{i}\boldsymbol{k}_0 \cdot \boldsymbol{r} + \mathrm{i}q_0 h(\boldsymbol{r})] \ . \tag{6.5.7}$$

This equation is easily solved with the Fourier transform. Substituting (2.1.24) into (6.5.7) (at $z = 0$) we have

$$-\frac{\mathrm{i}}{8\pi^2} \int \mu(\boldsymbol{r}') \exp(\mathrm{i}\boldsymbol{k} \cdot \boldsymbol{r} - \mathrm{i}\boldsymbol{k} \cdot \boldsymbol{r}') q_k^{-1} \, d\boldsymbol{k} \, d\boldsymbol{r}' = q_0^{-1/2} \exp[\mathrm{i}\boldsymbol{k}_0 \cdot \boldsymbol{r} + \mathrm{i}q_0 h(\boldsymbol{r})] \ .$$

Multiplying the latter by $\exp(-\mathrm{i}\boldsymbol{k}' \cdot \boldsymbol{r})$ and integrating over $\boldsymbol{r}$ using (2.1.20) yields

$$\int \mu(\boldsymbol{r}') \exp(-\mathrm{i}\boldsymbol{k}' \cdot \boldsymbol{r}') \, d\boldsymbol{r}' = 2\mathrm{i} q_0^{-1/2} q_{k'} \int \exp[\mathrm{i}\boldsymbol{k}_0 \cdot \boldsymbol{r} - \mathrm{i}\boldsymbol{k}' \cdot \boldsymbol{r} + \mathrm{i}q_0 h(\boldsymbol{r})] \, d\boldsymbol{r} \ .$$

After one more Fourier transform we find

$$\mu(\boldsymbol{r}) = 2\mathrm{i} q_0^{-1/2} \int q_{k'} \exp[-\mathrm{i}\boldsymbol{k}' \cdot (\boldsymbol{r}' - \boldsymbol{r}) + \mathrm{i}\boldsymbol{k}_0 \cdot \boldsymbol{r}' + \mathrm{i}q_0 h(\boldsymbol{r}')] \frac{d\boldsymbol{r}' \, d\boldsymbol{k}'}{(2\pi)^2} \ . \tag{6.5.8}$$

Substitution of (6.5.8) into (6.5.1) and using (2.1.24) at $z < \min h(\boldsymbol{r})$ gives

$$\Psi_{\mathrm{sc}}(\boldsymbol{R}_*) = \int S(\boldsymbol{k}, \boldsymbol{k}_0) q_k^{-1/2} \exp(\mathrm{i}\boldsymbol{k} \cdot \boldsymbol{r}_* - \mathrm{i}q_k z_*) \, d\boldsymbol{k} \ ,$$

where the formula for SA is

$$\begin{aligned} S(\boldsymbol{k}, \boldsymbol{k}_0) = -\frac{1}{(q_k q_0)^{1/2}} \int q_{k'} \exp[&-\mathrm{i}(\boldsymbol{k} - \boldsymbol{k}') \cdot \boldsymbol{r}_1 + \mathrm{i}q_k h(\boldsymbol{r}_1) \\ &- \mathrm{i}(\boldsymbol{k}' - \boldsymbol{k}_0) \cdot \boldsymbol{r}_2 + \mathrm{i}q_0 h(\boldsymbol{r}_2)] \frac{d\boldsymbol{r}_1 \, d\boldsymbol{r}_2 \, d\boldsymbol{k}'}{(2\pi)^4} \ . \end{aligned} \tag{6.5.9}$$

It can be easily shown that SA (6.5.9) satisfies both the reciprocity theorem (2.3.6) and the transformation properties (6.1.1, 2).

When passing from (6.5.4) to (6.5.5) we neglect the terms higher then the second order in h. Therefore, in the first order in small elevations when

$$\exp[iq_k h(\boldsymbol{r}_1) + iq_0 h(\boldsymbol{r}_2)] \approx 1 + iq_k h(\boldsymbol{r}_1) + iq_0 h(\boldsymbol{r}_2)$$

formula (6.5.9) transforms into (4.1.5, 6).

Now consider the high frequency limit for smooth roughness. Suppose that in integrating over $\boldsymbol{k}'$ in (6.5.9) one can restrict to the narrow region $\boldsymbol{k}' \approx \boldsymbol{k}_0$, then

$$q_{k'} \approx q_0 - \boldsymbol{k}_0 \cdot (\boldsymbol{k}' - \boldsymbol{k}_0)/q_0 \ . \tag{6.5.10}$$

Set

$$\begin{aligned}(\boldsymbol{k}' - \boldsymbol{k}_0)\exp[-i(\boldsymbol{k}' - \boldsymbol{k}_0)\cdot \boldsymbol{r}_2 + iq_0 h(\boldsymbol{r}_2)] \\ = i\nabla_{\boldsymbol{r}_2}\{\exp[-i(\boldsymbol{k}' - \boldsymbol{k}_0)\cdot \boldsymbol{r}_2 + iq_0 h(\boldsymbol{r}_2)]\} \\ + q_0 \nabla h(\boldsymbol{r}_2)\cdot \exp[-i(\boldsymbol{k}' - \boldsymbol{k}_0)\cdot \boldsymbol{r}_2 + iq_0 h(\boldsymbol{r}_2)] \ .\end{aligned}$$

In integrating over $\boldsymbol{r}_2$ in (6.5.9) the first term in this equality vanishes. In the rest of the terms it can be integrated over $\boldsymbol{k}'$ and then over $\boldsymbol{r}_1$. As a result,

$$\begin{aligned}S(\boldsymbol{k}, \boldsymbol{k}_0) = -(qq_0)^{-1/2} \int (q_0 - \boldsymbol{k}_0 \cdot \nabla h) \\ \times \exp[-i(\boldsymbol{k} - \boldsymbol{k}_0)\cdot \boldsymbol{r} + i(q + q_0)h(\boldsymbol{r})] \frac{d\boldsymbol{r}}{(2\pi)^2} \ .\end{aligned} \tag{6.5.11}$$

The term related to ∇h can be transformed as in Sect. 5.1, which leads to replacement (5.1.10). Hence, (6.5.11) transforms into (5.1.11) – the result obtained in the tangent plane approximation. One can verify that when using expansion $q_{k'}$ in the vicinity of the point $\boldsymbol{k}' = \boldsymbol{k}$ instead of $\boldsymbol{k}' = \boldsymbol{k}_0$ in (6.5.10) the result does not change.

Practical application of (6.5.9) is complicated owing to the rather high order of integration.

6.6 Relation Between Bahar's Full-Wave Approach and the Small Slope Approximation

The full-wave approach due to *Bahar* [6.8–10] has been elaborated for many problems. Here we shall not state this method completely but just compare it to the small slope approximation considering the two-scale acoustically soft surface $h = h(x)$ and assuming the slopes to be small:

$$\frac{dh}{dx} \ll 1 \ . \tag{6.6.1}$$

The two-dimensional Helmholtz equation can be represented as a set of the first order equations

$$\begin{cases} \dfrac{\partial \Psi}{\partial x} = \mathrm{i}\varphi \ , & (6.6.2a) \\[2ex] \dfrac{\partial \varphi}{\partial x} = \mathrm{i}\left(K^2\Psi + \dfrac{\partial^2 \Psi}{\partial z^2}\right) . & (6.6.2b) \end{cases}$$

Introduce functions dependent on some positive parameter u:

$$F_u(x, z) = \sin[u(z - h(x))] \ . \tag{6.6.3}$$

$F_u(x, z)$ can be considered at each fixed x as local modes of waveguide with one boundary situated at infinity. Then the fields Ψ and φ can be represented for given x as a superposition of functions of the kind (6.6.3):

$$\Psi(x, z) = \frac{2}{\pi} \int_0^\infty \tilde{\Psi}(x, u) \cdot (K^2 - u^2)^{-1/4} \exp(-\mathrm{i}uh) \cdot \sin[u(z - h(x))] \, du \ , \tag{6.6.4a}$$

$$\varphi(x, z) = \frac{2}{\pi} \int_0^\infty \tilde{\varphi}(x, u) \cdot (K^2 - u^2)^{1/4} \exp(-\mathrm{i}uh) \sin[u(z - h(x))] \, du \ . \tag{6.6.4b}$$

For convenience, the normalizing factors are introduced into (6.6.4). Inverse transformations are also sin-transforms:

$$\begin{aligned} \tilde{\Psi}(x, u) &= (K^2 - u^2)^{1/4} \exp(\mathrm{i}uh) \int_{-\infty}^{h(x)} \Psi(x, z) \sin[u(z - h(x))] \, dz \ , \\ \tilde{\varphi}(x, u) &= (K^2 - u^2)^{-1/4} \exp(\mathrm{i}uh) \int_{-\infty}^{h(x)} \varphi(x, z) \sin[u(z - h(x))] \, dz \ . \end{aligned} \tag{6.6.5}$$

Multiply (6.6.2a) by $F_u(x, z)$ and integrate the result over z. The obvious identity takes place

$$\begin{aligned} \int_{-\infty}^{h(x)} \frac{\partial \Psi}{\partial x} \sin[u(z - h(x))] \, dz &= \frac{\partial}{\partial x} \int_{-\infty}^{h(x)} \Psi(x, z) \sin[u(z - h(x))] \, dz \\ &\quad + u \frac{dh}{dx} \int_{-\infty}^{h(x)} \Psi(x, z) \cdot \cos[u(z - h(x))] \, dz \ . \end{aligned} \tag{6.6.6}$$

The first term in the right-hand side of this equality is proportional to $\partial \tilde{\Psi}(x, u)/\partial x$. Substituting into the second term in (6.6.6) instead of $\Psi(x, z)$ its representation (6.6.4a) we obtain a divergent integral over z being regularized with the help of factor $\exp \varepsilon(z - h)$, $\varepsilon \to +0$:

$$I(u,u') = \lim_{\varepsilon\to+0} \int_{-\infty}^{h} \cos[u\cdot(z-h)]\cdot\sin[u'\cdot(z-h)]\exp[\varepsilon(z-h)]\,dz$$
$$= \lim_{\varepsilon\to+0} 1/2[(u-u'+\mathrm{i}\varepsilon)^{-1} + (-u-u'+\mathrm{i}\varepsilon)^{-1} - (u'-u+\mathrm{i}\varepsilon)^{-1}$$
$$-(u+u'+\mathrm{i}\varepsilon)^{-1}] \ . \tag{6.6.7}$$

Using the known formula

$$\lim_{\varepsilon\to+0} (x\pm\mathrm{i}\varepsilon)^{-1} = P\left(\frac{1}{x}\right) \mp \mathrm{i}\pi\delta(x) \ ,$$

where P denotes principal value integral, we find

$$I(u,u') = P\cdot\frac{2u'}{(u^2-u'^2)} \ . \tag{6.6.8}$$

Taking into account (6.6.6 and 8), we obtain as a result, the equation

$$\frac{d}{dx}\tilde{\Psi}(x,u) - \mathrm{i}(K^2-u^2)^{1/2}\tilde{\varphi}(x,u) = \mathrm{i}u\frac{dh}{dx}\cdot\tilde{\Psi}(x,u)$$
$$+\frac{dh}{dx}\cdot\frac{1}{\pi}\int_0^\infty \exp[\mathrm{i}(u-u')h]\cdot\frac{2uu'}{(u'^2-u^2)}$$
$$\times\left(\frac{K^2-u^2}{K^2-u'^2}\right)^{1/4}\tilde{\Psi}(x,u')\,du' \tag{6.6.9a}$$

Equation (6.6.2b) is dealt with in quite a similar manner, and the term with the second derivative with respect to z is transformed by integration by parts:

$$\int_{-\infty}^{h}\frac{\partial^2\Psi}{\partial z^2}\cdot\sin[u(z-h)]\,dz = -\int_{-\infty}^{h}\frac{\partial\Psi}{\partial z}\cdot u\cdot\cos[u(z-h)]\,dz \ ,$$
$$= -u^2\int_{-\infty}^{h}\Psi(x,z)\sin[u\cdot(z-h)]\,dz \ .$$

We take into account also that the field Ψ satisfies the boundary condition

$$\Psi|_{z=h(x)} = 0 \ .$$

As a result (6.6.2b) takes the form

$$\frac{d}{dx}\tilde{\varphi}(x,u) - \mathrm{i}(K^2-u^2)^{1/2}\tilde{\Psi}(x,u) = \mathrm{i}u\frac{dh}{dx}\tilde{\varphi}(x,u) + \frac{dh}{dx}\cdot\frac{1}{\pi}\int_0^\infty \exp[\mathrm{i}(u-u')h]$$
$$\times\frac{2uu'}{(u'^2-u^2)}\cdot\left(\frac{K^2-u'^2}{K^2-u^2}\right)^{1/4}\tilde{\varphi}(x,u')\,du' \ . \tag{6.6.9b}$$

It is more convenient to introduce new variables instead of $\tilde{\Psi}$ and $\tilde{\varphi}$:

$$
\begin{aligned}
a(x,u) &= \tilde{\Psi}(x,u) + \tilde{\varphi}(x,u) \ , \\
b(x,u) &= \tilde{\Psi}(x,u) - \tilde{\varphi}(x,u) \ .
\end{aligned}
\tag{6.6.10}
$$

Equation (6.6.9) in terms of a, b takes the form of the so-called generalized telegraphists' equations

$$
\begin{aligned}
\frac{da}{dx} - \mathrm{i}(K^2 - u^2)^{1/2} a &= \mathrm{i}u\frac{dh}{dx}\cdot a + \frac{dh}{dx}\cdot\frac{1}{\pi}\int_0^\infty \exp[\mathrm{i}(u-u')h] \\
&\quad \times \frac{uu'}{(u'^2 - u^2)}[T_1(u,u')a(x,u') + T_2(u,u')b(x,u')]\,du' \ , \\
\frac{db}{dx} + \mathrm{i}(K^2 - u^2)^{1/2} b &= \mathrm{i}u\frac{dh}{dx}\cdot b + \frac{dh}{dx}\cdot\frac{1}{\pi}\int_0^\infty \exp[\mathrm{i}(u-u')h] \\
&\quad \times \frac{uu'}{(u'^2 - u^2)}[T_2(u,u')a(x,u') + T_1(u,u')b(x,u')]\,du' \ ,
\end{aligned}
\tag{6.6.11}
$$

where

$$
T_1(u,u') = \left(\frac{K^2 - u^2}{K^2 - u'^2}\right)^{1/4} + \left(\frac{K^2 - u'^2}{K^2 - u^2}\right)^{1/4} \ ,
$$

$$
T_2(u,u') = \left(\frac{K^2 - u^2}{K^2 - u'^2}\right)^{1/4} - \left(\frac{K^2 - u'^2}{K^2 - u^2}\right)^{1/4} \ .
$$

Equations (6.6.11) at $h(x) = \text{const}$ become independent and their solutions describe the waves propagating in positive and negative x directions, respectively. For small slopes of roughness (6.6.11) can be solved by the method of subsequent approximations over the slope (6.6.1).

The first approximation is

$$
\begin{cases}
a_0(x,u) = 2\pi\mathrm{i}(K^2 - u_0^2)^{1/4} u_0^{-1/2} \exp[\mathrm{i}(K^2 - u_0^2)^{1/2} x + 2\mathrm{i}u_0 h(x)]\delta(u - u_0) \ , \\
b_0(x,u) = 0 \ .
\end{cases}
\tag{6.6.12}
$$

Substituting (6.6.12) into (6.6.10) and (6.6.4a) we find

$$
\Psi_0(x,u) = u_0^{-1/2} \exp[\mathrm{i}(K^2 - u_0^2)^{1/2} x]\cdot\{\exp(\mathrm{i}u_0 z) - \exp[2\mathrm{i}u_0 h(x) - \mathrm{i}u_0 z]\} \ .
\tag{6.6.13}
$$

Thus (6.6.13) is a superposition of the incident wave of the kind (2.1.29), and the reflected wave adjusted to the local elevation. An equation for the first order correction results after substituting (6.6.12) into (6.6.11) and has the form $(u \neq u')$

$$\frac{da_1}{dx} - \mathrm{i}(K^2 - u^2)^{1/2} a_1 = Q(x, u) \ , \tag{6.6.14}$$

where

$$Q(x,u) = 2\pi \mathrm{i}(K^2 - u_0^2)^{1/4} u_0^{-1/2} \frac{dh}{dx} \cdot \{-\mathrm{i}u_0 \exp[2\mathrm{i}u_0 h(x) + \mathrm{i}(K^2 - u_0^2)^{1/2} x]$$
$$\times \delta(u - u_0) + \frac{1}{\pi} \frac{uu_0}{(u_0^2 - u^2)} \cdot T_1(u, u_0)$$
$$\times \exp[\mathrm{i}(u + u_0) h(x) + \mathrm{i}(K^2 - u_0^2)^{1/2} x] \ . \tag{6.6.15}$$

Suppose that roughness is concentrated at the finite interval $|x| < L$ (i.e., $h(x) = 0$, $|x| > L$). Since $a(x, u)$ is an amplitude of the wave propagating to positive x, then the solution for (6.6.14) at $x < -L$ should satisfy condition $a_1(x, u) = 0$. It is obvious that such a solution at $x > L$ is

$$a^1(x, u) = A(u) \exp[\mathrm{i}(K^2 - u^2)^{1/2} x] \ , \tag{6.6.16}$$

where

$$A(u) = \int_{-L}^{L} Q(x', u) \exp[-\mathrm{i}(K^2 - u^2)^{1/2} x'] \, dx' \tag{6.6.17}$$

(one can easily verify that the first term in (6.6.15) does not contribute into $A(u)$). The solution for correction b_1 is written down quite similarly. Due to the physical sense of b at $x > L$ condition $b_1(x, u) = 0$ should be fulfilled. Therefore, by (6.6.10) we have for $x > L$

$$\tilde{\Psi}_1(x, u) = a_1(x, u)/2 \ .$$

Substituting (6.6.16) into (6.6.4a) and introducing, instead of integration variable u (vertical wavenumber), a new variable k:

$$k = (K^2 - u^2)^{1/2}$$

(horizontal wavenumber), we obtain for $x \to +\infty$, $z \to -\infty$

$$\Psi_1(x, u) = \frac{\mathrm{i}}{\pi} \int_0^K k^{1/2} \cdot u_k^{-1/2} A(u) \exp(\mathrm{i}kx - \mathrm{i}u_k z) \cdot u_k^{-1/2} \, dk \ , \tag{6.6.18}$$

where

$$u_k = (K^2 - k^2)^{1/2} \ .$$

Comparing (6.6.18) to (2.2.14) we find

$$S_1(k, k_0) = \frac{i}{\pi} \cdot k^{1/2} \cdot u_k^{-1/2} A(u_k) \ ,$$

where $k_0 = (K^2 - u_0^2)^{1/2}$, $k, k_0 > 0$, and $k \neq k_0$. In virtue of (6.6.17, 15)

$$S_1(k, k_0) = -\frac{2(u_k u_0)^{1/2}}{u_0^2 - u_k^2} \cdot (k + k_0) \int_{-L}^{L} \frac{dh}{dx}$$

$$\times \exp[-i(k - k_0)x + i(u_k + u_0)h(x)] \frac{dx}{2\pi} \ .$$

Integration by parts in this formula leads to a replacement of the type (5.1.10). Taking into account the reflection from plane $z = 0$ at $|x| > L$, we obtain

$$S(k, k_0) = -\frac{2(u_k u_0)^{1/2}}{u_k + u_0} \int \exp[-i(k - \boldsymbol{k}_0)x + i(u_k + u_0)h(x)] \frac{dx}{2\pi} \ .$$

This formula coincides with the first order small slope approximation (6.1.22) (in the last formula one should set $B = -1$ and bear in mind the two-dimensional character of the problem). Thus, for the Dirichlet problem the first approximation of the full-wave approach and small slope approximation give coincident results.

6.7 Operator Expansion Method

The operator expansion (OE) method was introduced by *Milder* in 1991 [6.25]. It was checked numerically for the Dirichlet problem and proved to be remarkably accurate. Recently, it has been generalized to the case of EM waves scattering from a perfect conductor [6.26] and a dielectric surface [6.27]. In this section we describe the OE method for the Dirichlet problem basically following [6.25, 34].

The value of the scattered field at the boundary for an arbitrary incident field Ψ_{in} immediately follows from the Dirichlet boundary condition (2.1.15):

$$\Psi_{\text{sc}} = -\Psi_{\text{in}} \qquad (6.7.1)$$

If the value of normal derivative of the scattered field at the surface were also known, one could find the scattered field at any point and determine the scattering amplitude. This motivates one to introduce a linear operator which expresses the normal derivative of the field at the boundary through the value of the scattered field. Let us introduce the value $\partial\Psi_{\text{sc}}/\partial\tilde{n}$ which differs from $\partial\Psi_{\text{sc}}/\partial\boldsymbol{n}$ by the factor $[1 + (\nabla h)^2]^{1/2}$, see (2.1.10):

$$\frac{\partial \Psi_{\mathrm{sc}}}{\partial \tilde{n}} = \left(\frac{\partial}{\partial z} - \nabla h \cdot \nabla\right) \Psi_{\mathrm{sc}} \tag{6.7.2}$$

and define the operator $\hat{N}_+$ by the relation

$$\frac{\partial \Psi_{\mathrm{sc}}}{\partial \tilde{n}}\Big|_{z=h(\boldsymbol{r})-0} = \hat{N}_+ \left(\Psi_{\mathrm{sc}}|_{z=h(\boldsymbol{r})}\right) . \tag{6.7.3}$$

Here the term $h(\boldsymbol{r}) - 0$ indicates that the observation point tends to the boundary from below. Note that the operator $\hat{N}_+$ acts on functions of the horizontal coordinate $\boldsymbol{r}$. Let us recall that the solution of the Helmholtz equation with a known value of the field at the boundary was obtained in Sect. 3.4 and is given by (3.4.16), i.e.,

$$\Psi_{\mathrm{sc}}(\boldsymbol{R}_*) = \int \frac{\partial G(\boldsymbol{R}, \boldsymbol{R}_*)}{\partial \tilde{n}_{\boldsymbol{R}}} \Psi_{\mathrm{sc}}(\boldsymbol{R})|_{z=h(\boldsymbol{r})} d\boldsymbol{r} . \tag{6.7.4}$$

Here, $G(\boldsymbol{R}, \boldsymbol{R}_*)$ is the Green's function of the boundary problem (3.4.15 a, b). Applying to (6.7.4) the operator $\partial/\partial\tilde{n}_{\boldsymbol{R}*}$ defined in (6.7.2) one easily finds

$$\frac{\partial \Psi_{\mathrm{sc}}}{\partial \tilde{n}}\bigg|_{z_*\to h(\boldsymbol{r}_*)-0} = \int \frac{\partial G(\boldsymbol{R}, \boldsymbol{R}_*)}{\partial \tilde{n}_{\boldsymbol{R}} \partial \tilde{n}_{\boldsymbol{R}*}}\bigg|_{z=h(\boldsymbol{r})} \Psi_{\mathrm{sc}}(\boldsymbol{R})|_{z=h(\boldsymbol{r})} d\boldsymbol{r} \tag{6.7.5}$$

Thus, the kernel of the operator $\hat{N}_+$ can be expressed as the following (weak) limit

$$N_+(\boldsymbol{r}, \boldsymbol{r}_*) = \hat{N}_+ \delta(\boldsymbol{r} - \boldsymbol{r}_*) = \lim_{z_*\to h(\boldsymbol{r}_*)-0} \frac{\partial G(\boldsymbol{R}, \boldsymbol{R}_*)}{\partial \tilde{n}_{\boldsymbol{R}} \partial \tilde{n}_{\boldsymbol{R}_*}}\bigg|_{z=h(\boldsymbol{r})} . \tag{6.7.6}$$

Let us consider the case of the horizontal plane $h(\boldsymbol{r}) = 0$ with respect to the two-dimensional problem. Building up the Green's function $G(\boldsymbol{R}, \boldsymbol{R}_*)$ with the help of mirror images one obtains

$$G(\boldsymbol{R}, \boldsymbol{R}_*) = G_0(\boldsymbol{R} - \boldsymbol{R}_*) - G_0(\boldsymbol{R}^{(\mathrm{mirr})} - \boldsymbol{R}_*) .$$

In two dimensions the expression for the free-space Green's function G_0 is, see (2.1.23b)

$$G_0(\boldsymbol{R}, \boldsymbol{R}_*) = -\frac{i}{4} H_0^{(1)}(K|\boldsymbol{R} - \boldsymbol{R}_*|) .$$

Substituting this expression into (6.7.6) one finds

$$N_+(x,x_*) = -\frac{\mathrm{i}K}{2}\frac{H_1^{(1)}(K|x-x_*|)}{|x-x_*|} \ . \tag{6.7.7}$$

This operator was introduced by *Maystre* in 1983 [6.28]. Expression (6.7.7) can be used in (6.7.5) for an approximate calculation of $\partial\Psi_{\mathrm{sc}}/\partial\boldsymbol{n}$ for the case of non-planar surfaces as well. One can see that this approach is, in fact, similar to the Meecham-Lysanov method considered in Sect. 6.5. If instead of coordinate x in (6.7.7) the arc length s is used, one reaches the approxiimation which is called the PDBO (plane approximation delta function boundary operator) method [6.29]. A generalization of this approach for the case of a dielectric surface is also available [6.30].

It was assumed in (6.7.4) that the observation point belongs to the "basic" region $z < h(\boldsymbol{r})$. It is convenient, however, to consider a solution of the Helmholtz equation with respect to the "non-physical" half-space $z > h(\boldsymbol{r})$, too. It is obvious that the upper half-space $z > h(\boldsymbol{r})$ is congruent to the lower half-space with the shape of the boundary given by the equation $z = -h(\boldsymbol{r})$. Taking into account that the sign of the normal also changes, we obtain

$$\left.\frac{\partial\Psi_{\mathrm{sc}}}{\partial\tilde{n}}\right|_{z=h(\boldsymbol{r})+0} = -\hat{N}_-\left(\Psi_{\mathrm{sc}}|_{z=h(\boldsymbol{r})}\right)$$

with

$$\hat{N}_-([h]) = \hat{N}_+([-h]) \ . \tag{6.7.8}$$

(Here, $[h]$ indicates the functional dependence of the operator on the shape of the boundary). Now, we can determine the scattered field for any point $\boldsymbol{R}_* = (\boldsymbol{r}_*, z_*)$, $z_* < h(\boldsymbol{r}_*)$ using the Helmholtz formulae (3.1.27). In the present case

$$[f] = \Psi_{\mathrm{sc}}(\boldsymbol{R})|_{z=h(\boldsymbol{r})+0} - \Psi_{\mathrm{sc}}(\boldsymbol{R})|_{z=h(\boldsymbol{r})-0} = 0 \ , \tag{6.7.9}$$

$$\left[\frac{\partial f}{\partial \boldsymbol{n}}\right] = [1+(\nabla h)^2]^{-1/2}\left(\left.\frac{\partial\Psi_{\mathrm{sc}}}{\partial\tilde{n}}\right|_{z=h(\boldsymbol{r})+0} - \left.\frac{\partial\Psi_{\mathrm{sc}}}{\partial\tilde{n}}\right|_{z=h(\boldsymbol{r})-0}\right) \tag{6.7.10}$$

and one obtains

$$\Psi_{\mathrm{sc}}(\boldsymbol{R}_*) = -\int G_0(\boldsymbol{R}-\boldsymbol{R}_*)\cdot(\hat{N}_- + \hat{N}_+)\Psi_{\mathrm{sc}}|_{z=h(\boldsymbol{r})}d\boldsymbol{r} \ .$$

Introducing the operator

$$\hat{N}_s = \frac{1}{2}(\hat{N}_- + \hat{N}_+) \tag{6.7.11}$$

and using (6.7.1) one finds

$$\Psi_{sc}(\boldsymbol{R}_*) = 2\int G_0(\boldsymbol{R} - \boldsymbol{R}_*) \cdot \hat{N}_s \Psi_{sc}|_{z=h(\boldsymbol{r})} d\boldsymbol{r} \ .$$

Assuming now that the incident field is a plane wave (2.1.29)

$$\Psi_{in} = q_0^{-1/2} \exp(\mathrm{i}\boldsymbol{k}_0\boldsymbol{r} + \mathrm{i}q_0 z)$$

and representing G_0 as a superposition of plane waves according to Weyl's formula (2.1.24) we represent the scattered field for $z_* < \min h(\boldsymbol{r})$ according to (2.2.14). In our case the scattering amplitude is given by

$$S(\boldsymbol{k}, \boldsymbol{k}_0) = -\frac{\mathrm{i}}{\sqrt{q_k q_0}} \int \frac{d\boldsymbol{r}}{(2\pi)^2} \exp[-\mathrm{i}\boldsymbol{k}\boldsymbol{r} + \mathrm{i}q_k h(\boldsymbol{r})]\hat{N}_s \exp[\mathrm{i}\boldsymbol{k}_0\boldsymbol{r} + \mathrm{i}q_0 h(\boldsymbol{r})] \tag{6.7.12}$$

or in the explicit form

$$S(\boldsymbol{k}, \boldsymbol{k}_0) = -\frac{\mathrm{i}}{\sqrt{q_k q_0}} \int \frac{d\boldsymbol{r}d\boldsymbol{r}'}{(2\pi)^2} N_s(\boldsymbol{r}, \boldsymbol{r}') \exp[-\mathrm{i}\boldsymbol{k}\boldsymbol{r} + \mathrm{i}\boldsymbol{k}_0\boldsymbol{r}' + \mathrm{i}q_k h(\boldsymbol{r}) + \mathrm{i}q_0 h(\boldsymbol{r}')] \tag{6.7.13}$$

Note that according to (6.7.6, 6.7.11) the operators $\hat{N}_+$, $\hat{N}_-$ and $\hat{N}_s$ are symmetric

$$N_s(\boldsymbol{r}, \boldsymbol{r}') = N_s(\boldsymbol{r}', \boldsymbol{r}) \ .$$

Hence, the reciprocity theorem

$$S(\boldsymbol{k}, \boldsymbol{k}_0) = S(-\boldsymbol{k}_0, -\boldsymbol{k})$$

immediately follows from (6.7.13).

Let us expand the operator $\hat{N}_+$ into a power series with respect to the elevations $h(\boldsymbol{r})$:

$$\hat{N}_+ = \hat{N}_0 + \hat{N}_1 + \hat{N}_2 + \dots \ . \tag{6.7.14}$$

According to (6.7.8) for the operator $\hat{N}_-$ one obtains

$$\hat{N}_- = \hat{N}_0 - \hat{N}_1 + \hat{N}_2 - \dots \ .$$

Hence, expansion of the operator $\hat{N}_s$ consists of even terms only:

$$\hat{N}_s = \hat{N}_0 + \hat{N}_2 + \dots \ . \tag{6.7.15}$$

We emphasize that according to the physical meaning of the operators $\hat{N}_+, \hat{N}_-$ they should remain invariant when the boundary is shifted as a whole in the vertical direction by some constant value H:

$$\hat{N}_+([h+H]) = \hat{N}_+([h]) \ .$$

Hence, the expansions (6.7.14, 15) take place with respect to spatial derivatives of h rather than with respect to the elevation itself.

To calculate the operator $\hat{N}_+$ we will proceed as follows. Let us introduce the two-dimensional operator $\hat{\partial}$ which acts on smooth functions defined in three-dimensional space $f(\boldsymbol{r}, z)$ according to the following rule

$$\hat{\partial} f(\boldsymbol{r}, z) = \nabla f(\boldsymbol{r}, h(\boldsymbol{r})) = (\nabla f(\boldsymbol{r}, z))|_{z=h(\boldsymbol{r})} + \nabla h \cdot \left. \frac{\partial f(\boldsymbol{r}, z)}{\partial z} \right|_{z=h(\boldsymbol{r})} \tag{6.7.16}$$

Here, as usual, $\nabla = (\partial/\partial x, \partial/\partial y)$ is a standard horizontal gradient operator. Hence, $\hat{\partial}$ represents the horizontal gradient of the functions that result from the "cross-section" of functions defined in three dimensions by the boundary $z = h(\boldsymbol{r})$. Hence, a "usual" horizontal component of the gradient of any function $f(\boldsymbol{r}, z)$ calculated on the surface $z = h(\boldsymbol{r})$ can be expressed in terms of the operators $\hat{\partial}$ and $\hat{Z}$

$$\hat{Z} = \frac{\partial}{\partial z} \tag{6.7.17}$$

as follows

$$\nabla = \hat{\partial} - \nabla h \cdot \hat{Z} \ . \tag{6.7.18}$$

Thus, the operator $\partial/\partial\tilde{n}$ becomes

$$\hat{N}_+ = \frac{\partial}{\partial \tilde{n}} = \frac{\partial}{\partial z} - \nabla h \cdot \nabla = (1 + (\nabla h)^2)\hat{Z} - \nabla h \cdot \hat{\partial} \ . \tag{6.7.19}$$

When the boundary value of the solution of Helmholtz's equation is known (in the present case of a Dirichlet problem it is given by (6.7.1)), the result of the action of operator $\hat{\partial}$ on it is immediately known as well. Thus, if one could express the operator $\hat{Z}$ in terms of the operator $\hat{\partial}$, a normal derivative of the solution at the boundary could be calculated according to (6.7.19), and the scattering problem would be solved.

Of course, for an arbitrary smooth function, the operators $\hat{\partial}$ and $\hat{Z}$ are quite independent and cannot be related to each other. However, they become related when considered with respect to solutions of Helmholtz's equation satisfying the radiation condition. In this case after squaring equation (6.7.18) one obtains

$$-\hat{Z}^2 = K^2 + (\hat{\partial} - \nabla h \cdot \hat{Z})^2 \, . \tag{6.7.20}$$

This relation can be considered as an equation with respect to the operator $\hat{Z}$. For a surface roughness with small slopes this equation explicitly contains the small parameter ∇h, and its formal solution can be represented by the following recursion [6.31]

$$\hat{Z} = -\mathrm{i}\sqrt{K^2 + \left[\hat{\partial} + \mathrm{i}\nabla h\sqrt{K^2 + (\hat{\partial} + \mathrm{i}\nabla h \ldots)^2}\right]^2} \, .$$

Or, one can seek a solution of (6.7.20) in terms of an expansion with respect to the slope

$$\hat{Z} = \hat{Z}_0 + \hat{Z}_1 + \hat{Z}_2 + \ldots \tag{6.7.21}$$

To the lowest order one neglects ∇h on the right-hand side of (6.7.20) and gets

$$\hat{Z}_0 = -\mathrm{i}\hat{q} \, , \tag{6.7.22}$$

where $\hat{q}$ represents "vertical wave number" operator

$$\hat{q} = \sqrt{K^2 + \hat{\partial}^2} \, . \tag{6.7.23}$$

The sign of the square root was chosen according to the radiation condition. When acting on $\exp(\mathrm{i}\boldsymbol{k} \cdot \boldsymbol{r})$, which represent an eigenfunction of the operator $\hat{\partial}$, the operator $\hat{q}$ obviously reduces to a multiplication by q_k. To calculate (6.7.21) explicity let us introduce matrix elements of the operator $\hat{Z}$ on the basis of exponentials, i.e.,

$$Z(\boldsymbol{k}_1, \boldsymbol{k}_2) = \int \exp(-\mathrm{i}\boldsymbol{k}_1 \cdot \boldsymbol{r})\hat{Z} \exp(\mathrm{i}\boldsymbol{k}_2 \cdot \boldsymbol{r}) \frac{d\boldsymbol{r}}{(2\pi)^2} \, . \tag{6.7.24}$$

The matrix elements of the operator $\hat{N}_s$ are defined by the similar relation (of course, they can be equivalently defined as an appropriate Fourier transform of $N_s(\boldsymbol{r},\boldsymbol{r}')$ from (6.7.6) as well). In terms of N_s, SA given by (6.7.12) is expressed as

$$\begin{aligned} S(\boldsymbol{k},\boldsymbol{k}_0) = & -\frac{\mathrm{i}}{\sqrt{q_k q_0}} \int \frac{d\boldsymbol{r}_1 d\boldsymbol{r}_2 d\boldsymbol{k}_1 d\boldsymbol{k}_2}{(2\pi)^4} \\ & \times \exp[-\mathrm{i}(\boldsymbol{k}-\boldsymbol{k}_1)\cdot\boldsymbol{r}_1 + \mathrm{i}q_k h(\boldsymbol{r}_1)] N_s(\boldsymbol{k}_1,\boldsymbol{k}_2) \\ & \times \exp[\mathrm{i}(\boldsymbol{k}_0-\boldsymbol{k}_2)\cdot\boldsymbol{r}_2 + \mathrm{i}q_0 h(\boldsymbol{r}_2)] \,. \end{aligned} \tag{6.7.25}$$

For the operator $\hat{Z}_0$ according to the rule of calculation of operators (6.4.5) one finds

$$\begin{aligned} Z_0(\boldsymbol{k}_1,\boldsymbol{k}_2) = & -\mathrm{i} \int \exp(-\mathrm{i}\boldsymbol{k}_1\cdot\boldsymbol{r}) \sqrt{K^2+\hat{\partial}^2} \exp(\mathrm{i}\boldsymbol{k}_2\cdot\boldsymbol{r}) \frac{d\boldsymbol{r}}{(2\pi)^2} \\ = & -\mathrm{i}q_{k_1}\delta(\boldsymbol{k}_1-\boldsymbol{k}_2) \,. \end{aligned} \tag{6.7.26}$$

To the lowest order $\hat{N}_0 = \hat{Z}_0$, and substituting (6.7.26) into (6.7.25) one immediately obtains an expression which exactly coincides with (6.5.9) for the scattering amplitude obtained in the framework of the Meecham-Lysanov approach. Hence, conclusions drawn with respect to the latter in Sect. 6.5 are directly applicable to the lowest order of the OE approach. In particular, in the high-frequency limit OE reduces to the Kirchhoff approximation, and in the small roughness (perturbation) limit one obtains a result accurate through the first power of elevations.

To calculate higher-order corrections, $\hat{Z}_1, \hat{Z}_2, \ldots$, let us write down the equation with respect to matrix elements. In terms of $Z(\boldsymbol{k}_1,\boldsymbol{k}_2)$ relation (6.7.20) reads

$$\begin{aligned} -\int Z(\boldsymbol{k}_1,\xi)Z(\xi,\boldsymbol{k}_2)d\xi = & \; q_{k_1}^2\delta(\boldsymbol{k}_1-\boldsymbol{k}_2) \\ & + (\boldsymbol{k}_1+\boldsymbol{k}_2)\int(\boldsymbol{k}_1-\xi)h(\boldsymbol{k}_1-\xi)Z(\xi,\boldsymbol{k}_2)d\xi \\ & - \int(\boldsymbol{k}_1-\xi_1)h(\boldsymbol{k}_1-\xi_1)Z(\xi_1,\xi_2)(\xi_2-\xi_3) \\ & \times h(\xi_2-\xi_3)Z(\xi_3,\boldsymbol{k}_2)d\xi_1 d\xi_2 d\xi_3 \,. \end{aligned} \tag{6.7.27}$$

To the lowest order in slopes only the first term on the right-hand side survives, and this equation immediately reproduces (6.7.26). Using this relation in (6.7.27) one easily finds

$$(q_{k_1}+q_{k_2})Z_1(\boldsymbol{k}_1,\boldsymbol{k}_2) = (\boldsymbol{k}_2^2-\boldsymbol{k}_1^2)q_{k_1}h(\boldsymbol{k}_2-\boldsymbol{k}_1)$$

and

$$Z_1(\boldsymbol{k}_1,\boldsymbol{k}_2) = (q_{k_1}q_{k_2} - K^2 + k_2^2)h(\boldsymbol{k}_1 - \boldsymbol{k}_2)\ . \tag{6.7.28}$$

For the operator $\nabla h \cdot \hat{\partial}$ in (6.7.19) we find

$$\int \exp(-\mathrm{i}\boldsymbol{k}_1 \cdot \boldsymbol{r})\nabla h \cdot \hat{\partial} \exp(\mathrm{i}\boldsymbol{k}_2 \cdot \boldsymbol{r}) \frac{d\boldsymbol{r}}{(2\pi)^2} = (\boldsymbol{k}_2 - \boldsymbol{k}_1) \cdot \boldsymbol{k}_2 h(\boldsymbol{k}_1 - \boldsymbol{k}_2)\ . \tag{6.7.29}$$

Combining this result with (6.7.28) yields

$$N_1(\boldsymbol{k}_1,\boldsymbol{k}_2) = (q_{k_1}q_{k_2} + \boldsymbol{k}_1 \cdot \boldsymbol{k}_2 - K^2)h(\boldsymbol{k}_1 - \boldsymbol{k}_2)\ . \tag{6.7.30}$$

Similarly to (6.7.29) one can easily make sure that the operator $\hat{N}_1$ has the following representation in terms of the operator $\hat{q}$

$$\hat{N}_1 = \hat{q}h\hat{q} - \hat{\partial}h\hat{\partial} - K^2 h\ . \tag{6.7.31}$$

Here h is understood as a plain multiplication by $h(\boldsymbol{r})$.

Calculation of the matrix element $N_2(\boldsymbol{k}_1,\boldsymbol{k}_2)$ is quite similar and straightforward, and the result is

$$N_2(\boldsymbol{k}_1,\boldsymbol{k}_2) = -\frac{\mathrm{i}}{2}q_{k_1}q_{k_2}\int (q_{k_1} - 2q_\xi + q_{k_2})h(\boldsymbol{k}_1 - \xi)h(\xi - \boldsymbol{k}_2)d\xi\ . \tag{6.7.32}$$

In terms of the $\hat{q}$ operators this relation reads

$$\hat{N}_2 = -\frac{\mathrm{i}}{2}\hat{q}[h(h\hat{q} - \hat{q}h) - (h\hat{q} - \hat{q}h)h]\hat{q} = -\frac{\mathrm{i}}{2}\hat{q}[h,[h,\hat{q}]]\hat{q} \tag{6.7.33}$$

where $[\hat{a},\hat{b}] = \hat{a}\hat{b} - \hat{b}\hat{a}$ is a commutator.

Let us compare now the OE and the small slope approximation (SSA) considered in Sect. 6.1. If one "passes" the second exponential in (6.7.12) through the operator $\hat{N}_\mathrm{s}$ one arrives at the basic ansatz of the SSA (6.1.4) with

$$\varphi(\boldsymbol{k},\boldsymbol{k}_0;\boldsymbol{r};[h]) = -\frac{\mathrm{i}}{\sqrt{q_k q_0}} \cdot \exp(-\mathrm{i}\boldsymbol{k}_0 \cdot \boldsymbol{r} - \mathrm{i}q_0 h(\boldsymbol{r}))\hat{N}_\mathrm{s} \exp[\mathrm{i}\boldsymbol{k}_0 \cdot \boldsymbol{r} + \mathrm{i}q_0 h(\boldsymbol{r})]\ . \tag{6.7.34}$$

At this point the functional φ appears to be unambiguously determined. The reason is that using Helmholtz's formula with the surface source densities defined in (6.7.9, 10) corresponds to a particular choice of the "external" field in the "non-physical" region $z > h(\boldsymbol{r})$. This choice fixes the gauge arbitrariness of φ. When the "external" sources are chosen differently, the expression for φ also changes (the "antisymmetric" version of the OE represents an example of this ambiguity; see [6.25] for details).

To reach the SSA one needs now to expand (6.7.34) into a power series of the elevation and to allow gauge transformations of φ according to (6.1.15). However, for numerical simulations there is no need to make a power expansion. Let us assume that the series expansion (6.7.15) is truncated at some finite order, and appropriate terms are explicitly calculated similar to $\hat{N}_{0,1,2}$ above. As a result, the operator $\hat{N}_{\rm s}$ is approximated by a polynomial with respect to the operators $\hat{q}$ and h. In this case the action of $\hat{N}_{\rm s}$ on the incident field can be conveniently calculated with the help of the fast Fourier transform technique: multiplication by h is performed in the coordinate space, and the action of $\hat{q}$ is calculated in the Fourier domain. The results of numerical simulations, a detailed comparison with exact numerical solutions and other techniques for a broad range of parameters, which clearly demonstrates great potentials of the method, can be found in [6.25, 27, 32]. A review of the method was also published in [6.33, 34].

7. Waveguide with Statistically Rough Boundary in Noncorrelated Successive Reflections Approximation

The present chapter deals with wave propagation in the stratified medium bounded by a rough surface when refraction returns radiation scattered from roughness back to the boundary.

Wave propagation in the waveguides with rough boundaries is of great interest and have been discussed many times, see e.g. [7.1–4]. The present chapter suggests the approach with the following features.

The terminal purpose of the theory given in this chapter is calculating the correlation function of the field. For this we first assume that the waveguide is layered and all its properties depend only upon vertical coordinate z and statistics of roughness is horizontally homogeneous. As previously we suppose that excursions of the boundary points are near the level $z = 0$. The medium is supposed to be homogeneous in the layer $-H < z < h(\boldsymbol{r})$ which adjoins the boundary and whose thickness H exceeds amplitude of roughness. Thus, we neglect the refraction effect immediately near the roughness which is true for many practical tasks. The field source and receivers are deployed inside this homogeneous layer as well. The last supposition is not important and can be omitted. But this leads to using more cumbersome notation and to less clear statements. Therefore, we only emphasize modifications which should be done in the general case. For simplicity, we limit ourselves to the scalar problem concerning sound propagation.

Assuming that the homogeneous layer adjoins to the boundary we can consider the wave propagation as consisting of two processes. Let us analyze the sound field below the roughness and inside the homogeneous layer which can be uniquely represented as the superposition of upgoing and downgoing plane waves. The upgoing waves arrive at the rough boundary and are scattered. Previous chapters considered in detail this process. The downgoing waves are refracted in the stratified lower medium. In this case some wave beams arriving at the lower stratified medium leave it at the point shifted horizontally with respect to the arriving point by distance L. This distance in WKB-approximation is equal to the cycle length of the proper ray. In many practical cases such a shift greatly exceeds the correlation radius of boundary roughness. The appropriate small parameter often arose previously in considering this problem, however, at rather late stages when an approximate solution of the problem was constructed and proper limitations owing to the method used had already arisen. These limitations can be essential and not fulfilled in reality. However, statistical independence of successive reflections can be taken as a principle of analysis allowing us to progress to a solution without concretization of scattering properties at the rough boundary.

Hence, the approach stated in this chapter is fundamentally based on the assumption of noncorrelativity of successive reflections at the boundary. This assumption must be fulfilled for many problems of practical interest and its validity can be easily controlled. The advantage of this theory is also that quantities dealt with in it enable direct local measurements. This chapter is based on the work [7.5].

7.1 Directional Source in the Waveguide with Rough Boundary: General Solution

We shall consider the monochromatic sound fields and describe them using potential Ψ so that acoustic speed and pressure are determined by formulas

$$V = \mathrm{Re}\{\exp(-\mathrm{i}\omega t)\nabla\Psi\} , \qquad p = -\omega\varrho_0 \,\mathrm{Im}\{\exp(-\mathrm{i}\omega t)\Psi\} . \tag{7.1.1}$$

Discuss the following situation: Let the homogeneous liquid layer with sound speed $c = c_0 = \mathrm{const}$ be at $z > -H$ and bounded from above by the rough surface $z = h(\boldsymbol{r})$ (situated near the level $z = 0$). Let an arbitrary medium with properties independent of horizontal coordinates be at $z < -H$. This assumption enables us to introduce the plane-wave reflection coefficient from this medium

$$\mathscr{V}(\boldsymbol{k}) = \mathscr{V}_{\boldsymbol{k}} = |\mathscr{V}(\boldsymbol{k})| \exp[2\mathrm{i}\alpha(\boldsymbol{k})] . \tag{7.1.2}$$

Then solutions of the wave equation for $z > -H$ take the form

$$y(\boldsymbol{k}, z) = \exp(\mathrm{i}\boldsymbol{k}\cdot\boldsymbol{r} - \mathrm{i}qz) + \mathscr{V}_{\boldsymbol{k}} \exp(\mathrm{i}\boldsymbol{k}\cdot\boldsymbol{r} + \mathrm{i}qz) . \tag{7.1.3}$$

Here it is supposed that at $z > -H$ the fictitious isovelocity half-space is situated in a way in which the sound speed coincides with that in the homogeneous subsurface layer. As was mentioned above we shall assume that the field sources and the receivers are in the area $-H < z < 0$.

The reflection coefficient $\mathscr{V}$ is an exaustive characteristic of the lower medium ($z < -H$). For the cases not concerning the field structure outside the homogeneous layer, the physical nature of the lower medium is of no consequence. For example, it can consist of liquids or solids moving, or motionless layers, and so on. The requirement of horizontal homogeniety of the lower medium is the only important fact for introducing the reflection coefficient. For instance, if at $z < -H$ some layered liquid half-space occurs, then in the WKB-approximation we obtain $\mathscr{V}_k = 0$ for the continuous sound speed profile $c(z)$ without turning point and $|\mathscr{V}_k| = 1$ for available turning point. In the latter case

$$\alpha(k) = \int_{z_t}^{0} [\omega^2/c^2(z) - k^2]^{1/2}\, dz + \text{const} \;, \qquad c(z_t) = \omega/k \;,$$

and the vector

$$\boldsymbol{L}_{\boldsymbol{k}} = -2\nabla_{\boldsymbol{k}}\alpha \tag{7.1.4}$$

corresponds to the previously mentioned horizontal shift. It coincides in absolute value with the cycle length of the ray leaving the surface at the angle $\theta_0 = \sin^{-1}(k/K)$ to the vertical:

$$|\boldsymbol{L}_{\boldsymbol{k}}| = 2\int_{z_t}^{0} k\cdot[\omega^2/c^2(z) - k^2]^{-1/2}\, dz \;.$$

Vector $\boldsymbol{L}_{\boldsymbol{k}}$ is directed along $\boldsymbol{k}$. We keep the name "cycle length" or "skip distance" for value $|\boldsymbol{L}_{\boldsymbol{k}}|$ in the general case, when phase $\alpha(\boldsymbol{k})$ in (7.1.4) results from the exact solution of the motion equations in the lower medium, and, generally speaking, WKB- approximation is inadequate.

In the homogeneous layer below the boundary excursions we expand the field in the upgoing ($b_{\boldsymbol{k}}$) and downgoing ($a_{\boldsymbol{k}}$) plane waves:

$$\Psi = \int a(\boldsymbol{k})\exp(\mathrm{i}\boldsymbol{k}\cdot\boldsymbol{r} - \mathrm{i}q_k z)\, d\boldsymbol{k} + \int b(\boldsymbol{k})\exp(\mathrm{i}\boldsymbol{k}\cdot\boldsymbol{r} + \mathrm{i}q_k z)\, d\boldsymbol{k} \;. \tag{7.1.5}$$

Then for the amplitudes of these waves two equations can be written

$$q_k^{1/2} a(\boldsymbol{k}) = \int S(\boldsymbol{k},\boldsymbol{k}')q_{k'}^{1/2} b(\boldsymbol{k}')\, d\boldsymbol{k}' \;, \tag{7.1.6a}$$

$$b(\boldsymbol{k}) = \mathscr{V}(\boldsymbol{k})a(\boldsymbol{k}) + d(\boldsymbol{k}) \;. \tag{7.1.6b}$$

Equation (7.1.6a) is, in fact, a definition of SA S, see (2.2.14), and the term $d(\boldsymbol{k})$ in (7.1.6b) describes the field incident on the boundary immediately after its radiation by the source, i.e., the field which has not interacted with the boundary yet. Equations (7.1.6) with specified form of function d, see (7.1.10) following, completely describe the problem of sound propagation in a waveguide with a rough upper boundary. From these equations we obtain

$$q_k^{1/2} a(\boldsymbol{k}) = \int F(\boldsymbol{k},\boldsymbol{k}')q_{k'}^{1/2}\, d(\boldsymbol{k}')\, d\boldsymbol{k}' \;, \tag{7.1.7}$$

where

$$F(\boldsymbol{k},\boldsymbol{k}_0) = S(\boldsymbol{k},\boldsymbol{k}_0) + \int S(\boldsymbol{k},\boldsymbol{k}_1)\mathscr{V}_{k_1}S(\boldsymbol{k}_1,\boldsymbol{k}_0)\, d\boldsymbol{k}_1$$
$$+ \int S(\boldsymbol{k},\boldsymbol{k}_1)\mathscr{V}_{k_1}S(\boldsymbol{k}_1,\boldsymbol{k}_2)\mathscr{V}_{k_2}S(\boldsymbol{k}_2,\boldsymbol{k}_0)\, d\boldsymbol{k}_1\, d\boldsymbol{k}_2 + \cdots \;. \tag{7.1.8}$$

It is easily seen that function F satisfies the equation

$$F(\boldsymbol{k},\boldsymbol{k}_0) = S(\boldsymbol{k},\boldsymbol{k}_0) + \int S(\boldsymbol{k},\boldsymbol{k}')\mathscr{V}_{k'}F(\boldsymbol{k}',\boldsymbol{k}_0)\, d\boldsymbol{k}' \tag{7.1.9}$$

and describes multiple interactions of sound with the boundary owing to refraction in a layered medium. Interpretations of certain terms in series (7.1.8) is clear: the factors S and $\mathscr{V}$ describe wave scattering at the surface and passing through the lower medium, correspondingly.

Now consider the structure of the initial field $d(\boldsymbol{k})$. Let a directional source be at the point $(\boldsymbol{r}_0, z_0)(-H < z_0 < 0)$. Suppose the amplitudes of emitted upward and downward plane waves with a horizontal component of the wave vector equal to $\boldsymbol{k}$ are proportional to $\mathscr{D}(\boldsymbol{k}, +1)$ and $\mathscr{D}(\boldsymbol{k}, -1)$. The downgoing plane waves before arriving at the boundary are refracted in the lower medium. Hence, we obtain the following representation for function $d(\boldsymbol{k})$ in (7.1.6b):

$$d(\boldsymbol{k}) = \frac{d_0}{Kq_k}[\mathscr{D}(\boldsymbol{k}, +1)\exp(-\mathrm{i}\boldsymbol{k}\cdot\boldsymbol{r}_0 - \mathrm{i}q_k z_0) + \mathscr{D}(\boldsymbol{k}, -1)\mathscr{V}_k\exp(-\mathrm{i}\boldsymbol{k}\cdot\boldsymbol{r}_0 + \mathrm{i}q_k z_0)] , \tag{7.1.10}$$

where

$$d_0 = [W_0/2\pi^2\varrho_0 c_0]^{1/2} \tag{7.1.11}$$

and W_0 is the total power emitted by the sound source. It can be easily verified that power emitted at the solid angle $d\boldsymbol{n}$ in direction $\boldsymbol{n} = (\boldsymbol{k}, \varepsilon q_k)/K$, $\varepsilon = \pm 1$ is

$$dW = W_0\cdot|\mathscr{D}(\boldsymbol{k}, \varepsilon)|^2\, d\boldsymbol{n} .$$

Thus for $F(\boldsymbol{k}, \boldsymbol{k}_0)$ known $a(\boldsymbol{k})$ is determined in virtue of (7.1.7) and $b(\boldsymbol{k})$ in virtue of (7.1.6b). Substituting the result into (7.1.5) gives the following representation for the directional source field in the homogeneous layer:

$$\Psi = \Psi_0 + \Psi_{\mathrm{sc}} , \tag{7.1.12}$$

where

$$\Psi_0(\boldsymbol{r}, z) = d_0\cdot\int\exp[\mathrm{i}\boldsymbol{k}\cdot(\boldsymbol{r}-\boldsymbol{r}_0)]\cdot\{\mathscr{D}(\boldsymbol{k}, \mathrm{sign}(z-z_0))\exp[\mathrm{i}q|z-z_0|) + \mathscr{D}(\boldsymbol{k}, -1)\mathscr{V}(k)\cdot\exp[\mathrm{i}q(z+z_0)]\}\frac{d\boldsymbol{k}}{Kq_k} \tag{7.1.13}$$

is the field generated by the source in a medium without account of the boundary, that is, for a homogeneous layer extending to $z = +\infty$. The value Ψ_{sc} is

$$\Psi_{\mathrm{sc}}(\boldsymbol{r}, z) = d_0\cdot\int\exp(\mathrm{i}\boldsymbol{k}\cdot\boldsymbol{r} - \mathrm{i}\boldsymbol{k}_0\cdot\boldsymbol{r}_0)\cdot F(\boldsymbol{k}, \boldsymbol{k}_0)\cdot\sum_{\varepsilon,\varepsilon_0=\pm1} \times\mathscr{D}(\boldsymbol{k}_0, \varepsilon_0)\exp(-\mathrm{i}\varepsilon_0 q_0 z_0 + \mathrm{i}\varepsilon q z)\cdot\mathscr{V}_{k_0}^{(1-\varepsilon_0)/2}\mathscr{V}_k^{(1+\varepsilon)/2}\frac{d\boldsymbol{k}\, d\boldsymbol{k}_0}{K(qq_0)^{1/2}} . \tag{7.1.14}$$

When substituting into (7.1.14) the expression for F in the form of series (7.1.8), each term will be obviously interpreted: in the high-frequency limit it is the field corresponding to the ray with a trajectory determined by indices ε_0 and ε and by the sequence of factors $\mathscr{V}$ and S. In particular, ε_0 characterizes the ray emitted upward ($\varepsilon_0 = +1$) or downward ($\varepsilon_0 = -1$) from the source and, ε characterizes the ray arriving at the receiver from below ($\varepsilon = +1$) or from above ($\varepsilon = -1$); and $\mathscr{V}$ and S describe passing through the lower medium and scattering at the surface, respectively.

We emphasize that formulas (7.1.12–14) give an exact solution of the problem which is not related to the ray approximation, although we use the appropriate language to interpret various values. In order to go over to the ray approximation it is necessary to substitute into (7.1.14) expressions for F in the form of series (7.1.8) and then to calculate each of the resulting integrals by the stationary phase method. However, it will not be done in what follows.

7.2 Average Field

To determine the average field experimentally precise measurement of the signal phase is needed and this is sometimes a complicated problem. But in any case, it is of interest to calculate the average field because it contributes to the correlation function of the total field, see Sect. 7.3.

As was seen in Sect. 2.4 relation (2.4.6) takes place in virtue of the spatial homogeneity of a statistical emsemble of irregularities:

$$\overline{S(\boldsymbol{k}, \boldsymbol{k}_0)} = \bar{V}_{\boldsymbol{k}} \delta(\boldsymbol{k} - \boldsymbol{k}_0) \ . \tag{7.2.1}$$

The similar equation is valid for value $\bar{F}$:

$$\overline{F(\boldsymbol{k}, \boldsymbol{k}_0)} = \bar{F}_{\boldsymbol{k}} \cdot \delta(\boldsymbol{k} - \boldsymbol{k}_0) \ . \tag{7.2.2}$$

Relation (7.2.2) readily follows from (7.1.8) when using the same considerations by which relation (7.2.1) was obtained in Sect. 2.4. Assume notation

$$\Psi_{\boldsymbol{k}}(z) = \int \Psi(\boldsymbol{r}, z) \cdot \exp(-\mathrm{i}\boldsymbol{k} \cdot \boldsymbol{r}) \frac{d\boldsymbol{r}}{(2\pi)^2} \ . \tag{7.2.3}$$

Then we have from (7.1.5)

$$\begin{cases} \Psi_{\boldsymbol{k}}|_{z=0} = a_{\boldsymbol{k}} + b_{\boldsymbol{k}} \ , \\ \left.\dfrac{\partial \Psi_{\boldsymbol{k}}}{\partial z}\right|_{z=0} = -\mathrm{i}q_k(a_{\boldsymbol{k}} - b_{\boldsymbol{k}}) \ . \end{cases} \tag{7.2.4}$$

It goes without saying that equations (7.2.4) are fulfilled for mean values $\bar{\Psi}_{\boldsymbol{k}}$, $\bar{a}_{\boldsymbol{k}}$, and $\bar{b}_{\boldsymbol{k}}$ also. Averaging (7.1.7) with account of (7.2.2) yields

$$\bar{a}_{\boldsymbol{k}} = \bar{F}_{\boldsymbol{k}} \cdot d_{\boldsymbol{k}}$$

and we obtain from (7.1.6b)

$$\bar{b}_{\boldsymbol{k}} = \mathscr{V}_k \cdot \bar{a}_{\boldsymbol{k}} + d_{\boldsymbol{k}} \ .$$

Substitution of these relations into (7.2.4) gives, for average values

$$\bar{\Psi}_{\boldsymbol{k}}|_{z=0} = [(1 + \mathscr{V}_k)\cdot \bar{F}_{\boldsymbol{k}} + 1]\, d_{\boldsymbol{k}} \ ,$$

$$\left.\frac{\partial \bar{\Psi}_{\boldsymbol{k}}}{\partial z}\right|_{z=0} = -\mathrm{i}q_k[(1 - \mathscr{V}_k)\bar{F}_{\boldsymbol{k}} + 1]\, d_{\boldsymbol{k}} \ ,$$

whence relation

$$[(1 + \mathscr{V}_k)\bar{F}_{\boldsymbol{k}} + 1]\frac{\partial \bar{\Psi}_{\boldsymbol{k}}}{\partial z} + \mathrm{i}q_k[(1 - \mathscr{V}_k)\bar{F}_{\boldsymbol{k}} - 1]\Psi_{\boldsymbol{k}} = 0 \ , \qquad z = 0 \ , \qquad (7.2.5)$$

follows.

The average field $\bar{\Psi}$ satisfies the same Helmholtz equation as the initial field does. Thus the problem of determining the average field reduces to calculating the value $\bar{F}_{\boldsymbol{k}}$ and subsequently integrating the Helmholtz equation with impedance boundary condition (7.2.5).

Let us apply developed formalism to the case of an acoustically soft boundary with small irregularities when, for SA, representation (4.1.10) is valid

$$S(\boldsymbol{k}, \boldsymbol{k}_0) = -\delta(\boldsymbol{k} - \boldsymbol{k}_0) - 2\mathrm{i}(qq_0)^{1/2}h(\boldsymbol{k} - \boldsymbol{k}_0)$$
$$+ 2(qq_0)^{1/2}\int q_{k'}h(\boldsymbol{k} - \boldsymbol{k}')h(\boldsymbol{k}' - \boldsymbol{k}_0)\, d\boldsymbol{k}' + \mathrm{O}(h^3).$$

Let us seek solution of (7.1.9) in the form of power series in h:

$$F = F_0 + F_1 + F_2 + \cdots \ .$$

Straightforward calculations yield

$$F_0(\boldsymbol{k}, \boldsymbol{k}_0) = -(1 + \mathscr{V}_k)^{-1}\delta(\boldsymbol{k} - \boldsymbol{k}_0) \ ,$$

$$F_1(\boldsymbol{k}, \boldsymbol{k}_0) = -2\mathrm{i}(qq_0)^{1/2}(1 + \mathscr{V}_k)^{-1}(1 + \mathscr{V}_{k_0})^{-1}h(\boldsymbol{k} - \boldsymbol{k}_0) \ ,$$

$$F_2(\boldsymbol{k}, \boldsymbol{k}_0) = \frac{2(qq_0)^{1/2}}{(1 + \mathscr{V}_k)(1 + \mathscr{V}_{k_0})}\int q_{k'}\left(\frac{1 - \mathscr{V}_{k'}}{1 + \mathscr{V}_{k'}}\right)h(\boldsymbol{k} - \boldsymbol{k}')h(\boldsymbol{k}' - \boldsymbol{k}_0)\, d\boldsymbol{k}' \ .$$

As a result of averaging we obtain

$$\bar{F}_{\boldsymbol{k}} = -\frac{1}{(1 + \mathscr{V}_k)} + \frac{2q_k}{(1 + \mathscr{V}_k)^2}\int q_{k'}\frac{(1 - \mathscr{V}_{k'})}{(1 + \mathscr{V}_{k'})}W(\boldsymbol{k} - \boldsymbol{k}')\, d\boldsymbol{k}' + \mathrm{O}(W^2) \ , \qquad (7.2.6)$$

where $W(\boldsymbol{k})$ is the roughness spectrum, see (4.1.23). Substituting (7.2.6) into (7.2.5) we represent the last equation in the form

$$\overline{\Psi}_{\boldsymbol{k}}\cdot\left(\frac{\partial\overline{\Psi}_{\boldsymbol{k}}}{\partial z}\right)^{-1}\Bigg|_{z=0} = -\mathrm{i}\int W(\boldsymbol{k}-\boldsymbol{k}')\, q_{k'}\frac{(1-\mathscr{V}_{k'})}{(1+\mathscr{V}_{k'})}d\boldsymbol{k}' + \mathrm{O}(W^2)\ . \qquad (7.2.7)$$

This result has been obtained by many researchers [7.1, 7.4]. Replacement of the unperturbed boundary condition $\Psi(0) = 0$ by the impedance boundary condition (7.2.7) gives rise to decaying normal modes in the waveguide whose decrement is easily expressed through the right-hand side of (7.2.7).

Assumption of noncorrelated successive reflections was not applied in the previous calculations. Now we introduce an assumption of statistical independence of scattering amplitudes appearing in (7.1.8), since the latter depends, in fact, on roughness divided by the ray cycle length which is supposed to greatly exceed the correlation radius of roughness.

Here it is reasonable to give some comment. Since $S(\boldsymbol{k}, \boldsymbol{k}_0)$ is the scattering amplitude of plane waves having infinite extent over horizontal coordinates, then to substantiate the statistical independence of S in the series terms in (7.1.8) it is more accurate to say about SA of quasiplane waves which are tempered horizontally at distances less then cycle length. However, if this distance greatly exceeds the correlation radius of roughness and the wavelength then the statistical characteristics of scattering of such quasiplane waves in corresponding spatial regions will not practically differ from appropriate characteristics of plane waves.

Thus in the noncorrelated successive reflections approximation used in averaging series (7.1.8), we split correlators consisting of products of SA. With account of (7.2.1, 2) we find

$$\overline{F}_{\boldsymbol{k}} = \overline{V}_{\boldsymbol{k}} + \overline{V}_{\boldsymbol{k}}^2\cdot\mathscr{V}_k + \overline{V}_{\boldsymbol{k}}^3\mathscr{V}_k^2 + \cdots = \overline{V}_{\boldsymbol{k}}\cdot(1-\overline{V}_{\boldsymbol{k}}\mathscr{V}_k)^{-1}\ . \qquad (7.2.8)$$

Substituting this expression for $\overline{F}_{\boldsymbol{k}}$ in (7.2.5) we obtain

$$\frac{\partial\overline{\Psi}_k}{\partial z} - \mathrm{i}q_k\left(\frac{1-\overline{V}_{\boldsymbol{k}}}{1+\overline{V}_{\boldsymbol{k}}}\right)\overline{\Psi}_k = 0\ , \qquad z = 0\ . \qquad (7.2.9)$$

This relation corresponds to the impedance boundary condition in which value $\overline{V}_{\boldsymbol{k}}$ plays the role of reflection coefficient. The latter is quite natural and the expected result directly following from the assumption of noncorrelated successive reflections.

Now apply this approximation to (7.2.7) and express it in the following form:

$$\overline{\Psi}_{\boldsymbol{k}}\cdot\left(\frac{\partial\overline{\Psi}_{\boldsymbol{k}}}{\partial z}\right)^{-1} = -\mathrm{i}\int W(\boldsymbol{k}-\boldsymbol{k}')q_{k'}\,d\boldsymbol{k}'$$

$$+\ 2\mathrm{i}\sum_{n=1}^{\infty}\int W(\boldsymbol{k}-\boldsymbol{k}')q_{k'}|\mathscr{V}_{k'}|^n\exp[2\mathrm{i}n\alpha(k')]\,d\boldsymbol{k}'\ . \qquad (7.2.10)$$

Using (4.1.24) we can write down

$$\int W(\boldsymbol{k}-\boldsymbol{k}')q_{k'}|\mathscr{V}_{k'}|^n \exp[2in\alpha(k')]\,d\boldsymbol{k}'$$

$$= (2\pi)^{-2}\int d\boldsymbol{\varrho}\tilde{W}(\boldsymbol{\varrho})\exp(-i\boldsymbol{k}\boldsymbol{\varrho})\int|\mathscr{V}_{k'}|^n\exp[i\boldsymbol{k}'\boldsymbol{\varrho}+2in\alpha(k')]\,d\boldsymbol{k}' \ .$$

To estimate this expression one can use the stationary phase method. The equation on stationary point $\boldsymbol{k}_s'$ for the integral over $\boldsymbol{k}'$ is

$$\boldsymbol{\varrho} = -2n\nabla\alpha_{\boldsymbol{k}_s'} = n\cdot \boldsymbol{L}(\boldsymbol{k}_s') \ ,$$

see (7.1.4). Since the correlation radius of roughness is supposed to be less than the typical ray cycle this equation has no real solutions. Hence, the sum in (7.2.10) can be neglected. Taking into account (4.1.22) we can represent (7.2.10) as

$$\bar{\Psi}_{\boldsymbol{k}}\cdot\left(\frac{\partial\bar{\Psi}_{\boldsymbol{k}}}{\partial z}\right)^{-1} \approx -i\frac{1+\bar{V}_{\boldsymbol{k}}}{2q_k} \ .$$

This formula coincides in fact with (7.2.9) to an accuracy of $O(W)$ terms.

Note that (7.2.8) also directly follows from (7.1.9) after splitting the correlator $\overline{S\cdot F} = \bar{S}\cdot\bar{F}$ in the integral and taking into account (7.2.1). In the terms of the theory of wave propagation in random media, relation (7.2.8) is the Dyson's equation in the Bourett approximation [7.6].

Substituting (7.2.8) into (7.1.14) gives the explicit integral representation for average field $\bar{\Psi}$. Hence, calculating $\bar{\Psi}$ in the noncorrelated successive reflections approximation reduces to calculating the appropriate wave field in the layered medium, under replacement of the reflection coefficient from the plane boundary $V_0(k)$ by $\bar{V}(\boldsymbol{k})$ in the boundary condition (7.2.9).

7.3 Correlation Function of the Field

Consider calculating the correlation function of the field. Proceeding from (7.1.12–14) we write down

$$\Psi(\boldsymbol{r},z) - \bar{\Psi}(\boldsymbol{r},z) = \left[\frac{W_0}{2\pi^2\varrho_0 c_0}\right]^{1/2}\int \exp(i\boldsymbol{k}\cdot\boldsymbol{r} - i\boldsymbol{k}_0\cdot\boldsymbol{r}_0)\Delta F(\boldsymbol{k},\boldsymbol{k}_0)$$

$$\times \sum_{\varepsilon,\varepsilon_0=\pm1}\mathscr{D}(\boldsymbol{k}_0,\varepsilon_0)\exp(-i\varepsilon_0 q_0 z_0 + i\varepsilon q z)\mathscr{V}_{\boldsymbol{k}}^{(1+\varepsilon)/2}$$

$$\times \mathscr{V}_{\boldsymbol{k}_0}^{(1-\varepsilon_0)/2}\frac{d\boldsymbol{k}\,d\boldsymbol{k}_0}{K(qq_0)^{1/2}} \ , \qquad (7.3.1\text{a})$$

$$\Psi^*(\boldsymbol{r}', z') - \bar{\Psi}^*(\boldsymbol{r}', z') = \left[\frac{W_0}{2\pi^2 \varrho_0 c_0}\right]^{1/2} \int \exp(-\mathrm{i}\boldsymbol{k}' \cdot \boldsymbol{r}' + \mathrm{i}\boldsymbol{k}_0' \cdot \boldsymbol{r}_0) \Delta F^*(\boldsymbol{k}', \boldsymbol{k}_0')$$

$$\times \sum_{\varepsilon', \varepsilon_0' = \pm 1} \mathscr{D}^*(\boldsymbol{k}_0', \varepsilon_0') \cdot \exp(\mathrm{i}\varepsilon_0' q_0' z_0 - \mathrm{i}\varepsilon' q' z')$$

$$\times \mathscr{V}_{\boldsymbol{k}'}^{*(1+\varepsilon')/2} \mathscr{V}_{\boldsymbol{k}_0'}^{*(1-\varepsilon_0')/2} \frac{d\boldsymbol{k}'\, d\boldsymbol{k}_0'}{K(q' q_0')^{1/2}} . \qquad (7.3.1\mathrm{b})$$

Here

$$\Delta F(\boldsymbol{k}, \boldsymbol{k}_0) = F(\boldsymbol{k}, \boldsymbol{k}_0) - \overline{F(\boldsymbol{k}, \boldsymbol{k}_0)}$$

and $q' = q(k')$ and $q_0' = q(k_0')$ as usual. The receivers' coordinates are $\boldsymbol{R} = (\boldsymbol{r}, z)$ and $\boldsymbol{R}' = (\boldsymbol{r}', z')$, and the source is assumed at the point $(\boldsymbol{r}_0, z_0)$.

We set

$$C_{\mathrm{ch}}(\boldsymbol{R}, \boldsymbol{R}') = \bar{\Psi}(\boldsymbol{R}) \cdot \bar{\Psi}^*(\boldsymbol{R}') ,$$
$$C_{\mathrm{nch}}(\boldsymbol{R}, \boldsymbol{R}') = \overline{[\Psi(\boldsymbol{R}) - \bar{\Psi}(\boldsymbol{R})] \cdot [\Psi^*(\boldsymbol{R}') - \bar{\Psi}^*(\boldsymbol{R}')]} , \qquad (7.3.2)$$

and represent the total correlation function of the field as the sum of coherent and noncoherent parts:

$$\overline{\Psi(\boldsymbol{R}) \cdot \Psi^*(\boldsymbol{R}')} = C_{\mathrm{ch}}(\boldsymbol{R}, \boldsymbol{R}') + C_{\mathrm{nch}}(\boldsymbol{R}, \boldsymbol{R}') . \qquad (7.3.3)$$

Determining C_{ch} reduces to calculating $\bar{\Psi}$; this problem was considered in the previous section. It is seen from (7.3.1) that to calculate C_{nch} the correlator $\overline{\Delta F(\boldsymbol{k}, \boldsymbol{k}_0) \cdot \Delta F^*(\boldsymbol{k}', \boldsymbol{k}_0')}$ should be found. Supposed spatial homogeneity of undulations statistics can be expressed in the form

$$\overline{F(\boldsymbol{k} - \boldsymbol{a}/2, \boldsymbol{k}_0 - \boldsymbol{a}_0/2) \cdot F^*(\boldsymbol{k} + \boldsymbol{a}/2, \boldsymbol{k}_0 + \boldsymbol{a}_0/2)}$$
$$- \bar{F}(\boldsymbol{k} - \boldsymbol{a}/2, \boldsymbol{k}_0 - \boldsymbol{a}_0/2) \cdot \bar{F}^*(\boldsymbol{k} + \boldsymbol{a}/2, \boldsymbol{k}_0 + \boldsymbol{a}_0/2)$$
$$= \overline{\Delta F(\boldsymbol{k} - \boldsymbol{a}/2, \boldsymbol{k}_0 - \boldsymbol{a}_0/2) \cdot \Delta F^*(\boldsymbol{k} + \boldsymbol{a}/2, \boldsymbol{k}_0 + \boldsymbol{a}_0/2)}$$
$$= \Delta\varphi(\boldsymbol{k}, \boldsymbol{k}_0; \boldsymbol{a}) \delta(\boldsymbol{a} - \boldsymbol{a}_0) . \qquad (7.3.4)$$

Presence of factor $\delta(\boldsymbol{a} - \boldsymbol{a}_0)$ in (7.3.4) can be proved by the same approach as in (2.4.9).

Multiplying (7.3.1a, 1b), averaging the result over the ensemble of elevations, and replacing the integration variables yield

$$C_{\mathrm{nch}} = C_{\mathrm{nch}}^{(-)} + C_{\mathrm{nch}}^{(+)} , \qquad (7.3.5)$$

where

$$\begin{aligned}
&C_{\mathrm{nch}}^{(-)}(\boldsymbol{R},\boldsymbol{R}';\boldsymbol{r}_0,z_0)\\
&= \frac{W_0}{2\pi^2\varrho_0 c_0}\int \exp[\mathrm{i}\boldsymbol{k}\cdot(\boldsymbol{r}-\boldsymbol{r}') - \mathrm{i}\boldsymbol{a}\cdot(\boldsymbol{r}+\boldsymbol{r}')/2 + \mathrm{i}\boldsymbol{a}\cdot\boldsymbol{r}_0]\\
&\quad\times \Delta\varphi(\boldsymbol{k},\boldsymbol{k}_0;\boldsymbol{a})\sum_{\varepsilon_0=\pm1}\mathscr{D}(\boldsymbol{k}_0-\boldsymbol{a}/2,\varepsilon_0)\cdot\mathscr{D}^*(\boldsymbol{k}_0+\boldsymbol{a}/2,\varepsilon_0)\\
&\quad\times \mathscr{V}_{\boldsymbol{k}_0-\boldsymbol{a}/2}^{(1-\varepsilon_0)/2}\mathscr{V}_{\boldsymbol{k}_0+\boldsymbol{a}/2}^{*(1-\varepsilon_0)/2}\exp[-\mathrm{i}\varepsilon_0(q_{\boldsymbol{k}_0-\boldsymbol{a}/2}-q_{\boldsymbol{k}_0+\boldsymbol{a}/2})z_0]\sum_{\varepsilon=\pm1}\\
&\quad\times \mathscr{V}_{\boldsymbol{k}-\boldsymbol{a}/2}^{(1+\varepsilon)/2}\mathscr{V}_{\boldsymbol{k}+\boldsymbol{a}/2}^{*(1+\varepsilon)/2}\exp[\mathrm{i}\varepsilon(q_{\boldsymbol{k}-\boldsymbol{a}/2}z-q_{\boldsymbol{k}+\boldsymbol{a}/2}z')]\,d\boldsymbol{k}\,d\boldsymbol{k}_0\,d\boldsymbol{a}\\
&\quad\times [K^2(q_{\boldsymbol{k}-\boldsymbol{a}/2}q_{\boldsymbol{k}+\boldsymbol{a}/2}q_{\boldsymbol{k}_0-\boldsymbol{a}/2}q_{\boldsymbol{k}_0+\boldsymbol{a}/2})^{1/2}]\ ,
\end{aligned}\tag{7.3.6}$$

$$\begin{aligned}
&C_{\mathrm{nch}}^{(+)}(\boldsymbol{R},\boldsymbol{R}';\boldsymbol{r}_0,z_0)\\
&= \frac{W_0}{2\pi^2\varrho_0 c_0}\int \exp[\mathrm{i}\boldsymbol{k}\cdot(\boldsymbol{r}-\boldsymbol{r}') - \mathrm{i}\boldsymbol{a}\cdot(\boldsymbol{r}+\boldsymbol{r}')/2 + \mathrm{i}\boldsymbol{a}\boldsymbol{r}_0]\\
&\quad\times \Delta\varphi(\boldsymbol{k},\boldsymbol{k}_0;\boldsymbol{a})\sum_{\varepsilon_0\cdot\varepsilon_0'+\varepsilon\cdot\varepsilon'<2}\mathscr{D}(\boldsymbol{k}_0-\boldsymbol{a}/2,\varepsilon_0)\mathscr{D}^*(\boldsymbol{k}_0+\boldsymbol{a}/2,\varepsilon_0')\\
&\quad\times \mathscr{V}_{\boldsymbol{k}_0-\boldsymbol{a}/2}^{(1-\varepsilon_0)/2}\cdot\mathscr{V}_{\boldsymbol{k}_0+\boldsymbol{a}/2}^{*(1-\varepsilon_0')/2}\exp[-\mathrm{i}(\varepsilon_0 q_{\boldsymbol{k}_0-\boldsymbol{a}/2}-\varepsilon_0' q_{\boldsymbol{k}_0+\boldsymbol{a}/2})z_0]\\
&\quad\times \mathscr{V}_{\boldsymbol{k}-\boldsymbol{a}/2}^{(1+\varepsilon)/2}\mathscr{V}_{\boldsymbol{k}+\boldsymbol{a}/2}^{(1+\varepsilon')/2}\exp[\mathrm{i}(\varepsilon q_{\boldsymbol{k}-\boldsymbol{a}/2}z-\varepsilon' q_{\boldsymbol{k}+\boldsymbol{a}/2}z')]\,d\boldsymbol{k}\,d\boldsymbol{k}_0\,d\boldsymbol{a}\\
&\quad\times [K^2(q_{\boldsymbol{k}-\boldsymbol{a}/2}q_{\boldsymbol{k}+\boldsymbol{a}/2}q_{\boldsymbol{k}_0-\boldsymbol{a}/2}q_{\boldsymbol{k}_0+\boldsymbol{a}/2})^{1/2}]\ .
\end{aligned}\tag{7.3.7}$$

Contribution of inhomogeneous waves is neglected in relations (7.3.6, 7) and vertical wave numbers q are all considered real. C_{nch} is divided into two parts as follows: the terms corresponding to $\varepsilon_0 = \varepsilon_0'$ and $\varepsilon = \varepsilon'$ entered $C_{\mathrm{nch}}^{(-)}$, and the remaining combinations of values ε_0, ε_0', ε, and ε' entered $C_{\mathrm{nch}}^{(+)}$. The term $C_{\mathrm{nch}}^{(+)}$ corresponds to the correlation of signals either outgoing from the source or coming to the receiver, one above, another below. When setting $\boldsymbol{a} = \boldsymbol{0}$ in the integrands in (7.3.6, 7), $C_{\mathrm{nch}}^{(-)}$ depends only on the differences of the vertical coordinates of the receiving points $z - z'$, while the factors of the form $\exp[\pm\mathrm{i}q(z+z')]$ and/or $\exp(\pm 2\mathrm{i}qz_0)$ will enter $C_{\mathrm{nch}}^{(+)}$. Therefore, after averaging over $(z+z')/2$ and/or z_0, contribution of the term $C_{\mathrm{nch}}^{(+)}$ vanishes. However, when approaching the waveguide boundaries this term essentially contributes.

The expressions (7.3.6, 7) are too awkward to be practically used. One of the important simplifications of these formulas can be achieved in the following way: Average spatially the function $C_{\mathrm{nch}}(\boldsymbol{R},\boldsymbol{R}';\boldsymbol{r}_0,z_0)$ over some horizontal area centered at point $\bar{\boldsymbol{r}}_0$:

$$C_{\mathrm{nch}}^{(\Omega)}(\boldsymbol{R},\boldsymbol{R}';\bar{\boldsymbol{r}}_0,z_0) = \int C_{\mathrm{nch}}(\boldsymbol{R},\boldsymbol{R}';\boldsymbol{r}_0,z_0)\tilde{\Omega}(\boldsymbol{r}_0-\bar{\boldsymbol{r}}_0)\,d\boldsymbol{r}_0\ . \tag{7.3.8}$$

Here $\tilde{\Omega}$ is some smooth weight function (Gaussian, for instance). In particular, if $\tilde{\Omega}(\boldsymbol{x}) = \delta(\boldsymbol{x})$ then $C_{\text{nch}}^{(\Omega)} = C_{\text{nch}}$. It is obvious that such averaging in (7.3.6, 7) results in the appearance of an additional factor $\Omega(\boldsymbol{a})$ in the integrands

$$\Omega(\boldsymbol{a}) = \int \tilde{\Omega}(\boldsymbol{x}) \exp(\mathrm{i}\boldsymbol{a} \cdot \boldsymbol{x}) d\boldsymbol{x} \tag{7.3.9}$$

and to replacement $\boldsymbol{r}_0 \to \bar{\boldsymbol{r}}_0$. When supposing that horizontal scale $\bar{l}$ of function $\tilde{\Omega}(\boldsymbol{x})$ is sufficiently large, function $\Omega(\boldsymbol{a})$ differs significantly from zero only at rather small $\boldsymbol{a}$: $|\boldsymbol{a}| \lesssim \bar{l}^{-1}$. Therefore, linearization with respect to parameter $\boldsymbol{a}$ can be performed in the phase factors in (7.3.6, 7), and $\boldsymbol{a} = \boldsymbol{0}$ can be set in the rest of the factors. After averaging write down (7.3.6) in the form

$$\begin{aligned} C_{\text{nch}}^{(-)(\Omega)}(\boldsymbol{R}, \boldsymbol{R}', \bar{\boldsymbol{r}}_0, z_0) &= \frac{W_0}{2\pi^2 \varrho_0 c_0} \int \Omega(\boldsymbol{a}) \Delta\varphi(\boldsymbol{k}, \boldsymbol{k}_0; \boldsymbol{a}) \\ &\times \exp[\mathrm{i}\boldsymbol{k} \cdot (\boldsymbol{r} - \boldsymbol{r}') - \mathrm{i}\boldsymbol{a} \cdot (\boldsymbol{r} + \boldsymbol{r}')/2 + \mathrm{i}\boldsymbol{a} \cdot \bar{\boldsymbol{r}}_0] \\ &\times \exp(\mathrm{i}\beta) \sum_{\varepsilon_0 = \pm 1} \mathscr{D}(\boldsymbol{k}_0 - \boldsymbol{a}/2, \varepsilon_0) \cdot \mathscr{D}^*(\boldsymbol{k}_0 + \boldsymbol{a}/2, \varepsilon_0) \\ &\times |\mathscr{V}_{\boldsymbol{k}_0 - \boldsymbol{a}/2} \mathscr{V}_{\boldsymbol{k}_0 + \boldsymbol{a}/2}|^{(1-\varepsilon_0)/2} \sum_{\varepsilon = \pm 1} \\ &\times \frac{|\mathscr{V}_{\boldsymbol{k} - \boldsymbol{a}/2} \mathscr{V}_{\boldsymbol{k} + \boldsymbol{a}/2}|^{(1+\varepsilon)/2} d\boldsymbol{k}\, d\boldsymbol{k}_0\, d\boldsymbol{a}}{[K^2(q_{\boldsymbol{k}-\boldsymbol{a}/2} q_{\boldsymbol{k}+\boldsymbol{a}/2} q_{\boldsymbol{k}_0-\boldsymbol{a}/2} q_{\boldsymbol{k}_0+\boldsymbol{a}/2})^{1/2}]} , \end{aligned} \tag{7.3.10}$$

where

$$\begin{aligned} \beta = &-\varepsilon_0(q_{\boldsymbol{k}_0-\boldsymbol{a}/2} - q_{\boldsymbol{k}_0+\boldsymbol{a}/2})z_0 + \varepsilon(q_{\boldsymbol{k}-\boldsymbol{a}/2} \cdot z - q_{\boldsymbol{k}+\boldsymbol{a}/2} \cdot z') + (1 + \varepsilon) \\ &\times [\alpha(\boldsymbol{k} - \boldsymbol{a}/2) - \alpha(\boldsymbol{k} + \boldsymbol{a}/2)] + (1 - \varepsilon_0)[\alpha(\boldsymbol{k}_0 - \boldsymbol{a}/2) - \alpha(\boldsymbol{k}_0 + \boldsymbol{a}/2)] \end{aligned}$$

and $\alpha(\boldsymbol{k})$ is the phase of the reflection coefficient from the lower medium, see (7.1.2). Linearization of β gives

$$\beta = \varepsilon q_k(z - z') + a\{\boldsymbol{\varrho}[\varepsilon, \boldsymbol{k}, (z + z')/2] + \boldsymbol{\varrho}(-\varepsilon_0, \boldsymbol{k}_0, z_0)\} + \mathrm{O}(a^2) , \tag{7.3.11}$$

where

$$\boldsymbol{\varrho}(\varepsilon, \boldsymbol{k}, z) = \varepsilon \cdot \boldsymbol{k} \cdot z/q + (1 + \varepsilon)\boldsymbol{L}_{\boldsymbol{k}}/2 = -\varepsilon\boldsymbol{k} \cdot |z|/q + (1 + \varepsilon)\boldsymbol{L}_{\boldsymbol{k}}/2 \tag{7.3.12}$$

(recall that z, $z_0 < 0$). When considering the ray characterized by parameters $(\varepsilon, \boldsymbol{k})$ and outgoing from the point $(0, z)$, value $\varrho(\varepsilon, \boldsymbol{k}, z)$ has a sense of horizontal displacement which the ray undergoes before the first contact with surface $z = 0$ occurs, see Fig. 7.1. We ignore the weak (as compared to rapidly oscillating phase factor) dependence of values $|\mathscr{V}_{\boldsymbol{k} \pm \boldsymbol{a}/2}|$, $q_{\boldsymbol{k} \pm \boldsymbol{a}/2}$, $\mathscr{D}(\boldsymbol{k}_0 \pm \boldsymbol{a}/2)$, etc. on param-

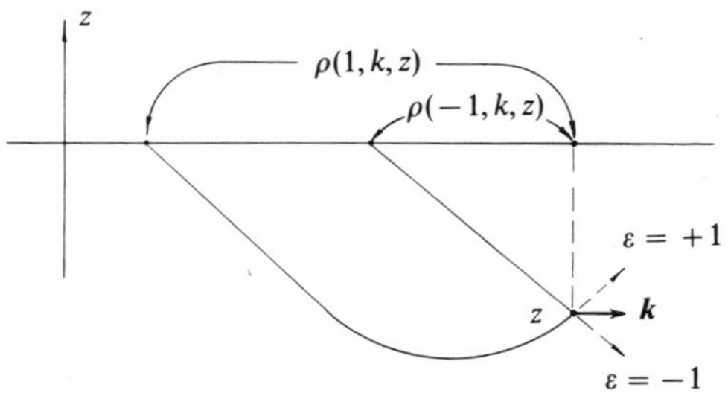

Fig. 7.1. Meaning of functions $\varrho(\varepsilon, \boldsymbol{k}, z)$: let the ray with parameters $(\varepsilon, \boldsymbol{k})$ outgo from the point situated at the depth z. Then $\rho(\varepsilon, \boldsymbol{k}, z)$ is a horizontal displacement which the ray undergoes before the first contact with average boundary $z = 0$

eter $\boldsymbol{a}$, set $\boldsymbol{a} = \boldsymbol{0}$ in them, and supress the term $O(a^2)$ in (7.3.11). Implication of spatial averaging is in substantiation of the reference linearization with respect to $\boldsymbol{a}$. Note that another way to substantiate linearization is the transition to the high-frequency limit, however, we do not use it.

As a result, we obtain the following algorithm for calculating $C_{\mathrm{nch}}^{(-)(\Omega)}$:

$$C_{\mathrm{nch}}^{(-)(\Omega)}(\boldsymbol{R}, \boldsymbol{R}'; \bar{\boldsymbol{r}}_0, z_0) = \int \Delta C(\boldsymbol{R}, \boldsymbol{R}'; \boldsymbol{r}_0, z_0) \tilde{\Omega}(\boldsymbol{r}_0 - \bar{\boldsymbol{r}}_0)\, d\boldsymbol{r}_0 \ . \tag{7.3.13}$$

Here

$$\Delta C(\boldsymbol{R}, \boldsymbol{R}'; \boldsymbol{r}_0, z_0) = \frac{2W_0}{\varrho_0 c_0} \sum_{\varepsilon = \pm 1} \int \frac{d\boldsymbol{k}}{K q_k} \Delta I(\boldsymbol{k}, \varepsilon; \bar{\boldsymbol{R}}) \times \exp[\mathrm{i}\boldsymbol{k} \cdot (\boldsymbol{r} - \boldsymbol{r}') + \mathrm{i}\varepsilon q_k (z - z')] \ , \tag{7.3.14}$$

$\bar{\boldsymbol{R}} = (\boldsymbol{r} + \boldsymbol{r}', z + z')/2$ and

$$\Delta I(\boldsymbol{k}, \varepsilon, \boldsymbol{R}) = |\mathscr{V}_k|^{1+\varepsilon} \sum_{\varepsilon_0 = \pm 1} \int \frac{d\boldsymbol{k}_0}{K q_0} \cdot |\mathscr{V}_{k_0}|^{1-\varepsilon_0} \cdot |\mathscr{D}(k_0, \varepsilon_0)|^2 \times g\{\boldsymbol{k}, \boldsymbol{k}_0; (\boldsymbol{r} + \boldsymbol{r}')/2 - \boldsymbol{r}_0 - \varrho[\varepsilon, \boldsymbol{k}, (z + z')/2] - \varrho(-\varepsilon_0, \boldsymbol{k}_0, z_0)\} \ . \tag{7.3.15}$$

Finally, dimensionless function g is

$$g(\boldsymbol{k}, \boldsymbol{k}_0; \boldsymbol{r}) = \int \Delta\varphi(\boldsymbol{k}, \boldsymbol{k}_0; \boldsymbol{a}) \exp(-\mathrm{i}\boldsymbol{a} \cdot \boldsymbol{r})\, d\boldsymbol{a}/(2\pi)^2 \ . \tag{7.3.16}$$

Taking into account that according to (2.4.20)

$$d\boldsymbol{k}/K q_k = d\boldsymbol{n} \ ,$$

where $d\boldsymbol{n}$ is the solid angle element, we see that (7.3.14) and (2.4.21) are coincident in form. Thus, the noncoherent component of the wave field has the structure corresponding to the radiative transport theory. In virtue of (7.3.14) the value ΔI is the appropriate Fourier transform of the smoothed correlation function $C_{\text{nch}}^{(-)(\Omega)}$ and can be measured experimentally. For the given source theoretical estimation of ΔI can be done by (7.3.15). Function $g(\boldsymbol{k}, \boldsymbol{k}_0; \boldsymbol{r})$ should be known for this. Let us discuss its physical meaning.

The vector of energy flux in the plane wave is [see Sect. 2.1 and (7.1.1)]:

$$\mathscr{I} = \tfrac{1}{2}\omega\varrho_0 \operatorname{Im}\{\Psi^* \nabla \Psi\} \ .$$

Therefore, for the noncoherent component of the energy flux we find from (7.3.14)

$$\Delta\mathscr{I} = \tfrac{1}{2}\omega\varrho_0 \cdot \nabla|_{\boldsymbol{R}=\boldsymbol{R}'} \Delta C(\boldsymbol{R}, \boldsymbol{R}; \boldsymbol{r}_0, z_0) = W_0 \cdot K^2 \int \boldsymbol{n} \Delta I(\boldsymbol{k}, \varepsilon; \boldsymbol{R})\, d\boldsymbol{n} \ , \qquad (7.3.17)$$

where $\boldsymbol{n} = (\boldsymbol{k}, \varepsilon q_{\boldsymbol{k}})/K$ is the unit vector in the direction of propagation of the corresponding wave. Let the source be at $z_0 = 0$ and radiate only in upward direction ($\varepsilon_0 = +1$); the receiving system is also near the boundary $z = 0$ and selects downgoing radiation ($\varepsilon = -1$). Then, according to (7.3.15)

$$\Delta I(\boldsymbol{k}, -1; \boldsymbol{R}) = \int |\mathscr{D}(\boldsymbol{n}_0, +1)|^2 \cdot g(\boldsymbol{k}, \boldsymbol{k}_0; \boldsymbol{r})\, d\boldsymbol{n}_0$$

and (7.3.17) is written in the form

$$\begin{aligned}\Delta\mathscr{I} &= W_0 K^2 \int \boldsymbol{n} \cdot |\mathscr{D}(\boldsymbol{n}_0, +1)|^2 \cdot g(\boldsymbol{k}, \boldsymbol{k}_0; \boldsymbol{r})\, d\boldsymbol{n}_0\, d\boldsymbol{n} \\ &= \int \boldsymbol{n} \cdot K^2 \cdot g(\boldsymbol{k}, \boldsymbol{k}_0; \boldsymbol{r})\, d\boldsymbol{n} \cdot W_0 \cdot |\mathscr{D}(\boldsymbol{n}_0, +1)|^2\, d\boldsymbol{n}_0 \ .\end{aligned}$$

Taking into account that $W_0 \cdot |\mathscr{D}(\boldsymbol{n}_0, +1)|^2\, d\boldsymbol{n}_0$ is the power radiated from the source into the solid angle $d\boldsymbol{n}_0$, see the comment following (7.1.11), we conclude that $K^2 g(\boldsymbol{k}, \boldsymbol{k}_0; \boldsymbol{r})$ is the ratio of the noncoherent component of the energy flux per unit solid angle in $(\boldsymbol{k}, -q_{\boldsymbol{k}})$ direction at the point $(\boldsymbol{r}, 0)$ to the energy flux radiated into the unit solid angle in $(\boldsymbol{k}_0, q_0)$ direction from the source situated at the coordinate origin.

7.4 Radiative Transport Theory

Up to now in calculating the noncoherent component of the correlation function we did not do any principal approximations, and relations (7.3.6, 7) correspond to the exact solution of the problem. However, for explicit calculation of $\Delta\varphi$, some simplifying assumptions should be done and we again use the approximation of noncorrelated successive reflections.

Consider expression (7.1.8) for F. Recall that factor $\mathscr{V}_{\boldsymbol{k}_i}$ in this formula describe the shift of the corresponding wave packet by horizontal distance $L_{\boldsymbol{k}_i}$ (Sect. 7.1). Proceeding from (7.1.8)

$$F(\boldsymbol{k}-\boldsymbol{a}/2,\boldsymbol{k}_0-\boldsymbol{a}_0/2)=S(\boldsymbol{k}-\boldsymbol{a}/2,\boldsymbol{k}_0-\boldsymbol{a}_0/2)+\int S(\boldsymbol{k}-\boldsymbol{a}/2,\boldsymbol{k}_1)\mathscr{V}_{\boldsymbol{k}_1}S(\boldsymbol{k}_1,\boldsymbol{k}_0-\boldsymbol{a}_0/2)\,d\boldsymbol{k}_1+\cdots,$$

$$\updownarrow \qquad\qquad \updownarrow \qquad\qquad \updownarrow$$

$$F^*(\boldsymbol{k}+\boldsymbol{a}/2,\boldsymbol{k}_0+\boldsymbol{a}_0/2)=S^*(\boldsymbol{k}+\boldsymbol{a}/2,\boldsymbol{k}_0+\boldsymbol{a}_0/2)+\int S^*(\boldsymbol{k}+\boldsymbol{a}/2,\boldsymbol{k}_1')\mathscr{V}^*_{\boldsymbol{k}_1'}S(\boldsymbol{k}_1',\boldsymbol{k}_0+\boldsymbol{a}_0/2)\,d\boldsymbol{k}_1'+\cdots. \tag{7.4.1}$$

Approximation of noncorrelated successive reflections reduces in this case to the following: After multiplying these two series, averaging over the ensemble of roughness is performed as follows: in the products of the different order terms the correlator is completely split, for instance:

$$\overline{S(\boldsymbol{k}-\boldsymbol{a}/2,\boldsymbol{k}_0-\boldsymbol{a}_0/2)\cdot\int S^*(\boldsymbol{k}+\boldsymbol{a}/2,\boldsymbol{k}_1')\mathscr{V}^*_{\boldsymbol{k}_1'}S^*(\boldsymbol{k}_1',\boldsymbol{k}_0+\boldsymbol{a}_0/2)\,d\boldsymbol{k}_1'}$$
$$=\bar{S}(\boldsymbol{k}-\boldsymbol{a}/2,\boldsymbol{k}_0-\boldsymbol{a}_0/2)\cdot\int\bar{S}^*(\boldsymbol{k}+\boldsymbol{a}/2,\boldsymbol{k}_1')\cdot\mathscr{V}^*_{\boldsymbol{k}_1'}\bar{S}^*(\boldsymbol{k}_1',\boldsymbol{k}_0+\boldsymbol{a}_0/2)\,d\boldsymbol{k}_1'\,.$$

The product of the same order terms is split into the products of the pair of correlators of SA joined by arrows in (7.4.1), i.e., into these scattering amplitudes which correspond to interaction with the same portions of roughness. For instance,

$$\int\overline{S(\boldsymbol{k}-\boldsymbol{a}/2,\boldsymbol{k}_1)\mathscr{V}_{\boldsymbol{k}_1}S(\boldsymbol{k}_1,\boldsymbol{k}_0-\boldsymbol{a}_0/2)\cdot S^*(\boldsymbol{k}+\boldsymbol{a}/2,\boldsymbol{k}_1')\mathscr{V}^*_{\boldsymbol{k}_1'}S^*(\boldsymbol{k}_1',\boldsymbol{k}_0+\boldsymbol{a}_0/2)}\,d\boldsymbol{k}_1\,d\boldsymbol{k}_1'$$
$$=\int\overline{S(\boldsymbol{k}-\boldsymbol{a}/2,\boldsymbol{k}_1)\cdot S^*(\boldsymbol{k}+\boldsymbol{a}/2,\boldsymbol{k}_1')}\cdot\mathscr{V}_{\boldsymbol{k}_1}\mathscr{V}^*_{\boldsymbol{k}_1'}\overline{S(\boldsymbol{k}_1,\boldsymbol{k}_0-\boldsymbol{a}_0/2)\cdot S^*(\boldsymbol{k}_1',\boldsymbol{k}_0+\boldsymbol{a}_0/2)}$$
$$\times\, d\boldsymbol{k}, d\boldsymbol{k}_1'\,. \tag{7.4.2}$$

Setting in the last integral $\boldsymbol{k}_1\to\boldsymbol{k}_1-\boldsymbol{a}_1/2$, $\boldsymbol{k}_1'\to\boldsymbol{k}_1+\boldsymbol{a}_1/2$ and $d\boldsymbol{k}_1\,d\boldsymbol{k}_1'\to d\boldsymbol{k}_1\,d\boldsymbol{a}_1$ we can continue the previous equality in the following way:

$$(7.4.2)\ =\int\mathscr{E}(\boldsymbol{k},\boldsymbol{k}_1;\boldsymbol{a})\delta(\boldsymbol{a}-\boldsymbol{a}_1)\mathscr{V}_{\boldsymbol{k}_1-\boldsymbol{a}_1/2}\cdot\mathscr{V}^*_{\boldsymbol{k}_1+\boldsymbol{a}_1/2}\mathscr{E}(\boldsymbol{k}_1,\boldsymbol{k}_0;\boldsymbol{a}_1)\delta(\boldsymbol{a}_1-\boldsymbol{a}_0)\,d\boldsymbol{k}, d\boldsymbol{a}_1$$
$$=\int\mathscr{E}(\boldsymbol{k},\boldsymbol{k}_1;\boldsymbol{a})\mathscr{V}_{\boldsymbol{k}_1-\boldsymbol{a}/2}\mathscr{V}^*_{\boldsymbol{k}_1+\boldsymbol{a}/2}\mathscr{E}(\boldsymbol{k}_1,\boldsymbol{k}_0;\boldsymbol{a})\,d\boldsymbol{k}_1\,\delta(\boldsymbol{a}-\boldsymbol{a}_0)\,,$$

where in virtue of (2.4.8, 9)

$$\mathscr{E}(\boldsymbol{k},\boldsymbol{k}_0;\boldsymbol{a})=\Delta\mathscr{E}(\boldsymbol{k},\boldsymbol{k}_0;\boldsymbol{a})+\bar{V}_{\boldsymbol{k}-\boldsymbol{a}/2}\cdot\bar{V}^*_{\boldsymbol{k}+\boldsymbol{a}/2}\delta(\boldsymbol{k}-\boldsymbol{k}_0)\,. \tag{7.4.3}$$

Quite similar transformations are fulfilled in other terms. Taking into account (7.3.4) and (7.2.8), we obtain as a result

$$\Delta\varphi(\boldsymbol{k},\boldsymbol{k}_0;\boldsymbol{a})=\mathscr{E}(\boldsymbol{k},\boldsymbol{k}_0;\boldsymbol{a})+\int\mathscr{E}(\boldsymbol{k},\boldsymbol{k}_1;\boldsymbol{a})\mathscr{V}_{\boldsymbol{k}_1-\boldsymbol{a}/2}\mathscr{V}^*_{\boldsymbol{k}_1+\boldsymbol{a}/2}\mathscr{E}(\boldsymbol{k}_1,\boldsymbol{k}_0;\boldsymbol{a})\,d\boldsymbol{k}_1+\cdots$$
$$-\bar{V}_{\boldsymbol{k}-\boldsymbol{a}/2}\cdot\bar{V}^*_{\boldsymbol{k}+\boldsymbol{a}/2}\cdot(1-\bar{V}_{\boldsymbol{k}-\boldsymbol{a}/2}\bar{V}^*_{\boldsymbol{k}+\boldsymbol{a}/2}\cdot\mathscr{V}_{\boldsymbol{k}-\boldsymbol{a}/2}\mathscr{V}^*_{\boldsymbol{k}+\boldsymbol{a}/2})^{-1}\delta(\boldsymbol{k}-\boldsymbol{k}_0)\,. \tag{7.4.4}$$

According to (7.4.3) the kernel $\mathscr{E}(\boldsymbol{k},\boldsymbol{k}_0;\boldsymbol{a})$ contains the singular component related to the product of mean reflection coefficients. To separate the corresponding terms we set

$$T(\boldsymbol{k},\boldsymbol{k}_0;\boldsymbol{a}) = \mathscr{E}(\boldsymbol{k},\boldsymbol{k}_0;\boldsymbol{a}) + \int \mathscr{E}(\boldsymbol{k},\boldsymbol{k}_1;\boldsymbol{a})\mathscr{V}_{\boldsymbol{k}_1-\boldsymbol{a}/2}\mathscr{V}^*_{\boldsymbol{k}_1+\boldsymbol{a}/2}\mathscr{E}(\boldsymbol{k}_1,\boldsymbol{k}_0;\boldsymbol{a})\,d\boldsymbol{k}_1 + \cdots .$$

Quantity T obviously satisfies equation

$$T(\boldsymbol{k},\boldsymbol{k}_0;\boldsymbol{a}) = \mathscr{E}(\boldsymbol{k},\boldsymbol{k}_0;\boldsymbol{a}) + \int \mathscr{E}(\boldsymbol{k},\boldsymbol{k}_1;\boldsymbol{a})\mathscr{V}_{\boldsymbol{k}_1-\boldsymbol{a}/2}\mathscr{V}^*_{\boldsymbol{k}_1+\boldsymbol{a}/2}T(\boldsymbol{k}_1,\boldsymbol{k}_0;\boldsymbol{a})\,d\boldsymbol{k}_1 . \tag{7.4.5}$$

Now set

$$T(\boldsymbol{k},\boldsymbol{k}_0;\boldsymbol{a}) = T_0(\boldsymbol{k},\boldsymbol{a})\delta(\boldsymbol{k}-\boldsymbol{k}_0) + \Delta T(\boldsymbol{k},\boldsymbol{k}_0;\boldsymbol{a}) , \tag{7.4.6}$$

where ΔT does not contain singularities yet. Substituting relations (7.4.3) for $\mathscr{E}$ and (7.4.6) for T into (7.4.5), and separating the terms proportional to $\delta(\boldsymbol{k}-\boldsymbol{k}_0)$ we have

$$T_0(\boldsymbol{k},\boldsymbol{a}) = \overline{V}_{\boldsymbol{k}-\boldsymbol{a}/2}\cdot\overline{V}^*_{\boldsymbol{k}+\boldsymbol{a}/2}/M(\boldsymbol{k},\boldsymbol{a}) ,$$

where

$$M(\boldsymbol{k},\boldsymbol{a}) = 1 - \overline{V}_{\boldsymbol{k}-\boldsymbol{a}/2}\overline{V}^*_{\boldsymbol{k}+\boldsymbol{a}/2}\mathscr{V}_{\boldsymbol{k}-\boldsymbol{a}/2}\cdot\mathscr{V}^*_{\boldsymbol{k}+\boldsymbol{a}/2} . \tag{7.4.7}$$

Thus in virtue of (7.4.5, 7) the last term with δ-function in (7.4.4) cancels with $T_0\delta(\boldsymbol{k}-\boldsymbol{k}_0)$ and we obtain $\Delta T = \Delta\varphi$. The equation for ΔT following from (7.4.5) gives as a result

$$M(\boldsymbol{k},\boldsymbol{a})\cdot\Delta\varphi(\boldsymbol{k},\boldsymbol{k}_0;\boldsymbol{a}) = \frac{\Delta\mathscr{E}(\boldsymbol{k},\boldsymbol{k}_0;\boldsymbol{a})}{M(\boldsymbol{k}_0,\boldsymbol{a})} + \int \Delta\mathscr{E}(\boldsymbol{k},\boldsymbol{k}';\boldsymbol{a})\mathscr{V}_{\boldsymbol{k}'-\boldsymbol{a}/2}\mathscr{V}^*_{\boldsymbol{k}'+\boldsymbol{a}/2}$$
$$\times\,\Delta\varphi(\boldsymbol{k}',\boldsymbol{k}_0;\boldsymbol{a})\,d\boldsymbol{k}' . \tag{7.4.8}$$

The values $\boldsymbol{k}_0$ and $\boldsymbol{a}$ in (7.4.8) are parameters. In terms of the theory of wave propagation in random media (7.4.8) is the Bethe-Salpeter equation in the ladder approximation [7.6].

To obtain the equation of the radiative transport theory for function g one should use the presence of function $\Omega(\boldsymbol{a})$ in (7.3.10) and perform in (7.4.8) similarly to the previous linearization with respect to $\boldsymbol{a}$. We have

$$\mathscr{V}_{\boldsymbol{k}'-\boldsymbol{a}/2}\cdot\mathscr{V}^*_{\boldsymbol{k}'+\boldsymbol{a}/2} \approx |\mathscr{V}_{\boldsymbol{k}'}|^2\cdot\exp(\mathrm{i}\boldsymbol{a}\cdot\boldsymbol{L}_{\boldsymbol{k}'}) ,$$

$$M(\boldsymbol{k},\boldsymbol{a}) \approx 1 - |\overline{V}_{\boldsymbol{k}}|^2|\mathscr{V}_{\boldsymbol{k}}|^2\cdot\exp(\mathrm{i}\boldsymbol{a}\cdot\boldsymbol{L}_{\boldsymbol{k}}) ,$$

and (7.4.8) can be represented in the form

$$[1 - |\overline{V}_{\boldsymbol{k}}|^2|\mathscr{V}_{\boldsymbol{k}}|^2\exp(\mathrm{i}\boldsymbol{a}\cdot\boldsymbol{L}_{\boldsymbol{k}})]\cdot\Delta\varphi(\boldsymbol{k},\boldsymbol{k}_0;\boldsymbol{a}) = \sigma(\boldsymbol{k},\boldsymbol{k}_0)\sum_{n=0}^{\infty}|\overline{V}_{\boldsymbol{k}_0}\mathscr{V}_{\boldsymbol{k}_0}|^{2n}$$
$$\times\,\exp[\mathrm{i}\boldsymbol{a}\cdot(n\boldsymbol{L}_{\boldsymbol{k}_0})] + \int \sigma(\boldsymbol{k},\boldsymbol{k}')|\mathscr{V}_{\boldsymbol{k}'}|^2\exp(\mathrm{i}\boldsymbol{a}\cdot\boldsymbol{L}_{\boldsymbol{k}'})$$
$$\times\,\Delta\varphi(\boldsymbol{k}',\boldsymbol{k}_0;\boldsymbol{a})\,d\boldsymbol{k}' . \tag{7.4.9}$$

Here, according to (2.4.12) we set $\sigma(\boldsymbol{k}, \boldsymbol{k}_0) = \Delta\mathscr{E}(\boldsymbol{k}, \boldsymbol{k}_0; 0)$. Then, multiplying (7.4.9) by $\exp(-\mathrm{i}\boldsymbol{a}\cdot\boldsymbol{r})/(2\pi)^2$ and integrating over $\boldsymbol{a}$ leads to the radiative transport equation in the usual coordinate representation:

$$g(\boldsymbol{k}, \boldsymbol{k}_0; \boldsymbol{r}) = |\bar{V}_k \mathscr{V}_k|^2 g(\boldsymbol{k}, \boldsymbol{k}_0; \boldsymbol{r} - \boldsymbol{L}_{\boldsymbol{k}}) + \sigma(\boldsymbol{k}, \boldsymbol{k}_0) \sum_{n=0}^{\infty} |\bar{V}_{k_0} \mathscr{V}_{k_0}|^{2n}$$

$$\times\, \delta(\boldsymbol{r} - n\boldsymbol{L}_{\boldsymbol{k}_0}) + \int \sigma(\boldsymbol{k}, \boldsymbol{k}') |\mathscr{V}_{k'}|^2 g(\boldsymbol{k}', \boldsymbol{k}_0; \boldsymbol{r} - \boldsymbol{L}_{\boldsymbol{k}'})\, d\boldsymbol{k}' \ . \quad (7.4.10)$$

Equations (7.4.9, 10) are completely equivalent. Equation (7.4.10) has an obvious physical sense. In particular, the first term on the right-hand side describes the transmission of the noncoherent radiation component through the "coherent channel", i.e., by the specular reflection with coefficient $|\bar{V}_k|^2$. The second term describes the arising of the noncoherent component as a result of the scattering processes from the coherent component; and the third term describes the result of rescattering of the noncoherent component. The equation of the radiative transport theory (7.4.10) contains values of the mean reflection coefficient $\bar{V}$ and the scattering cross-sections σ whose calculation was outlined in Chaps. 4–6. It is important to emphasize that the equation was obtained only under the assumption on the statistical independence of successive reflections. The possibility of a geometric-optical description of the field was not used; it was replaced by the procedure of spatial averaging of the correlation function.

The developed theory allows one to calculate the correlation characteristics of the sound field in a homogeneous layer $-H < z < 0$. The simplest way to take into account stratification of the waveguide below the level $z = -H$ is to modify (7.3.14) according to the WKB-approximation. The latter reduces to the following substitutions in this formula:

$$\varrho_0, c_0, K \to \varrho(\bar{z}), \quad c(\bar{z}), K(\bar{z}) \ ; \qquad \bar{z} = (z + z')/2 \ ;$$

$$q_k \to q_k(\bar{z}) = [K^2(\bar{z}) - k^2]^{1/2}$$

(it is assumed that $|z - z'|$ is sufficiently small). Function $\varrho(\varepsilon, \boldsymbol{k}, z)$ in (7.3.15), see (7.3.12), and Fig. 7.1, has the previous meaning and is merely the horizontal distance covered by the appropriate ray from the point $(0, z)$ to the first contact with surface $z = 0$; modifications which should be done in (7.3.12) are obvious.

7.5 Problem of Noise Surface Sources: Diffusion Approximation

In the present section we consider the problem which is of interest in ocean acoustics, for instance [7.7]. Suppose that there are some noise sound sources situated at the upper boundary of the waveguide $z = h(\boldsymbol{r})$. Physically such sources can be turbulent pressure fluctuations, caused by wind, the processes of wave breaking, etc. This noise field can propagate in an oceanic waveguide over large distances and undergo scattering at the rough boundary while propagating. The question arises as to what kind of noise field forms, as a result, in the ocean.

When considering this problem we use the following simplified model: Assume that the sources are situated not at the rough boundary but are uniformly distributed at the level $z = 0$ and radiate sound only into the halfspace $z < 0$ [i.e., in (7.1.10) and the next formulas we should set $\mathscr{D}(\boldsymbol{k}_0, +1) = 0$]. This sound is refracted further in the lower medium and after that is scattered at the rough boundary. Fictitious sound sources at $z = 0$ are assumed transparent by that. Passing through the stratified lower medium and scattering at the rough boundary is repeated many times.

The noncoherent component of the correlation function of the field is calculated, as previously, by formula (7.3.14), however, ΔI is no more dependent on $\mathscr{R}$ due to the spatial homogeneity of the problem; ΔI is obtained as a result of averaging (7.3.15) over the horizontal coordinates of the field sources $\bar{\boldsymbol{r}}_0$:

$$\Delta I(\boldsymbol{k}, \varepsilon) = |\mathscr{V}_{k}|^{1+\varepsilon} \int \frac{d\boldsymbol{k}_0}{K q_0} |\mathscr{D}(\boldsymbol{k}_0, -1)|^2 \cdot |\mathscr{V}_{k_0}|^2 \Delta\varphi(\boldsymbol{k}, \boldsymbol{k}_0) \ , \tag{7.5.1}$$

where

$$\Delta\varphi(\boldsymbol{k}, \boldsymbol{k}_0) = \Delta\varphi(\boldsymbol{k}, \boldsymbol{k}_0; \boldsymbol{a})|_{\boldsymbol{a}=0} = \int g(\boldsymbol{k}, \boldsymbol{k}_0; \boldsymbol{r})\, d\boldsymbol{r} \ . \tag{7.5.2}$$

Now the product $W_0 |\mathscr{D}(\boldsymbol{k}_0, -1)|^2$ in (7.3.14) has the meaning of the energy flux radiated into the unit solid angle from the unit horizontal area, cf. (7.1.11). Integrating (7.4.10) over all $\boldsymbol{r}$ we obtain the following equation for function $\Delta\varphi(\boldsymbol{k}, \boldsymbol{k}_0)$:

$$(1 - |\bar{V}_k \mathscr{V}_k|^2) \Delta\varphi(\boldsymbol{k}, \boldsymbol{k}_0) = \frac{\sigma(\boldsymbol{k}, \boldsymbol{k}_0)}{1 - |\bar{V}_{k_0} \mathscr{V}_{k_0}|^2} + \int \sigma(\boldsymbol{k}, \boldsymbol{k}') |\mathscr{V}_{k'}|^2 \Delta\varphi(\boldsymbol{k}', \boldsymbol{k}_0)\, d\boldsymbol{k}' \ . \tag{7.5.3}$$

Recall that the contribution of inhomogeneous waves was neglected when deriving (7.4.10) for function g, so that integration over $\boldsymbol{k}'$ in (7.5.3) is related in fact only to region $|\boldsymbol{k}'| < K$.

Integrating (7.5.3) over $\boldsymbol{k}$ leads, in virtue of (2.4.11), to the relation

$$\int\limits_{|\boldsymbol{k}| < K} (1 - |\mathscr{V}_k|^2) \Delta\varphi(\boldsymbol{k}, \boldsymbol{k}_0)\, d\boldsymbol{k} = \frac{1 - |\bar{V}_{k_0}|^2}{1 - |\bar{V}_{k_0} \mathscr{V}_{k_0}|^2} < 1$$

whence it is clear that the solution of (7.5.3) exists only at the presence of losses: $|\mathscr{V}_k| \not\equiv 1$. When considering the homogeneous equation (7.5.3), one can be convinced by using the same approach that the positive solution of this equation is unique. It can be easily shown that replacement

$$\Delta\tilde{\varphi}(\boldsymbol{k},\boldsymbol{k}_0) = |\mathscr{V}_k \mathscr{V}_{k_0}| P_k P_{k_0} \Delta\varphi(\boldsymbol{k},\boldsymbol{k}_0) \ ,$$
$$\tilde{\sigma}(\boldsymbol{k},\boldsymbol{k}_0) = |\mathscr{V}_k \mathscr{V}_{k_0}|^{-1} P_k^{-1} P_{k_0}^{-1} \sigma(\boldsymbol{k},\boldsymbol{k}_0) \ , \tag{7.5.4}$$

where

$$P(k) = (1 - |\overline{V}_k \mathscr{V}_k|^2)^{1/2}$$

symmetrizes (7.5.3):

$$\Delta\tilde{\varphi}(\boldsymbol{k},\boldsymbol{k}_0) = \tilde{\sigma}(\boldsymbol{k},\boldsymbol{k}_0) + \int \tilde{\sigma}(\boldsymbol{k},\boldsymbol{k}')\Delta\varphi(\boldsymbol{k}',\boldsymbol{k}_0)\,d\boldsymbol{k}' \ . \tag{7.5.5}$$

In the isotropic case $\sigma(\boldsymbol{k},\boldsymbol{k}_0) = \sigma(\boldsymbol{k}_0,\boldsymbol{k})$, see (2.4.16). It follows from (7.5.5) that here $\Delta\varphi(\boldsymbol{k},\boldsymbol{k}_0) = \Delta\varphi(\boldsymbol{k}_0,\boldsymbol{k})$ also. In the general case (7.5.5) is solved by diagonalization of the integral operator having a symmetric real (and positive) kernel. It should be pointed out that despite the trivial character of transformation (7.5.4), (7.5.5) is much more convenient for numerical analysis than (7.5.3) due to the kernel symmetry in the former. A lot of efficient numerical algorithms were developed for analysis of such equations. Proceeding from (7.5.5) the problem of noise sources can be qualitatively classified in the following way: First, we note that the case $|\mathscr{V}| \ll 1$ corresponds to very rapid energy dissipation when sources more than the cycle length far from the given point contribute little into the sound field. Consequently, the scattering processes can be ignored in this case. Therefore, consider the case $|\mathscr{V}| \lesssim 1$. Introduce the parameter

$$N \sim (1 - |\mathscr{V}|^2)^{-1}$$

calculated for characteristic values of k; N is the number of cycles before the sound attenuates by $\mathrm{e} = 2.718$ times in energy. Consider the case of weak scattering described by the perturbation theory. Set

$$\varepsilon_k = 1 - |\overline{V}_k|^2 = \int\limits_{|\boldsymbol{k}'|<K} \sigma(\boldsymbol{k},\boldsymbol{k}')\,d\boldsymbol{k}' \ll 1 \ .$$

Quantity ε_k characterizes the energy transfer into the incoherent component in the single act of scattering. The character of the solution appeared to be determined by the product $N\varepsilon$ where ε is the characteristic value of ε_k. Let $N\varepsilon \ll 1$. Then

$$P = (1 - |\overline{V}|^2 |\mathscr{V}|^2)^{1/2} \approx (1 - |\mathscr{V}|^2)^{1/2} \sim N^{-1/2}$$

and, correspondingly, $\tilde{\sigma} \sim P^{-2}\sigma \sim N\sigma$. Therefore,

$$\int \tilde{\sigma}(\boldsymbol{k},\boldsymbol{k}')\,d\boldsymbol{k}' \sim N\varepsilon \ll 1$$

so that the integral term in (7.5.5) can be omitted, which means that the contribution of noncoherent component C_{nch} into the correlation function can be neglected. Thus, the scattering processes in this case are not essential. This result has a clear physical sense: during N cycles till attenuation scattering transfers into the noncoherent component only the $N\varepsilon$th (small) portion of its energy. Generally speaking, in the case $N\varepsilon \gtrsim 1$ we have

$$\int \tilde{\sigma}(\boldsymbol{k}, \boldsymbol{k}')\, d\boldsymbol{k}' \sim 1 \ .$$

Therefore, the integral term in (7.5.5) is essential and the contribution of the noncoherent component into the correlation function is not small. Thus, for $N\varepsilon \sim 1$ the scattering processes should be taken into account.

Now suppose that the Rayleigh parameter is large and it can be approximately assumed that $\overline{V} = 0$, see (6.1.34). In addition, let roughness have a large horizontal scale. Since in this case the scattering cross-section $\sigma(\boldsymbol{k}, \boldsymbol{k}_0)$ differs much from zero only at $\boldsymbol{k} \approx \boldsymbol{k}_0$, this allows one to go over in the integral equation to the so-called diffusion approximation. The corresponding procedure is the following: Equation (7.5.3) is rewritten in the form

$$\begin{aligned}(1 - |\mathscr{V}_k|^2)\Delta\varphi(\boldsymbol{k}, \boldsymbol{k}_0) \\ = \sigma(\boldsymbol{k}, \boldsymbol{k}_0) + \int \sigma(\boldsymbol{k}, \boldsymbol{k}')[|\mathscr{V}_{k'}|^2 \Delta\varphi(\boldsymbol{k}', \boldsymbol{k}_0) - |\mathscr{V}_k|^2 \Delta\varphi(\boldsymbol{k}, \boldsymbol{k}_0)]\, d\boldsymbol{k}' \ . \qquad (7.5.6)\end{aligned}$$

Here it is considered that, according to (2.4.11), from assumption $\overline{V} = 0$ follows:

$$\int \sigma(\boldsymbol{k}, \boldsymbol{k}')\, d\boldsymbol{k}' = 1 \ .$$

Multiply (7.5.6) by an arbitrary smooth function $f(\boldsymbol{k})$ and integrate over $\boldsymbol{k}$. Assuming that the kernel is symmetric, we represent the result as follows:

$$\begin{aligned}&\int (1 - |\mathscr{V}_k|^2)\Delta\varphi(\boldsymbol{k}, \boldsymbol{k}_0) f(\boldsymbol{k})\, d\boldsymbol{k} \\ &= \int \sigma(\boldsymbol{k}, \boldsymbol{k}_0) f(\boldsymbol{k})\, d\boldsymbol{k} + 1/2 \int \sigma(\boldsymbol{k}, \boldsymbol{k}')[|\mathscr{V}_{k'}|^2 \Delta\varphi(\boldsymbol{k}', \boldsymbol{k}_0) - |\mathscr{V}_k|^2 \Delta\varphi(\boldsymbol{k}, \boldsymbol{k}_0)] \\ &\quad \times [f(\boldsymbol{k}) - f(\boldsymbol{k}')]\, d\boldsymbol{k}\, d\boldsymbol{k}' \ . \qquad (7.5.7)\end{aligned}$$

Now take into account that function σ differs from zero in the vicinity of diagonal $\boldsymbol{k} \approx \boldsymbol{k}'$, and expand in a series the factors in the integrand term in (7.5.7):

$$f(\boldsymbol{k}) - f(\boldsymbol{k}') \approx \frac{\partial f}{\partial k_i}(k - k')^{(i)} \ ,$$

$$|\mathscr{V}_{k'}|^2 \cdot \Delta\varphi(\boldsymbol{k}', \boldsymbol{k}_0) - |\mathscr{V}_k|^2 \cdot \Delta\varphi(\boldsymbol{k}, \boldsymbol{k}_0) \approx \frac{\partial}{\partial k_j}[|\mathscr{V}_k|^2 \Delta\varphi(\boldsymbol{k}, \boldsymbol{k}_0)] \cdot (k' - k)^{(j)} \ .$$

Then we integrate by parts in (7.5.7) getting rid of derivatives of function f. The arbitrariness of this function enables one to omit integration over f ensuing:

$$(1 - |\mathscr{V}_k|^2)\Delta\varphi(\boldsymbol{k}, \boldsymbol{k}_0) - \frac{\partial}{\partial k_i}[D_{ij}(\boldsymbol{k}) \frac{\partial}{\partial k_j} |\mathscr{V}_k|^2 \Delta\varphi(\boldsymbol{k}, \boldsymbol{k}_0)] = \delta(\boldsymbol{k} - \boldsymbol{k}_0) \ , \qquad (7.5.8)$$

where

$$D_{ij} = \tfrac{1}{2} \int \sigma(\boldsymbol{k}, \boldsymbol{k}')(k' - k)^{(i)}(k' - k)^{(j)}\, d\boldsymbol{k}' \; ; \qquad i, j = 1, 2 \; . \tag{7.5.9}$$

Equation (7.5.8) should be solved with the boundary condition

$$D_{ij} \frac{\partial}{\partial k_j} (|V_k|^2 \Delta\varphi)|_{|\boldsymbol{k}|=K} = 0 \tag{7.5.10}$$

implying the absence of the energy flux into inhomogeneous waves. It is clear that since, in virtue of (7.5.9), the quadratic form D_{ij} is positively determined, then homogeneous equation (7.5.8) has only a trivial solution and the boundary problem, with respect to k-space, (7.5.8, 10) is uniquely solvable. It is this boundary problem that forms the diffusion approximation for the radiative transport equation. The boundary condition (7.5.10) is written down proceeding from physical considerations, although it can be obtained from the initial integral equation using the appropriate asymptotic procedure, as well.

Taking into account (7.5.1) it is easy to obtain the equation for value $\Delta I(\boldsymbol{k}, -1)$:

$$\left[1 - |\mathscr{V}_k|^2 - \frac{\partial}{\partial k_i}\left(D_{ij} \frac{\partial}{\partial k_j} |\mathscr{V}_k|^2\right)\right] \Delta I(\boldsymbol{k}, -1) = |\mathscr{D}(\boldsymbol{k}, -1)|^2 |\mathscr{V}_k|^2 / K q_k \; . \tag{7.5.11}$$

To calculate D_{ij} apply the Kirchhoff approximation (5.2.30) with $g^2 = B^2 = 1$. From (7.5.9) we obtain

$$D_{xx} = 2q^2 \delta_x^2 \; , \qquad D_{yy} = 2q^2 \delta_y^2 \; , \qquad D_{xy} = 0 \; ,$$

where $\delta_{x,y}^2$ are mean square slopes of roughness along x and y axes.

It is obvious from (7.5.8) that just as in the case of small roughness, the situation again depends qualitatively on the parameter

$$N \cdot \varepsilon \sim (1 - |V|^2)^{-1} \delta_{x,y}^2 \; .$$

For $N \cdot \varepsilon \ll 1$ the diffusion term in (7.5.11) can be neglected and scattering is insignificant:

$$\Delta I(\boldsymbol{k}, -1) = \frac{|\mathscr{D}(\boldsymbol{k}, -1)|^2}{K q_k} \frac{|V_k|^2}{1 - |V_k|^2} \; .$$

It can be shown in this case that the coherent component of correlation function C_{ch} is calculated by a formula of the form (7.3.14) with replacement of ΔI by I_{ch}. However, in virtue of condition $\overline{V} \approx 0$, only those noise sources contribute to I_{ch} whose radiation arrive at the observation point directly without scattering, see (7.1.13):

$$I_{\text{ch}}(\boldsymbol{k}, -1) = \frac{|\mathscr{D}(\boldsymbol{k}, -1)|^2}{K q_k} \; .$$

As a result

$$I(\boldsymbol{k}, -1) = I_{\mathrm{ch}} + \Delta I = \frac{|\mathscr{D}(\boldsymbol{k}, -1)|^2}{Kq_k \cdot (1 - |V_k|^2)} \ .$$

Thus the waveguide propagation leads to the simple modification of source function $\mathscr{D}$ as

$$|\mathscr{D}(\boldsymbol{k}, -1)|^2 \to |\mathscr{D}(\boldsymbol{k}, -1)|^2 \cdot (1 - |V_k|^2)^{-1} \ .$$

The last result corresponds to the obvious summation of geometrical progression originating while considering the infinite sequences of the sources contributing to the field at the given point due to the waveguide propagation effect.

In the inverse case: $N\varepsilon \sim 1$, the diffusion term in (7.5.11), should be taken into account and scattering becomes essential. The sense of this parameter is quite obvious in the case of large-scale irregularities also: this is the square mean value of the angle at which radiation deviates from its initial direction before attenuating. Thus the qualitative difference between situations $N\varepsilon \ll 1$ and $N\varepsilon \sim 1$ have a clear physical interpretation in this case as well.

References

Chapter 1

1.1 H. Hoinkes: Rev. Modern Phys. **52**, 933–970 (1980)
1.2 J.W. Vandersande: Phys. Rev. B **13**, 4560–4567 (1976)
1.3 A.A. Krokhin, N.M. Makarov, V.A. Yampolsky: Zh. Eksp. Teor. Fiz. **99**, 520–529 (1991) (in Russian)
1.4 J.M. Ziman: *Electrons and Phonons* (Claredon, Oxford 1960)
† 1.5 R. Petit (ed.): *Electromagnetic Theory of Gratings*, Topics Curr. Phys., Vol. 22 (Springer, Berlin, Heidelberg, 1980)
1.6 S.V. Gaponov, V.M. Genkin, N.N. Salaschenko, A.A. Fraerman: JETP Lett. **41** (2), 63–65 (1985)
1.7 N.N. Krupenio: *Radioissledovania Luny i Planet Zemnoi Gruppy*, in Itogi Nauki i Techniki, ser. Astronomia, Vol. 16 (Vses. Inst. Nauch.-Tech. Inform., 1980)
1.8 P. Beckmann, A. Spizzichino: *The Scattering of Electromagnetic Waves from Rough Surfaces* (McMillan, New York 1963)
1.9 F.G. Bass, I.M. Fuks: *Rasseyanie Voln na Statisticheski Nerovnoy Poverkhnosti* (Nauka, Moscow 1972) [English transl.: *Wave Scattering from Statistically Rough Surfaces* (Pergamon, New York 1978)]
† 1.10 S.M. Rytov, Yu.A. Kravtsov, V.I. Tatarskii: *Vvedenie v Statisticheskuyu Radiophysiku* (Nauka, Moscow 1978) [English transl.: *Introduction of Statistical Radiophysics* (Springer, Berlin, Heidelberg year?)]
† 1.11 L. Brekhovskikh, Yu. Lysanov: *Fundamentals of Ocean Acoustics* 2nd edn, Springer Ser. Wave Phenom, vol. 8 (Springer, Berlin, Heidelberg 1991)
1.12 L. Tsang, J.A. Kong, R.T. Shin: *Theory of Microwave Remote Sensing* (Wiley, New York 1985)
1.13 Lord Rayleigh, O.M.: Proc. Roy. Soc. A **79**, 399–416 (1907)
1.14 E.I. Thorsos: J. Acoust. Soc. Am. **88**, 335–349 (1990)
1.15 D.H. Berman: J. Acoust. Soc. Am. **89**, 623–636 (1991)
1.16 P.C. Waterman: J. Acoust. Soc. Am. **57**, 791–802 (1975)
1.17 L.M. Brekhovskikh: Zh. Eksp. Teor. Fiz. **23**, 275–304 (1952) (in Russian)
1.18 M.A. Isakovich: Zh. Eksp. Teor. Fiz. **23**, 305–314 (1952) (in Russian)
1.19 B.F. Kur'yanov: Akust. Zh. **8**, 325–333 (1962) [English Transl.: Sov. Phys. Acoust. **8**, 252–257 (1963)]
1.20 I.M. Fuks: Izv. Vuzov ser. Radio-Physica **28**, 177–183 (1985) (in Russian)
1.21 B.A. Lippman: J. Opt. Soc. Am. **43**, 408 (1953)
1.22 R. Petit, M. Cadilhac: C.R. Acad. Sc. Paris **262**, 468–471 (1966)
1.23 H.W. Marsh: J. Acoust. Soc. Am. **35**, 1835–1836 (1963)
1.24 R.F. Millar: Proc. Camb. Phil. Soc. **69**, 217–225 (1971)
1.25 E.G. Liszka, J.J. McCoy: J. Acoust. Soc. Am. **71**, 1093–1100 (1982)
1.26 G.G. Zipfel, J.A. DeSanto: J. Math. Phys. **13**, 1903–1911 (1972)
1.27 J.A. DeSanto: Coherent multiple scattering from rough surfaces, in *Multiple Scattering and Waves in Random Media*, ed. by P.L. Chow, W.E. Kohler, G.C. Papanicolau (North-Holland, Amsterdam 1981) pp. 123–141

1.28 J.A. DeSanto, G.S. Brown: Analytical techniques for multiple scattering from rough surfaces. *Progress in Optics* 1–62 (North-Holland, Amsterdam 1986)
1.29 A.G. Voronovich: Dokl. Akad. Nauk SSSR **273**, 830–834 (1983)
1.30 A.G. Voronovich: Zh. Eksp. Teor. Fiz. **89** (7), 116–125 (1985) [Engl. transl.: Sov. Phys. JETP **62** (1), 65–70 (1985)]
1.31 A. Hessel, A.A. Oliner: Appl. Opt. **4**, 1275–1297 (1965)
1.32 D. Winebrenner, A. Ishimaru: Radio Sci. **20**, 161–170 (1985)
1.33 D. Winebrenner, A. Ishimaru: J. Opt. Soc. Am. **2**, 2285–2294 (1985)
1.34 S.L. Broschat, E.I. Thorsos, A. Ishimaru: IEEE Trans. GRS. **28**, 202–205 (1990)
1.35 A.G. Voronovich: Dokl. Akad. Nauk SSSR **272**, 1351–1355 (1983) (in Russian)
1.36 W.C. Meecham: J. Acoust. Soc. Am. **28**, 370–377 (1956)
1.37 Yu.P. Lysanov: Akust. Zh. **2** (2), 182–187 (1956) (in Russian)
1.38 E. Bahar: Canad. J. Phys. **50**, 3132–3142 (1972)

Chapter 2

2.1 P.C. Waterman: J. Acoust. Soc. Am. **57**, 791–802 (1975)
2.2 L.D. Landau, E.M. Lifshitz: *Kwantovaya Mekhanika* (*Nerelativistskaya Teoriya*) (Nauka, Moscow 1974) [English transl.: *Quantum Mechanics* (Nonrelativistic Theory) (Pergamon, Oxford 1959)
†2.3 L. Brekhovskikh, Yu. Lysanov: *Fundamentals of Ocean Acoustics* 2nd edn. Springer Ser. Wave Phenom., Vol. 8 (Springer, Berlin, Heidelberg 1991)

Chapter 3

3.1 J.W.S. Rayleigh: *The Theory of Sound* (Dover, New York 1945) vol. 2
3.2 B.A. Lippman: J. Opt. Soc. Am. **43**, 408 (1953)
3.3 R. Petit, M. Cadilhac: C.R. Acad. Sc. Paris **262**, 468–471 (1966)
3.4 R.F. Millar: Proc. Camb. Phil. Soc. **65**, 773–791 (1969)
3.5 R.F. Millar: Proc. Camb. Phil. Soc. **69**, 217–225 (1971)
3.6 P.M. van den Berg, J. T. Fokkema: J. Opt. Soc. Am. **69**, 27–31 (1979)
3.7 H.W. Marsh: J. Acoust. Soc. Am. **35**, 1835–1836 (1963)
3.8 N.R. Hill, V. Celli: Phys. Rev. **B 17**, 2478–2481 (1978)
3.9 A.G. Voronovich: Dokl. Akad. Nauk. SSSR **273**, 85–89 (1983)
3.10 R.F. Millar: Radio Sci **8**, 785–796 (1973)
3.11 H. Ikuno, K. Yasuura: IEEE Trans. **AP-21**, 657–662 (1973)

Chapter 4

4.1 U. Fano: J. Opt. Soc. Am. **31**, 213–222 (1941)
4.2 S.O. Rice: Commun. Pure Appl. Math. **4**, 351–378 (1951)
4.3 R.K. Rosich, J.R. Wait: Radio Sci. **12**, 719–729 (1977)
4.4 M.L. Burrows: Electron. Lett. **5**, 277–278 (1969)
4.5 E.Toigo, A. Marvin, V. Celli, N.R. Hill: Phys. Rev. **B 15**, 5618–5626 (1977)
4.6 L. Kazandjian: Acoust. Lett. **13**, 187–191 (1990)
4.7 D.R. Jackson, D.P. Winebrenner, A. Ishimaru: J. Acoust. Soc. Am. **83**, 961–969 (1988)
4.8 A.G. Voronovich: Dokl. Akad. Nauk SSSR **273**, 85–89 (1983) (in Russian)

Chapter 5

5.1 L.M. Brekhovskikh: Zh. Eksp. Teor. Fiz. **23**, 275–288 (1952) (in Russian)
5.2 L.M. Brekhovskikh: Zh. Eksp. Teor. Fiz. **23**, 289–304 (1952) (in Russian)
5.3 M.A. Isakovich: Zh. Eksp. Teor. Fiz. **23**, 305–314 (1952) (in Russian)
5.4 G.G. Zipfel, J.A. DeSanto: J. Math. Phys. **13**, 1903–1911 (1972)
5.5 J.A. DeSanto, G.S. Brown: Analytical techniques for multiple scattering from rough surfaces. In *Progress in Optic* **23**, 1–62 (North-Holland, Amsterdam 1986)
5.6 A.G. Voronovich: Dokl. Akad. Nauk SSSR **273**, 830–834 (1983) (in Russian)
5.7 B.F. Kur'yanov: Akust. Zh. **8**, 325–333 (1962) [English Transl.: Sov. Phys. Acoust. **8**, 252–257 (1963)]
5.8 E.G. Liszka, J.J. McCoy: J. Acoust. Soc. Am. **71**, 1093–1100 (1982)
5.9 E.I. Thorsos: J. Acoust. Soc. Am **83**, 78–92 (1988)
5.10 S.T. McDaniel, A.D. Gorman: J. Geophys. Res. **87**, 4127–4136 (1982)

Chapter 6

6.1 S.Z. Dunin, G.A. Maksimov: Zh. Eksp. Teor. Fiz. **98**, 391–406 (1990 (in Russian)
6.2 A.G. Voronovich: Zh. Eksp. Teor. Fiz. **89** (7), 116–125 (1985) [English transl.: Sov. Phys. JETP **62** (1), 65–70 (1985)]
6.3 D. Winebrenner, A. Ishimaru: Radio Sci. **20**, 161–170 (1985)
6.4 D. Winebrenner, A. Ishimaru: J. Opt. Soc. Am. **2**, 2285–2294 (1985)
6.5 A.G. Voronovich: Dokl. Akad. Nauk SSSR **272**, 1351–1355 (1983) (in Russian)
6.6 W.C. Meecham: J. Acoust. Soc. Am. **28**, 370–377 (1956)
6.7 Yu.P. Lysanov: Akust. Zh. **2** (2), 182–187 (1956) (in Russian)
6.8 E. Bahar: Canad. J. Phys. **50**, 3132–3142 (1972)
6.9 E. Bahar: J. Geophys. res. **92**, 5209–5224 (1987)
6.10 E. Bahar: J. Acoust. Soc. Am. **89**, 19–26 (1991)
6.11 A.B. Kozin, S.D. Chuprov: Akust. Zh. **27**, 411–417 (1981) (in Russian)
6.12 R. Dashen, F.S. Henyey, D. Wurmser: J. Acoust. Soc. Am. **88**, 310–323 (1990)
6.13 L.A. Vainstein, A.I. Sykov: Radiot. Elektr. **29**, 1472–1478 (1984) (in Russian)
6.14 A.G. Mikheev: Chislennye i priblizhennye metody v praymoi i obratnoi zadachach difrakchii na periodicheskoy poverkhnosti. Dissertation, Moscow University (1991)
6.15 D.H. Berman: J. Acoust. Soc. Am. **89**, 623–636 (1991)
6.16 J. Shen, A.A. Maradudin: Phys. Rev. B **22**, 4234–4240 (1980)
6.17 S.L. Broschat, E.I. Thorsos, A. Ishimaru: J. Electromag. Waves Appl. **3**, 237–256 (1989)
6.18 S.L. Broschat, E.I. Thorsos, A. Ishimaru: IEEE Trans. GRS-**28**, 202–205 (1990)
†6.19 S.M. Rytov, Yu.A. Kravtsov, V.I. Tatarskii: *Vvedenie v Statisticheskuyu Radiophysiku* (Nauka, Moscow 1978) [English transl.: *Introduction of Statistical Radiophysics* (Springer, Berlin, Heidelberg year)]
6.20 E. Rodriguez: Radio Sci. **24**, 681–693 (1989)
6.21 D.H. Berman, D.K. Dacol: J. Acoust. Soc. Am. **87**, 2024–2032 (1990)
6.22 L. Fortuin: J. Acoust. Soc. Am. **47**, 1209–1228 (1970)
6.23 J.L. Uretsky: J. Acoust. Soc. Am. **35**, 1293–1294 (1963)
6.24 E.O. Lacasce, Yr., P. Tamarkin: J. Appl. Phys. **27**, 138–148 (1956)
6.25 D.M. Milder: J. Acoust. Soc. Am. **89**, 529–541 (1991)
6.26 D.M. Milder: Radio Sci. **31**, 1369–1376 (1996)
6.27 R.A. Smith: Radio Sci. **31**, 1377–1385 (1996)
6.28 D. Maystre: IEEE Trans. AP **31** (1983)
6.29 H. Faure-Geors, D. Maystre, A. Roger: J. Optics (Paris) **19**, 51–57 (1988)
6.30 H. Faure-Geors, D. Maystre, A. Roger: J. Optics (Paris) **19**, 221–229 (1988)

6.31 A.G. Voronovich: Trans. USSR Acad. Sciences, Earth Sci. Sec. **287**, 186–190 (March–April 1986)
6.32 P.J. Kaczkowski, E.I. Thorsos: J. Acoust. Soc. Am. **96**, 957–972 (1994)
6.33 D.M. Milder: Waves in Random Media **8**, 67–78 (1998)
6.34 D.M. Milder: J. Acoust, Soc. Am. **100**, 759–768 (1996)

Chapter 7

7.1 F.G. Bass, I.M. Fuks: *Rasseyanie Voln na Statisticheski Nerovnoy Poverkhnosti* (Nauka, Moscow 1972) [English transl.: *Wave Scattering from Statistically Rough Surfaces* (Pergamon, New York 1978)]
7.2 W. Meecham: J. Acoust. Soc. Am. **25**, 1012–1013 (1953)
7.3 A. Beilis, F.D. Tappert: J. Acoust. Soc. Am. **66**, 311–326 (1979)
7.4 F.I. Kryazhev, V.M. Kudryashov, N.A. Petrov: Akust. Zh. **22**, 377–384 (1976) [Engl. transl.: Sov. Phys. Acoust. **22**, 211–215 (1976)]
7.5 A.G. Voronovich: Akust. Zh. **33**, 19–30 (1987) (in Russian)
† 7.6 S.M. Rytov, Yu.A. Kravtsov, V.I. Tatarskii: *Vvedenie v Statisticheskuyu Radiophysiku* (Nauka, Moscow 1978) [English transl.: *Introduction of Statistical Radiophysics* (Springer, Berlin, Heidelberg year)]
† 7.7 L. Brekhovskikh, Yu. Lysanov: *Fundamentals of Ocean Acoustics* 2nd edn., Springer Ser. Wave Phenom., Vol. 8 (Springer, Berlin, Heidelberg 1991)

Subject Index